PAPERS PRESENTED AT THE INTERNATIONAL SYMPOSIUM ON

INDUSTRIAL APPLICATION OF HEAT PUMPS

held at the University of Warwick, UK
24-26 March, 1982

Sponsored and organised by

BHRA Fluid Engineering, Cranfield, Bedford MK43 0AJ, U.K.

Published by

Editors: H.S. Stephens
         Mrs. B. Jarvis

The Organisers are not responsible for statements or opinions
made in the papers.

The papers have been reproduced by offset printing from the
authors' original typescripts and illustrations to minimise
delay. Whilst every effort is made to produce a reasonable
volume of preprints the organisers are unable to accept
responsibility for the quality of the printing.

When citing papers from this volume the following reference
should be used:
Title, Author, Paper No., Pages, International Symposium
on the Industrial Application of Heat Pumps, Coventry, U.K.
BHRA Fluid Engineering, Cranfield, Bedford, U.K. 24-26
March, 1982.

Printed and published by
BHRA Fluid Engineering
Cranfield, Bedford MK43 0AJ, U.K.

ISBN 0 906085 65 9

ACKNOWLEDGEMENTS

The valuable assistance of the Organising Committee and
Panel of Referees is gratefully acknowledged.

ORGANISING COMMITTEE

Dr. H.J. Abrams	UMIST
J. Foster	Energy Research Support Unit
Prof. F.A. Holland	University of Salford
Mrs. B. Jarvis	BHRA Fluid Engineering
Dr. C. Lopez-Cacicedo	Electricity Council Research Centre
J. Masters	British Gas Corporation
Dr. M.J. McCall	Building Research Establishment
T.J.M. Moore	BHRA Fluid Engineering
Dr. D. Reay	International Research and Development Ltd
Prof. I.E. Smith	Cranfield Institute of Technology
H.S. Stephens	BHRA Fluid Engineering
D.F. Warne	ERA Technology Ltd

CORRESPONDING MEMBERS

J. Berghmans	Katholieke Universiteit Leuven, Belgium
J.R. Feuga	Electricité de France
Dr. D. Hodgett	Battelle Institute, FRG
J. Knobbout	TNO, Netherlands
Dr. B. Sternlicht	Mechanical Technology Inc., USA

PAPERS PRESENTED AT THE INTERNATIONAL SYMPOSIUM ON THE INDUSTRIAL APPLICATION OF HEAT PUMPS

CONTENTS

*   These papers were not received in time for publication.

The publishers have made every effort to print all papers presented at the Symposium but regret they cannot accept responsibility for not including papers which were not received in time to be printed.

THE PENDAR HEAT PUMP DEVELOPMENT AND APPLICATIONS GROUP

Background

Since 1976 Pendar has been at the forefront of heat pump technology in the U.K. and Europe. During this period the Company has built up a wealth of expertise and experience in all aspects of heat pump technology and has established a range of facilities for heat pump testing.

The heat pump design and applications work with which Pendar has been involved has, in the past, been treated as just one of many aspects of the business in design and development contract work. However, the current energy situation and the increasing importance of heat pumps in industry, commerce and in the home has resulted in the formation of this specific group within the company. The design and development capabilities in materials technology, mechanical engineering and electronics are all available to the Group as well as the in depth knowledge of heat pumps and energy technology which has been built up over the years.

What We Will Do

Pendar has full capability for research, design and development ranging from theoretical studies through to manufacture of prototypes or one off special systems. Pendar's engineers make sure they are up-to-date with worldwide developments in heat pumps and keep in contact with a number of important research centres in the UK and Europe.

The aim of Pendar's Heat Pump Group is to provide a commercial and practical base for the successful application of the technology. So given the bare bones of a prospective project, Pendar will:-

- Finalise the project specification.
- Highlight the objectives and requirements.
- Submit a detailed proposal for the work.
- Quote time scales and costs to complete
  and then, having received an order, Pendar will:-
- carry out the work with minimum of effort from the customer including control of subcontractors in the case of large installations.
- Liaise with the customer to ensure the project is in line with requirements.
- Be ever concious of the commercial pressures associated with the development and application of the technology.

What We Can Offer

Application:-
Identification of application areas.
Cost studies and economic assessment.

Component and System design:-
Recommendations on selection of components and equipment.
Organisation, supervision and commissioning of new installations.
Monitoring of new or existing installations.

Design and development:-
Development and testing of new components.
Design of new heat pumps including electrically driven and fossil fuel fired systems.

Testing:-
Performance testing of components.
Performance testing of systems.

Control and instrumentation:-
Development of electronic controls for various heat pump applications.
Optimisation of heat pump performance using microprocessor and other techniques.
Monitoring and measurement.

For further information contact:-

**Pendar Technical Associates Limited,
Hamp Industrial Estate, Old Taunton Rd.,
Bridgwater, Somerset TA6 3NT,
England.**

Telephone: (0278) 56888/9

Where We Will Go

Pendar engineers are always available at short notice to visit potential customers and discuss the possibilities for heat pump development and application. Normally the cost for the initial discussion and suggestions will be borne completely by Pendar as will the cost for preparing detailed proposals. Travel, especially in the U.K. and the rest of Europe, presents no problem and Pendar engineers will be pleased to visit customer's works and/or sites for potential applications as and when necessary.

What We Have Done

In one way or another Pendar has been involved with almost all aspects of heat pumps ranging from recommendations on the installation of electrically driven machines with many hundreds of kilowatts capacity to detailed research and development work on domestic gas fired systems.

Some of our past customers include:-

U.K. Department of Energy
Building Research Station (U.K. Department of the Environment)
Calor Group Ltd.,
National Coal Board

STAL VM 100 SERIES

A new range of STAL water chiller and heat pump units is now being introduced. The designs have been based on STAL's advanced technology and long experience in the refrigeration field which ensure compliance with most requirements from the market. The units are built in the company's modern workshops and are fully tested before delivery. Main features of the chillers are:

Increased heat transfer surfaces

Resulting in lower power consumption and reduced operating costs.

Compact assembly

Minimum space for transport and installation is required.

Electronic regulation

Giving a simple, accurate and safe operation.

Flexibility

A standard unit can easily be adapted to different refrigerants, chilled fluids and temperatures.

Capacity control

Stepless control from 25% to 100% duty.

Heat recovery, heat pumps

The VM 100 series also includes standard models for heat recovery. Heat pump units are available for heat carrier temperatures up to +110°C.

Fully tested

All units are test run and adjusted for the actual operating conditions before delivery. This means that commissioning work is kept to a minimum.

STAL liquid chiller units are built to give high service reliability, using advanced technology and well-proven components. Modern, safe and simple electronic controls are used for the operating and regulating equipment. Thus correct heat transfer medium temperature can be maintained, and it is simple to select other set-point values for adaption to different processes or outdoor temperatures. The compressors are equipped with stepless capacity control between 25 and 100% of nominal duty. Many of the features of the previous range of liquid chillers from STAL, such as independent refrigeration circuits for units including two or three compressors, have been retained. A wide range of optional accessories is available.

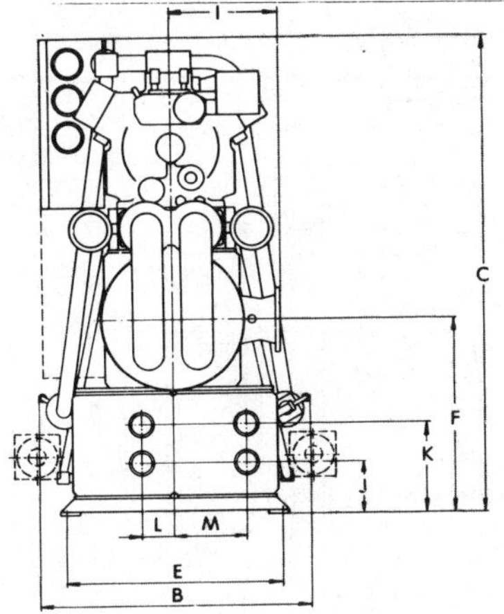

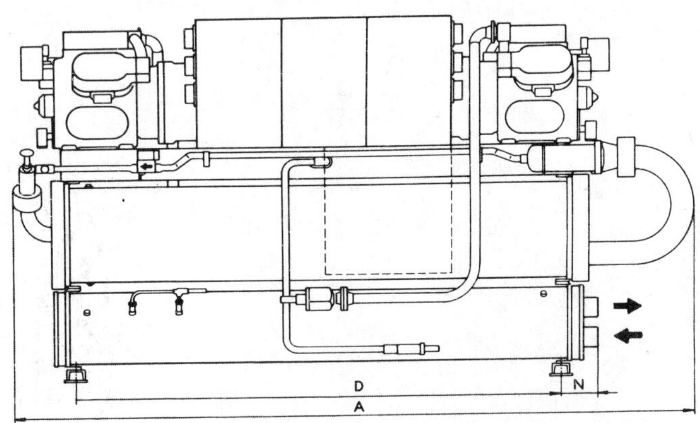

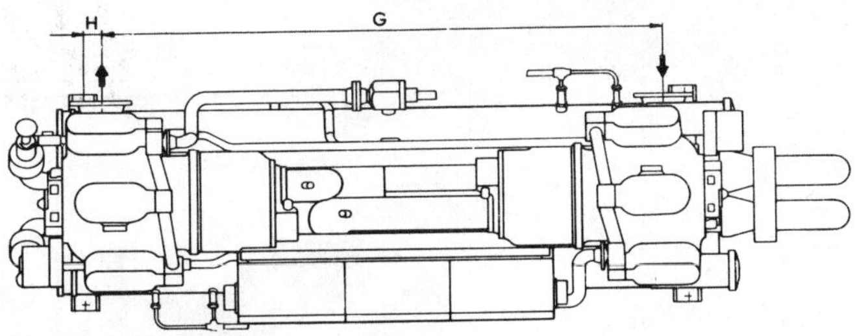

STAL-LEVIN LTD
YIEWSLEY HIGH ST · WEST DRAYTON · MIDDX · UB7 7TA
West Drayton (089 54) 46561

RIVER PINN WORKS

Telex: 262621

HEAT PUMPS
From P.A. Hilton

Manufacturers of Engineering Laboratory Teaching Equipment

The need to conserve energy has increased the interest in heat pumps. The illustrated units allow students to undertake a comprehensive analysis of the heat pump cycle. These units have been designed and manufactured specifically for teaching.

Air and Water
Heat Pump R830

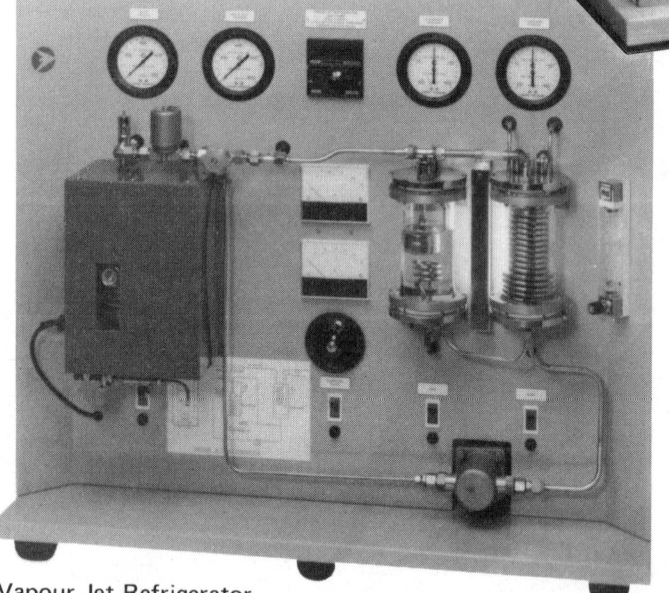

Vapour Jet Refrigerator
and Heat Pump R850

We manufacture a range of engineering laboratory teaching equipment which is used for studying:

Refrigeration	Fluid Flow
Air Conditioning	Combustion
Heat Transfer	Propulsion

P.A.Hilton Ltd.

Horsebridge Mill, King's Somborne, Hampshire SO20 6PX, England.
Telephone: 07947-382 or 450
Telex 477538

HEAT DRIVEN FREE PISTON HEAT PUMP

D. Vokaer and Y. Vandendael

Université Libre de Bruxelles, Belgium

Summary

The simple principle of the free piston has been used in a new way, combining two coupled Rankine cycles in a single free piston machine; the system presented here has been derived from the prototype of a solar cooling unit.

The technological simplicity and small dimensions of this machine succeed in combining optimum performance, reliability and economy. The design point is defined by the heat-sources temperatures:

- high temperature source (pressurized water boiler) $150^{\circ}C$
- useful heat (20 kW) output $40^{\circ}C$
- ambient (low grade input) $8^{\circ}C$

The thermodynamic cycle and the performance are presented both at the design point and under off-design operation.

It is shown that the proposed machine adapts itself naturally to the load, by a change in its stroke while the oscillation frequency remains almost constant.

An economical comparison with a conventional oil-fired boiler shows that short pay-back periods can be obtained, thanks to a C.O.P. (Primary energy) of 1.34.

By comparison with absorption systems, this machine has the great advantage that it can work at high-source temperatures, and at lower ambient temperatures than $LiBr/H_2O$ installations.

Held at the University of Warwick, U.K.
Symposium organised and sponsored by
BHRA Fluid Engineering
©BHRA Fluid Engineering, Cranfield, Bedford MK43 0AJ, England.

1. INTRODUCTION

It is somewhat unusual for a heat pump to be used to recover heat in the form of hot
gases for use in space heating.
In most instances, these gases can benefit very well from the application of conventional heat exchangers.
However, if they are at relatively high temperatures - say 150 °C - they can easily
drive a heat-actuated heat pump which then upgrades the ambient temperature (Fig. 1).

From a First Law point of view, the C.O.P. of the system will always be greater than
unity, and this will lead automatically to some energy savings.

In the case of domestic application, the hot source will consist of pressurised water
heated in an oil-fired boiler.
The use of such a system leads to substantial energy savings, which balance the higher
cost and complexity of the installation.

The results discussed in section 4 are based on the experience gained with an identical
machine built for a solar cooling unit, whose technology has been proven to be reliable
and cost-effective (Ref. 1 to 5).

2. THERMODYNAMIC DATA

The lay out of the installation and the corresponding thermodynamic cycle are illustrated Fig. 2 a,b. A common working fluid is used both for the power cycle and for the
vapour-compression cycle of the heat pump.

The useful heat is removed both from the power cycle and from the heat pump cycle in a
common condenser.

The critical temperature of most halocarbon compounds is lower than the maximum
temperature of the cycle. R11 and R113 remain the only candidates which can be used
in a Rankine process without superheating the vapour.

On the other hand, the high value of saturated vapour pressure of R11 (21 bar at 150°C)
makes its use problematic with respect to technological aspects of the installation.
These considerations led to the selection of R113 as the working fluid.

The evaporator can operate satisfactorily below atmospheric pressure (.17 bar) because,
as will be seen in the next section, the expander, compressor and pump are enclosed in
a sealed cylinder, so that no leakage of the working fluid or air infiltration can
occur.
The only penalty will be paid to the swept volume of the compressor, which of course
directly depends on the specific volume at compressor suction.

Table I details the most important operating characteristics at the design point.
The C.O.P. of the thermodynamic cycle (ratio of the useful heat over the thermal input
of the R113 boiler) is 1.68 or 50 % of the Carnot efficiency. This figure has been
obtained by making the assumption of an expansion and compression isentropic efficiency
of .8. If the water boiler is considered as a part of the installation, the
"Primary Energy Ratio" will be 1.34.
In other words, the complete heat pump system will save : 1. - 1./1.68 = 40 % of
primary energy, by comparison with a conventional space heating system.

3. PRACTICAL DESIGN

The major originality of the system lies in the way which has been choosen to design
the mechanical components : expander, compressor and pump. All these functions are
achieved at the same time by a single free piston, whose patented shape means that its
two ends dictate the annular and cylindrical shape of the chambers, each of which
performs a specific task (Fig. 3) :

    - chambers 1 and 1' : double acting motor
    - chambers 2 and 2' : double acting compressor
    - chambers 3 and 3' : double acting pump.

Looking at Fig. 3, it is essential to notice that the force exerted on the piston by
the driving fluid is directly transmitted to the fluids in the opposite chambers.

A standard jack tube is used for the cylinder, which is simply sealed with two cover-plates by means of so-called "O-rings". There are no other seals necessary, except the piston rings, which are made of a special low-wear PTFE compound.

In order to avoid vibration problems, the system includes two identical machines connected in parallel and oscillating in opposition. The main dimensions and characteristics of each machine are given in Table II. It can be seen that the overall dimensions (450x160 mm) are small with respect to the output of the system.

The low oscillation frequency of the piston, and technological simplicity lead to the following advantages :

- no need for a lubrication system (PTFE piston rings);
- low noise level;
- nearly no maintenance requirements.

4. OFF-DESIGN PERFORMANCES

When the ambient or condensing temperatures vary, the vapour pressure of the working fluid, of course, changes. This influences the dynamic behaviour of the piston, in terms of stroke and frequency of oscillation.

As can be seen in Fig. 4 a,b these parameters adapt themselves naturally until they match the new operating conditions. Fig. 4a shows that the stroke varies from 150 to 210 mm, while the frequency range is quite narrow (6 to 7 Hz), even for a wide variation of the temperature (Fig. 4b).

The heat output and C.O.P. of the system are plotted in Fig. 5 and 6, respectively, versus the ambient temperature and for different temperatures of the useful heat, T_U. The dotted line representing the characteristic of the load shows that both the C.O.P. and the thermal power increase favourably when the ambient temperature decreases.

It should be pointed out that decreasing the useful temperature T_U at the same time, can be accomplished easily in a Rankine cycle, whereas the situation would not have been the same in an absorption machine.

5. ECONOMIC ASSESSMENT

The economical evaluation presented in Table III has been calculated by making the following assumptions :

- operation of the heat pump at the design point (C.O.P.=1.34 ; output=20 kW)
- 2000 or 4000 hours operation per year
- fuel cost = 1 US cent / MJ
- efficiency of the oil-fired boiler = .8

The results indicate a quite reasonable pay-back period of 2 to 6 years, depending on the running time of the space heating system (respectively 4000 and 2000 hours). This is due to the relatively high value of the C.O.P. : 1.34 in terms of primary energy or 1.68 by comparison with the oil-fired boiler.

6. CONCLUSIONS

The aim of the present project has been to develop a new heat pump which combines all mechanical functions required by the system in a single free piston machine.

With its single moving part, the design of this machine reduces the cost and keeps maintenance to a minimum.

By comparison with absorption systems, the C.O.P. of this Rankine / Rankine heat pump remains high enough in a wide temperature range, to ensure substantial cost-savings and short pay-back periods.

7. REFERENCES

1. <u>Vokaer, D.</u> : "Etude et réalisation d'une machine frigorifique à piston libre, à trois sources de chaleur, fonctionnant suivant un cycle de Rankine". Commission of the European Communities - Report EUR 6786, 1980 (177 pp).

2. <u>Vokaer, D.</u> : "Development of an Autonomous Free Piston Refrigerating Unit Driven by Rankine Cycle". In : Proc. Contractors meeting, 2nd Solar Energy R/D Programme (Brussels : Jan 21-23, 1981). Report EUR 7343 EN.

3. <u>Vokaer, D. and Bougard, J.</u> : "Solar cooling unit using a single free piston Rankine Machine". In : Proc. Solar World Forum (Brighton : Aug. 23-28, 1981). International Solar Energy Society. (In press).

4. <u>Bougard, J. and Vokaer, D.</u> : "Machine frigorifique solaire autonome". Ecole d'été de Cargèse (Corsica, 1981). To be published in Revue Générale du Froid.

5. <u>Vokaer, D.</u> : "Develoment of an autonomous free piston refrigerating unit driven by Rankine cycle". In : Proc. Contractors meeting, 2nd Solar Energy R/D Programme (Athens : 11-13 Nov., 1981). (In press).

TABLE I - OPERATING CHARACTERISTICS AT THE DESIGN POINT (R113)

Design sources temperatures : 155°C - 40°C - 8°C		
Motor inlet temp. - pressure	°C - bar	150-12.22
Compressor outlet temp. - pressure	°C - bar	55- 1.03
Compressor suction temp. - pressure	°C - bar	8- .17
Mass and volume flow rates	kg/s - m3/s	
Motor (x 10^{-3})		65 - .76
Pump (x 10^{-3})		65 - .04
Compressor (x 10^{-3})		65 - 47
Heat fluxes	kW	
To the R113 - boiler		11.6
From the condenser (useful heat)		19.6
To the evaporator		8.0
Coefficients of performance		
Carnot efficiency		3.36
Thermodynamic cycle		1.68
In terms of primary energy		1.34
(efficiency of the water boiler = .8)		

TABLE II - CHARACTERISTICS OF THE FREE PISTON MACHINE

Outer diameter	mm	161
Compressor bore	mm	126
Overall length	mm	450
Stroke at the design point	mm	195
Oscillation frequency (design point)	Hz	6.6
Volumetric ratio / motor	-	17
Volumetric ratio / compressor	-	17

TABLE III - ECONOMICAL COMPARISON WITH A CONVENTIONAL BOILER (costs in US $)

	Conventional boiler		Heat pump system	
	2000h	4000h	2000h	4000h
Investment	500		3000	
Costs per annum				
Depreciation (over 10 years)	50	50	300	300
Maintenance	125	125	150	150
Running cost	1800	3600	1075	2150
Total cost per annum	1975	3775	1525	2600
Cost-savings	–	–	450	1175
Pay-back period (years)	–	–	5.6	2.2

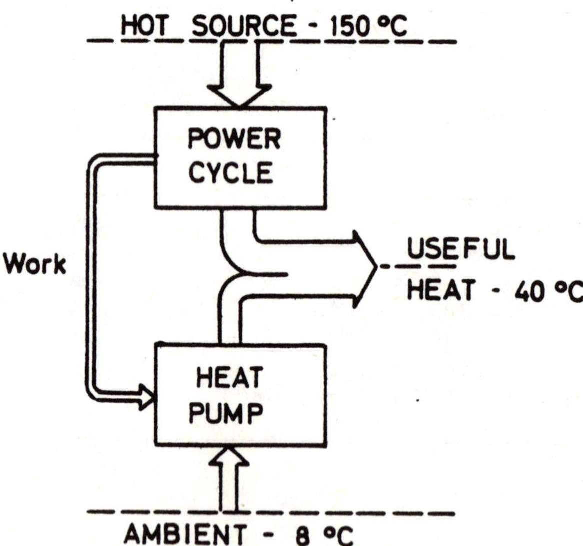

Fig. 1 - Energy flow chart of the system

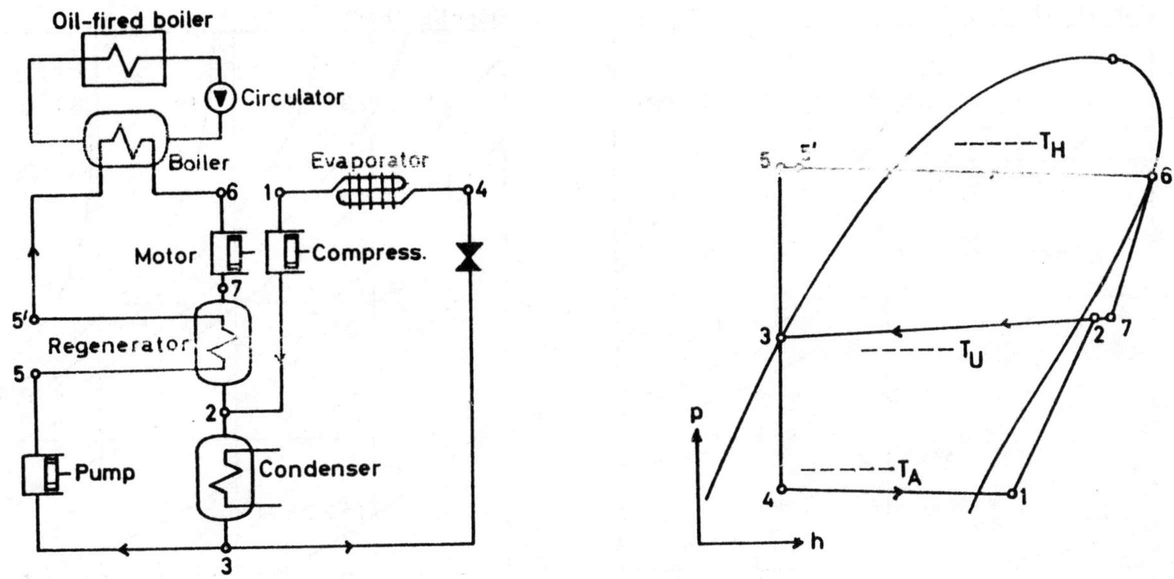

Fig. 2 - a - Lay out of the installation
- b - Thermodynamic cycle

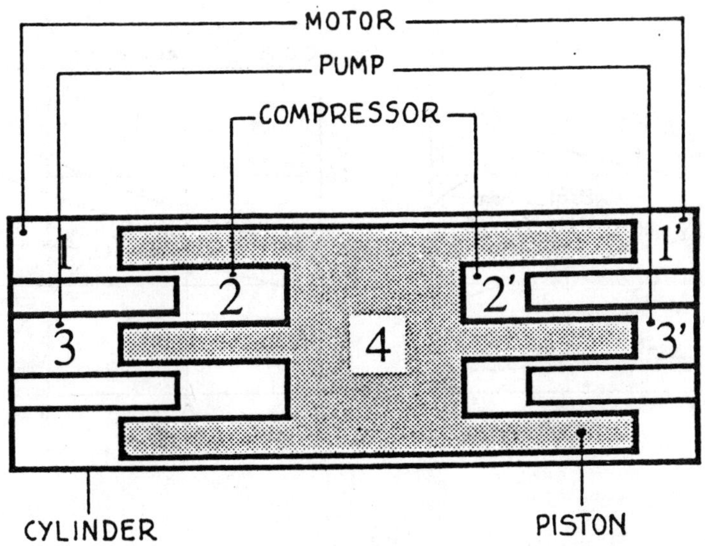

Fig. 3 - Free piston principle

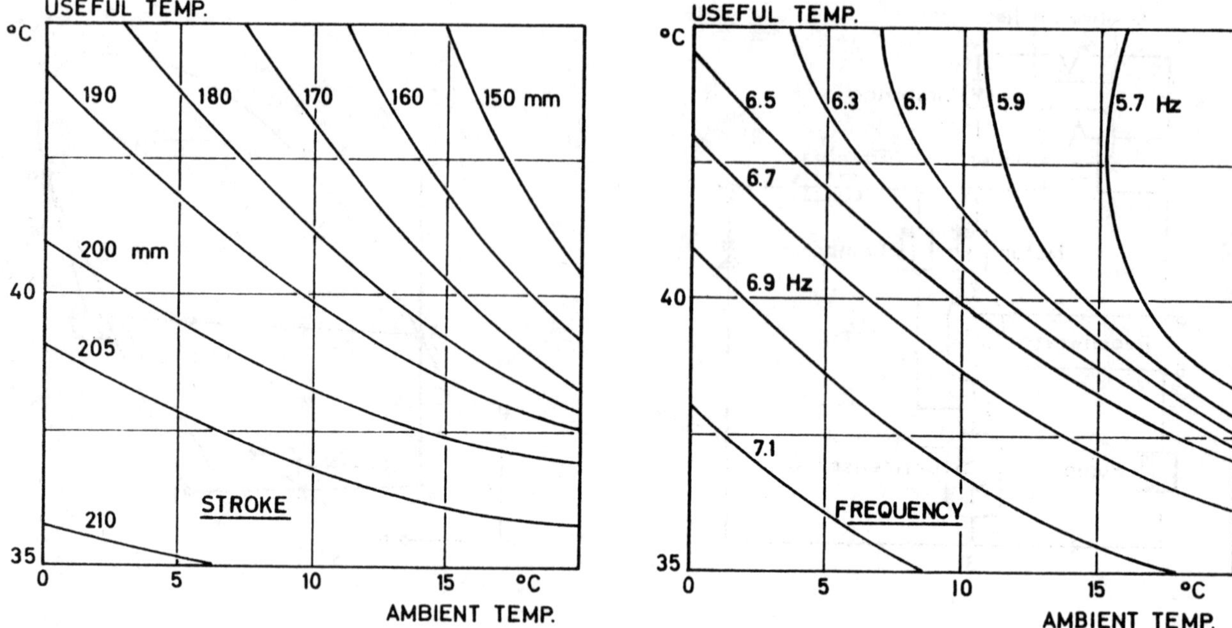

Fig. 4 – a – Stroke of the free piston
      – b – Frequency of oscillation of the free piston

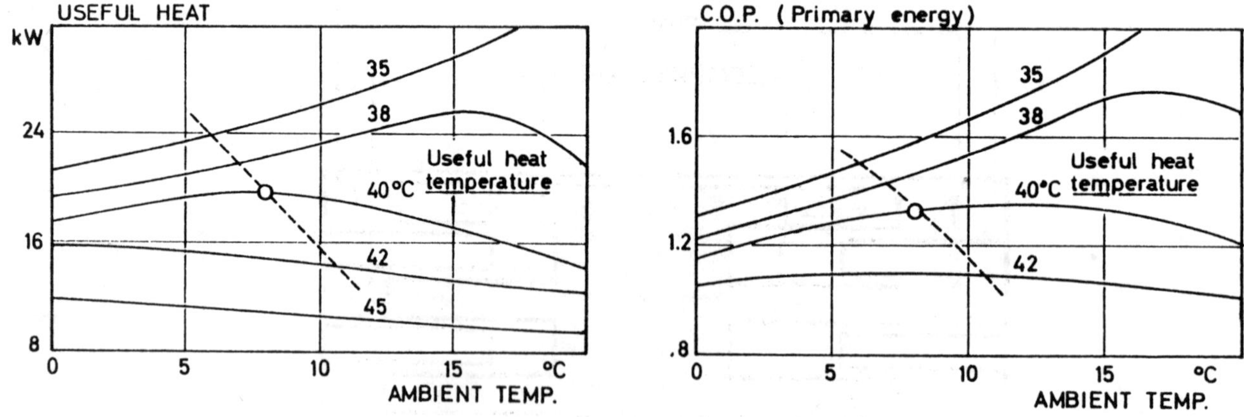

Fig. 5 – a – Heat output of the system
      – b – C.O.P. (primary energy) of the system

STIRLING ENGINE HEAT PUMPS

G. Walker, R. Fauvel, R. Gustafson, J. van Bentham

The University of Calgary, Canada

Summary

A Stirling engine is a thermal system that may be used to produce power from a high temperature heat source or as a refrigerator and heat pump to deliver energy at a higher temperature than abstracted from the source. A Stirling engine may therefore be used as the driver for natural gas heated air conditioning/heat pump Rankine cycle vapour compression systems or itself be used as the refrigerating/heat pump system requiring an input of work. Two Stirling systems, one acting as the driver, the other as the heat pump may be combined into the Stirling-Stirling or duplex Stirling arrangement. This paper touches briefly on a number of topics about fundamental aspects and recent developments in this field.

Held at the University of Warwick, U.K.
Symposium organised and sponsored by
BHRA Fluid Engineering

1. INTRODUCTION

The Stirling engine is a thermal system operating on a closed thermodynamic regenerative cycle with the working fluid successively and cyclically compressed and expanded at different temperature levels to achieve net conversion of heat to work or vice-versa.

Stirling engines may be used as heat engines or prime movers receiving heat at high temperature from an external source, converting some to mechanical work and rejecting the balance as waste heat at a low temperature. Alternatively, they may be used as refrigerating machines or heat pumps receiving heat at a low temperature and delivering heat at a higher temperature. Such machines require an input of mechanical work to accomplish their operation. In the refrigerating case heat is normally received at a temperature below ambient and rejected at the ambient atmospheric temperature. In the case of heat pump energy is normally abstracted from a source at ambient atmospheric temperature and rejected at a super-ambient value. There is no intrinsic reason for this and Stirling prime movers, refrigerators and heat pumps can operate in any temperature regimen within the capability of the material of construction.

Cryocooling is the principal established application of Stirling engines at present, particularly the miniature versions used for infra-red night vision equipment and missile guidance systems. Stirling cryocoolers are used extensively for local laboratory size gas liquefaction plants, cryopumping and superconducting systems in a variety of civil and military systems (Walker, Ref. 1).

Research and development on Stirling engines for power production and heat pumps has been in progress for many years with substantially elevated level of activity in the 1970's as consequence of the energy crisis. The omniverous fuel capability of the Stirling engine, the ready availability of combustible solid wastes and the substantial price differential between coal and oil are expected to result in extensive adoption of Stirling engines for stationary power, cogeneration and heat pump systems, locomotive and off-highway mining, forestry and agricultural mobile equipment. The first international conference on Stirling engines is sceduled to occur immediately following this meeting (I. Mech. E., Ref. 2).

Substantial research on Stirling heat pumps has been proceeding for over a decade in various parts of the world. Most work has been done with heat-activated systems using natural gas as the fuel, either as a) prime movers providing the work input to the compressors of vapour-compression heat pumps, or b) dual arrangements as duplex systems incorporating two Stirling systems, one acting as the prime mover driving the other acting as the heat pump or refrigerating system.

2. ELEMENTARY CONSIDERATIONS

Figure 1 is a cross-section of an idealized Stirling engine. It consists of a cylinder containing a displacer and a piston. The space above the displacer, called the expansion space, is heated to a high temperature, $T_{(max)}$. The space between the displacer and the piston is called the compression space and is maintained at a low temperature close to ambient, $T_{(min)}$. The volume of the two spaces varies cyclically with the motion of the piston and displacer as shown in Figure 2. The volume variations of the expansion space lead those of the compression space by about a quarter of a cycle, i.e. 90° of crank rotation.

The two spaces are coupled by a duct so the pressure is substantially the same in the two spaces. As the volumes of the spaces vary some of the working fluid passes from one space to the other. The connecting duct contains three heat exchangers termed the heater, the regenerator, the cooler. The heater is a recuperative heat exchanger with slots or fine tubes for the passage of the working fluid. It is the primary unit for the transfer of energy at high temperature to the working fluid from the energy source, a combustor, radioisotope, solar energy or thermal store.

The cooler is similarly equipped with slots or fine tubes for passage of the working fluid and also the cooling medium, usually water at ambient temperature.

In operation, the working fluid is concentrated in the hot expansion space and the pressure is high when the piston is descending (the working or expansion stroke). Conversely the displacer is at the top of its stroke with the working fluid concentrated in the compression space so the pressure is lower when the piston is ascending, (the compression stroke). The cyclic pressure variation is approximately sinusoidal and so phased with the piston motion to develop the kidney shaped pressure-volume or work diagrams shown in Figure 3.

There is a net transfer of energy at high temperatures through the heater to the engine. Part of the energy is converted to mechanical work at the piston for useful application and the remainder is rejected from the engine through the cooler at approximately ambient temperature.

The regenerator is another heat exchanger interposed between the heater and the cooler. It consists of a porous matrix of finely divided material, metal strips, wires, or sintered powder, and may be thought of as a thermodynamic sponge alternately accepting and rejecting heat from the working fluid.

The displacer and piston both reciprocate in the cylinder and appear to have much in common, but there are significant differences to justify the use of separate names. The open duct connecting the expansion and compression spaces assures that the pressure above and below the displacer is the same except for minor pressure losses in the regenerator and other heat exchangers. The displacer is therefore essentially a no-work element simply displacing fluid from the compressor to expansion spaces or vice-versa. The displacer seal is subject only to a small pressure difference and furthermore some leakage across the seal is tolerable although for proper operation of the engine the main flow of working fluid must always be directed through the heat exchangers.

Although the displacer has no significant pressure difference it experiences the full temperature difference between T_{max} at the topside and T_{min} at the bottom. To minimize thermal conduction the displacer is therefore a long thin-wall element of relatively low structural strength and light in weight.

The piston sustains the full pressure difference between the working space pressure on the top face and the crankcase or bounce space pressure on the bottom side. This pressure difference across the piston is very substantial and varies cyclically so the piston must be a strong structural element. At the same time there is no great temperature difference across the piston so that it can be designed without regard to thermal conduction between the upper and lower faces. In simple terms a displacer has 'high ΔT, zero ΔP' whereas a piston has 'high ΔP, zero ΔT'.

The rubbing seals of the piston have to contain the working fluid in the working space. They serve another, more demanding, function; to prevent egress of lubricating oil into the working space where it eventually degrades and blocks the fine interstices of the regenerator. There are two piston seals shown in Figure 1, one for the piston and cylinder, another for the piston and displacer rod.

Stirling refrigerators and heat pumps operate in precisely the same way as the prime mover described above with the essential difference that the expansion space is at a lower temperature than the compression space. As a consequence the work diagram for the compression space has a greater area than that of the expansion space and so a work input to the system is required.

3. MECHANICAL ARRANGEMENTS

The above discussion has introduced the essential elements of a Stirling engine as two spaces at different temperature levels, T_{max} and T_{min}, whose volumes are varied with the hot expansion space leading the cold compression space. The two spaces are coupled through a regenerator and two other heat exchangers, a heater and a cooler.

These elements can be arranged in endless variety but can be generally classified in two groups as:

    a)   single-acting,
    b)   double-acting.

Single-acting machines comprise two reciprocating elements, two pistons or a piston and displacer for each Stirling cycle system. Most versions can be further categorized into one of the three different types shown in Figure 4, where the piston and displacer are contained in the same cylinder, the tandem arrangement, or in separate

cylinders. Another family has a second piston instead of a displacer. We shall call
these 'two-piston machines'.

Small Stirling engines are always of the single-acting variety. Large engines can also be single-acting Stirling engines with multiple units arranged on a
common crankshaft.

The other major group of Stirling arrangements are the double-acting types
represented in Figure 5. Here multiple cylinders are connected successively so the
expansion space of one cylinder, above the reciprocating element, is coupled to the
compression space below the reciprocating element, of the adjacent cylinder. Each
Stirling system requires only one reciprocating element, a piston-displacer, so that
major economies are possible compared with ensembles of multiple single-acting
machines. Anything from 2 to 7 cylinders may be used. One favourite combination
called the Siemens-Stirling 'square four' coupling is shown in Figure 6. This was
mistakenly designated as the 'Rinia arrangement' by Walker in his 1973 book and unfortunately the term has entered into general usage. Sir William Siemens invented
this arrangement in 1863.

4. FREE-PISTON ENGINES

In many cases the displacers and pistons are coupled to some kind of drive
mechanism. Such arrangements are generically called 'kinematic drives' and include
the simple crank-slider (connecting rod/crank), rhombic drive, Scotch yoke and various
other ingenious mechanisms.

Other forms of Stirling engines have no kinematic drive but simply operate
by virtue of the balance of fluidic forces acting upon the reciprocating elements.
These are called Beale free-piston Stirling engines. A cross-section of an early
(1960's) Beale free-piston Stirling engine manufactured by Sunpower, Athens, Ohio, is
shown in Figure 7.

5. RINGBOM-STIRLING ENGINES

Yet other variations have the piston coupled to a kinematic drive and have
a free displacer as shown in Figure 8. These are called Ringbom-Stirling engines
after the originator Ossian Ringbom, a Russian living in Finland at the time of his
original U.S. patent in 1905.

6. THEORETICAL ASPECTS

Detailed discussion of theoretical aspects of Stirling engines is beyond
the scope of this paper but will be found in Walker (Ref. 1, Ref. 3) with further
references to advanced computer simulation methods. Comprehensive discussion of free
piston and Ringbom-Stirlings will be found in Walker (Ref. 4).

Design guideleins for elementary estimates of machine performance (given
in Ref. 4) include use of the Beale Number concept where:

$$\text{Beale Number} = \frac{P}{pVf} = 0.015 \text{ for typical power applications}$$

where
P = power output (watts)
p = mean pressure (bar)
V = piston displacement (cm^2)
f = engine frequency (hz)

A similar equation for the power input to Stirling heat pumps undoubtedly
exists but so far insufficient data exists to establish the range and temperature
dependence of the 'constant' (0.015 for typical power applications).

Estimates of engine efficiency or Coefficient of Performance may be made
by assuming values equivalent to half the Carnot value.

By way of illustrations, consider the Stirling heat pump system shown in
Figure 9 wherein heat is absorbed (in the expansion space) at T_e from a ground, air or
water source and discarded at the higher temperature T_c for use in space heating.

The Carnot coefficient of performance

$$COP_c = \frac{\text{Heat Rejected}}{\text{Work Done}} = \frac{T_c}{T_e - T_c}$$

Let the temperature of the source T_s be 10°C and allow a temperature potential of say 8°C in the heat exchanger designated the absorber in Figure 9. The expansion space temperature T_e is then 10-8 = 2°C. Now assume the temperature of hot water delivered by the heater is required to be 75°C and assume a temperature potential of 15°C in the heater. The compression space T_c is then 75+15 = 90°C.

The Carnot coefficient of performance is then

$$COP_c = \frac{90+273}{(90-2)}$$

$$= 4.125$$

Well designed systems are able to achieve about half the Carnot value so we may look to an actual coefficient of performance of 2. This implies an input of work W_h to drive the system having the same magnitude as the energy lifted from the source Q_s, to deliver to the heating system a combined total $Q_s + W_h = Q_h = 2Q_s$.

The work W_h to drive the Stirling heat pump may come from an electric motor or some other heat engine. Let us assume use of another Stirling engine acting as prime mover to drive the Stirling heat pump in the duplex arrangement shown in Figure 10. In this case we shall have perhaps the combustion system shown in Figure 11 wherein natural gas and air are burned to produce hot products of combustion to provide a high temperature heat source to the Stirling prime mover heater. A fraction (say 66%) of the heat released by combustion (Q_r) of the natural gas will be absorbed into the engine (Q_p) and the balance ($Q_r - Q_p$) can be incorporated into the heat supplied for space heating. Of the heat (Q_p) supplied to the engine a fraction will be converted to work, W_p, and the difference ($Q_p - W_p$) = Q_c will be rejected from the engine as waste heat in the engine cooler but this too can be added to the heat available for heating purposes.

Let us assume the prime mover heater is a modest tubular or finned exchanger constructed of relatively low cost stainless steels. This limits the maximum temperature T_p of the working fluid to about 650°C despite the fact the combustion products may be at a much higher temperature, say 2000°C.

We will further assume the temperature of heat rejection from the prime mover to be the same T_c = 90°C as that rejected from the heat pump. Knowing these temperatures, the Carnot thermal efficiency may be calculated as:

$$\eta_c = \frac{T_r - T_c}{T_p} = \frac{650-90}{(650+273)}$$

$$= 0.606$$

Well designed Stirling engines have overall brake thermal efficiencies about half the Carnot value or somewhat higher. Taking half the Carnot value and adjusting it slightly for convenience gives an actual thermal efficiency, η_{act}, of 33%. In other words, one third of the heat supplied *to the engine*, Q_p, is converted to work, W_p, and two thirds, Q_c, is rejected *from the engine* to the heating system. But the heat supplied to the engine, Q_p, is only 66% of the heat released by combustion, Q_r, with the balance, $Q_r - Q_p = 1/2\, Q_p$, supplied to the space heating system. Now the work output, W_p, of the Stirling prime mover must be exactly equal to the work input, W_h, to the heat pump and we can use the equality to relate the various heat transfer quantities in terms of the work transfer W from prime mover to heat pump. Summarizing, therefore

a)	Heat abstracted from source, Q_s	= 1 W
b)	Work input to heat pump, W_h	= 1 W
c)	Heat delivered to heating system from heat pump, Q_c	= 2 W
d)	Heat delivered to heating system from combustion gases	= 1/2 (3 W)
e)	Heat delivered to heating system from prime mover	= 2 W

Total heat delivered to heating system   5.5 W

Total heat release in combustion system (3/2 x 3 W)   = 4.5 W

$\therefore$ Primary energy ratio = 5.5/4.5 = 1.22

7. ALTERNATIVE ARRANGEMENTS

A Stirling engine system consists of two spaces at different temperatures and whose volumes are varied cyclically. The two spaces are coupled through heat exchangers for heating, cooling or regeneratively conserving heat. This seemingly simple system can be reduced to practice in literally hundreds of different manifestations. The principal ones are reviewed in Ref. 1, 3 and 4. Systems can be single-acting or double-acting, reciprocating or rotary, be composed of pistons or pistons and displacers, use kinematic mechanisms to couple reciprocating elements to a rotating shaft or operate as free-piston engines with entirely fluidic coupling of the piston and displacer with the piston driving a pump, compressor, or electric power generator.

The Stirling engine system may be externally heated to produce power, the power to drive a Rankine cycle vapour-compression heat pump or refrigerator. Alternatively, it may be driven and itself operate as the heat pump or refrigerating machine.

Free-piston Stirling engines have few moving parts, have no piston side forces induced by the kinematic mechanism and therefore have great potential for the long-lived, low maintenance characteristics desired in heat pump systems. Furthermore, they can be made self-starting for when the normal operating temperature regimes are established they are in fact tuned fluidic oscillators and any slight natural environmental disturbance is sufficient to upset their delicate stationary equilibrium and to start them oscillating.

8. RECENT DEVELOPMENTS

At the present time, there are no commercially available Stirling engines for power generation or for heat pumps. They have long (1953) been available as cryocoolers for small capacity, very low temperature refrigeration and indeed, Stirling engines and their thermocompressor cousin, the Vuilleumier engine, completely dominate the field (Ref. 1).

The high costs of diesel fuel compared with coal, the escalating price of electric power and the relative abundance of natural gas and solid wastes as alternate fuels are all forcing the pace of research and development of Stirling engine systems to higher levels so that commercial applications are foreseen well within the decade. Research and development on Stirling engines is in progress in the U.S.A., Sweden, England, Germany, Denmark, Holland, Japan, South Africa, Spain, Italy, India, China and the U.S.S.R.

William Martini (Ref. 5) publishes a directory of Stirling engine activities worldwide and also (Ref. 6) a very extensive bibliography of Stirling engine references. Martini is the editor of the Stirling Engine Newsletter published quarterly (Ref. 7) with up to a dozen pages of news about recent developments.

The proceedings of the Intersociety Energy Conversion Engineering Conference (IECEC) held every August in the United States have become the major forum for publication of current Stirling engine open literature publications. 1982 saw the 16th conference held in Atlanta, Georgia, with approximately 20 papers on different Stirling engine topics. This represents a quarter of the present annual crop of about 80 papers a year published in the field.

In the specialized area of Stirling engine heat pumps, most work appears to have been done with free-piston Stirling engine and particularly by Sunpower Inc. of Athens, Ohio, the company headed by William Beale. Beale invented the free-piston Stirling engine in the early 1960's while a professor at the University of Ohio and later founded Sunpower to carry the free-piston Stirling engine forward to a commercial stage. The American Gas Association funded the development of a 3 kW gas fired free-piston Stirling engine driving the inertia compressor of a Rankine vapour compression system. At about the fourth generation prototype of the very advanced system, A.G.A. apparently decided the system was ready for commercial

development, and with a logic that is hard to understand, switched the funding to a
new team at the General Electric Space Power Division. Little further progress with
the engine hardware appears to have been made judging from the various reports published by G.E. in the IECEC and elsewhere, and there appears little prospect of the
impending commercial introduction of this machine.

Later, Sunpower worked as the engine development design and test laboratory
for Mechanical Technology Inc. of Latham, N.Y. on a variety of free-piston Stirling
engines including heat pumps. This association terminated several years ago and
little further hardware development appears to have occurred at M.T.I. Meanwhile,
Sunpower has continued free-piston Stirling engine hardware development on a variety
of projects for other sponsors including the prototypes of commercial heat pump
systems. Sunpower staff contribute regularly to the IECEC proceedings.

Another company, E.R.G. Inc. of Oakland, California, is also believed to
be active in the free-piston Stirling engine. They are better known for the in-
genuity of their paper engines and sweeping claims than for demonstrations of
operating hardware. Some time ago, Dr. Glen Benson of E.R.G. Inc., gave an impres-
sive and fascinating *tour de force* (Ref. 8) of free-piston heat pumps that is highly
recommended.

The Philips Company of Eindhoven pursued Stirling engine research for
over 40 years (Ref. 3) and were master of the field until their recent withdrawal.
Few of their many published contributions were devoted to heat pumps but two papers
on the specific topic were given by Hermans and Asselman (Ref. 9) and by van
Eekelen (Ref. 10).

9. RESEARCH TOPICS

The entire field of Stirling engines is bristling with unknowns and areas
of uncertainty to provide a rich and fruitful area for research. Important new
discoveries and concepts for improved design, mechanical arrangements, bearings
seals, heat exchangers, materials and working fluids are regularly being announced
by workers in the Stirling engine field.

Topics of current research and development interest to the author include
the use of air as a working fluid and water as the lubricant with the prospect of
water injection to the hot space to provide a power boost during maximum power demand.
With a two-phase two-component working fluid very substantial gains in the maximum
to minimum pressure variation can be had with consequent improvement in specific
power output. The use of the liquid phase component as the bearing/seal lubricant
avoids the complexities of liquid/liquid separation of oil and water mixtures.
Furthermore, oil and compressed air are a dangerous combination it is best to avoid.
Air is attractive as the working fluid for it resolves the problem of containing
the light gases (hydrogen and helium) used in high specific output engines. Air can
be readily replenished and is more readily contained than hydrogen or helium but
air engines are larger, heavier and ran more slowly than systems with light gas
working fluids. Air in combination with one of the refrigerants used in Rankine
cycle systems appears worth investigating for heat pump/refrigerator applications of
the 2-phase, 2-component working fluid with supplementary lubricating functions.
Lubrication with water is not feasible in refrigeration systems because it freezes.
Preliminary work along these lines was reported by Walker (Ref. 11) but no extensive
studies in the literature are known.

10. <u>REFERENCES</u>

1. <u>Walker, G.</u>: "Cryocoolers". International Series on Cryogenics, Plenum Press, New York, (in press - scheduled 1982) (750 pp.).

2. ___________: "Stirling engines: progress towards reality". Inst. Mech. Eng. Conference, Reading, March 1982.

3. <u>Walker, G.</u>: "Stirling engines". Oxford Uniersity Press, 1980, 550 pp.

4. <u>Walker, G.</u>: "Free-piston Stirling engines". (in preparation) (information from G. Walker, Department of Mechanical Engineering, University of Calgary, Alberta, Canada.

5. <u>Martini, W.</u>: "Directory of Stirling engine activity". Martini Engineering, 2303 Harris, Richland, Washington, 99352, (88 entries, 44 pages).

6. <u>Martini, W.</u>: "Index to the Stirling engine literature". Martini Engineering, 2303 Harris, Richland, Washington, 99352, (1300 references, 132 pages).

7. <u>Martini, W.</u>: "Stirling engine newsletter". Martini Engineering, 2303 Harris, Richland, Washington,, 99352.

8. <u>Benson, G.M.</u>: "Free-piston heat pumps". Paper No. 779068, Proc. 12th IECEC, (Soc. Auto. Eng., Warren, Pa.), pp. 416-425.

9. <u>Hermans, M.L. and Asselman, G.A.A.</u>: "A Stirling engine heat pump system". Paper No. 789274, Proc. 13th IECEC, 1978, pp. 1830-1833.

10. <u>van Eekelen</u>: "State of a Stirling powered heat activated heat pump development". Paper No. 799254, Proc. 14th IECEC, pp. 1186-1190.

11. <u>Walker, G.</u>: "Stirling cycle cooling engine with two-phase, two-component working fluid". Cryogenics, Vol. 14, No. 8, August 1974, pp. 459-462.

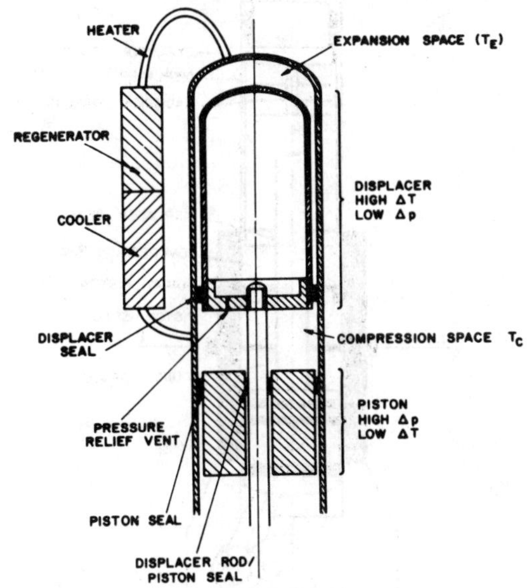

Figure 1   Elements of a Stirling Engine System

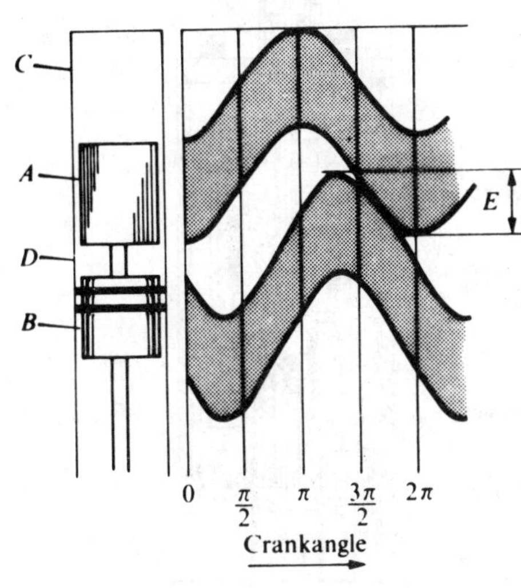

Figure 2   Volume Variation in a Stirling Engine

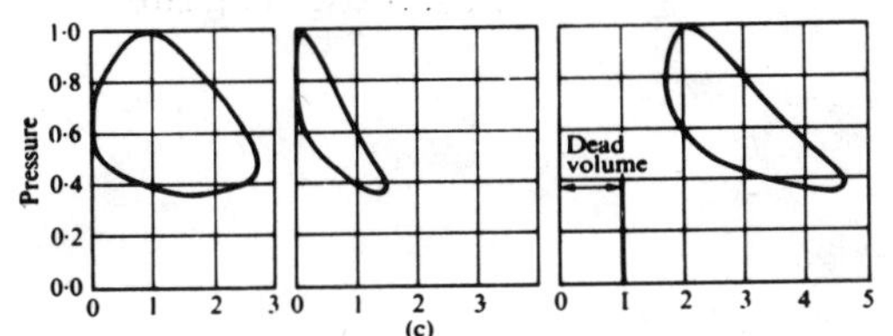

Figure 3   Work Diagrams for a Stirling Engine

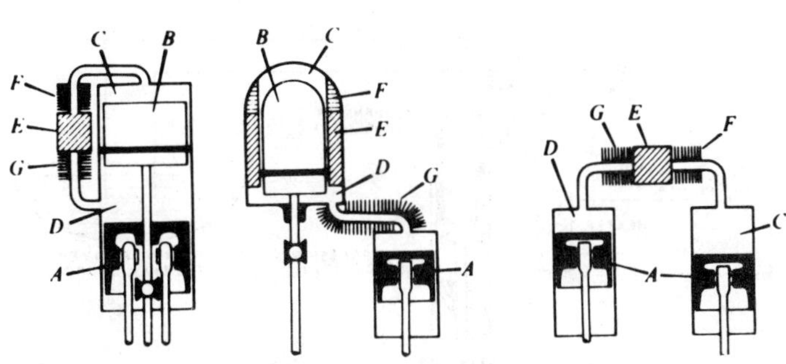

Figure 4   Mechanical Arrangements of a Single Acting Stirling Engine

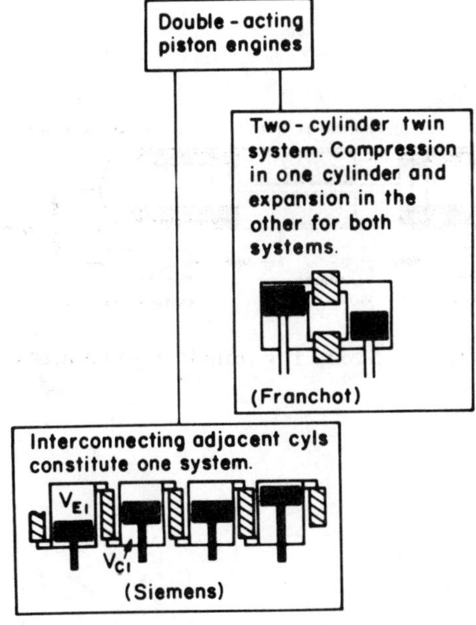

Figure 5   Double-Acting Stirling Engine

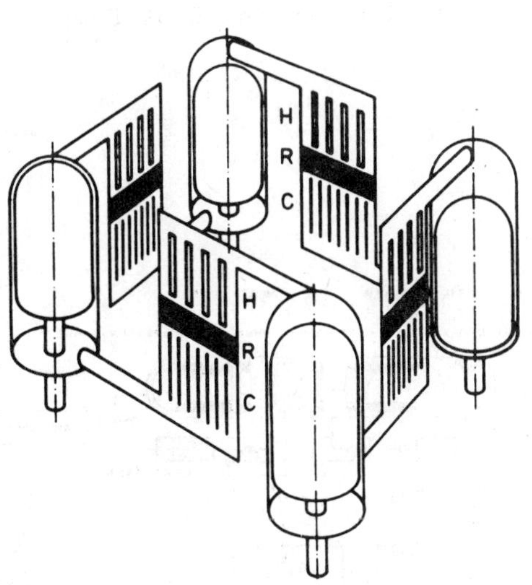

Figure 6   The Square-Four Siemens Stirling Engine Arrangement

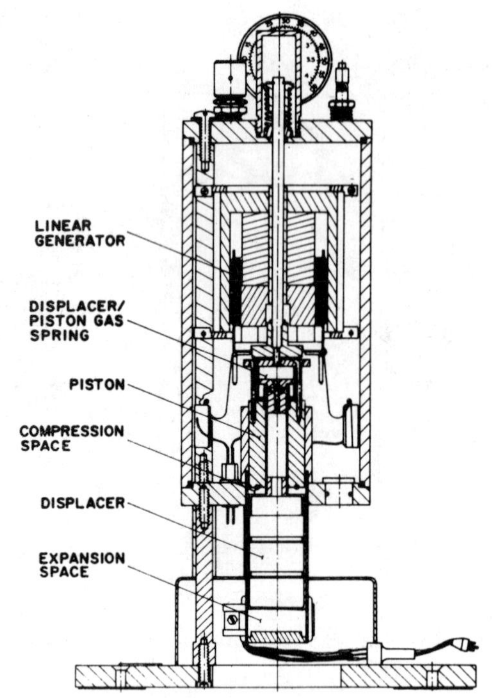

Figure 7   Cross-Section of Early Sunpower Free-Piston Stirling Engine, (courtesy Sunpower Inc., Ohio)

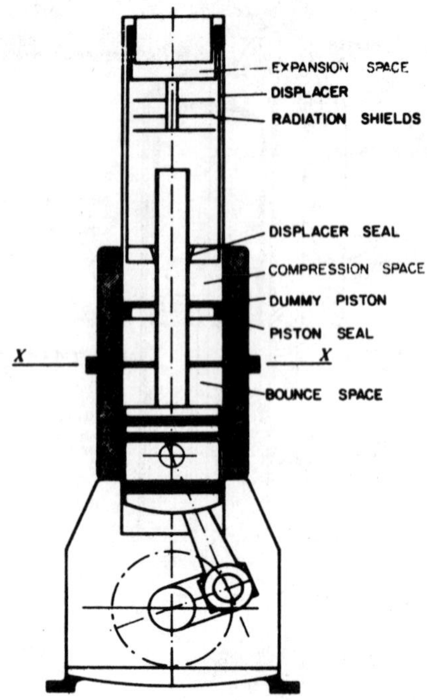

Figure 8   Hybrid-Ringbom Stirling Engine

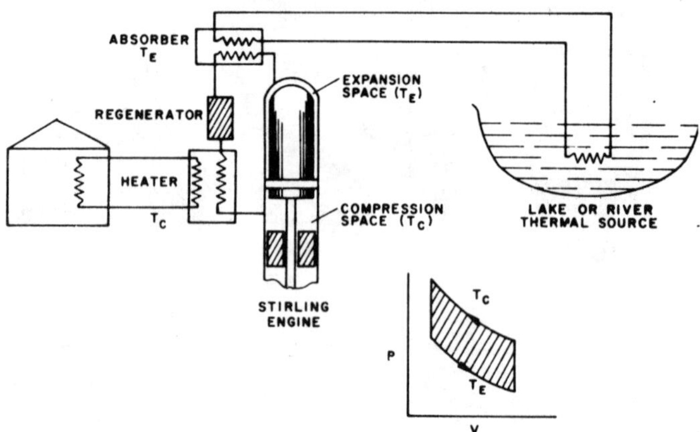

Figure 9   Stirling Engine Heat Pump

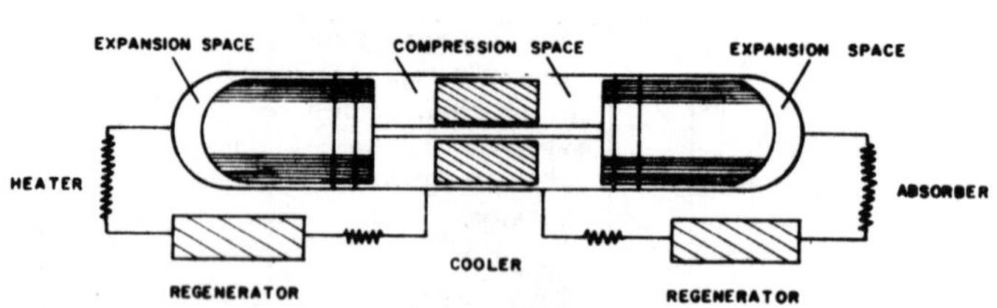

Figure 10   Duplex Stirling Engine Arrangement

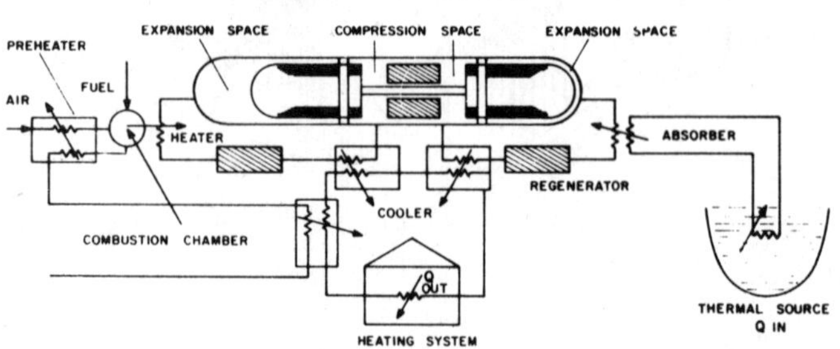

Figure 11   Duplex Stirling Heat Pump System

THE THERMAL STABILITY OF R11 AND R12B1 AS HIGH TEMPERATURE HEAT PUMP WORKING FLUIDS

F. A. Watson, S. P. Abbas, P. Srinivasan and S. Devotta

University of Salford, U.K.

Summary

The results of experimental studies of the stability of R11 and R12B1 in the presence of metals and of lubricants under static test conditions are reported. It is shown that, at temperatures up to 220°C, R11 is more stable than R12B1 in all combinations of lubricant and metals studied. It is also shown that the metal, particularly copper, is less subject to attack by R11 than by R12B1. However, the lubricant effectiveness of naphthenic oils, as measured by the viscosity, is less impaired by R12B1 than by R11. In the absence of lubricant R11 is acceptably stable even up to 220°C but is less stable in the presence of alkylbenzene lubricants and still less stable in the presence of naphthenic lubricants.

A new equipment, designed to decompose the working fluid under cyclic conditions similar to those experienced in vapour compression heat pumps, is described. Some preliminary results from the equipment are reported with R11 as the working fluid.

Held at the University of Warwick, U.K.
Symposium organised and sponsored by
BHRA Fluid Engineering
© BHRA Fluid Engineering, Cranfield, Bedford MK43 0AJ, England.

1. INTRODUCTION

Increasing interest in the potential use of heat pump systems operating at temperatures in excess of 100°C has made stability studies of the materials used in such systems of immediate importance. In vapour compression heat pumps the working fluid will be evaporated in a heat exchanger, will often be slightly superheated in an auxilliary heat exchanger by hot liquid condensate, will be compressed, will be desuperheated and condensed at a higher temperature and the hot condensate will then return to the evaporator via an expansion valve or a capillary restriction. In addition to these hardware items there will be valves, thermocouples, pressure gauges and piping so that the working fluid will be in contact with a variety of metal and plastic combinations over a wide range of temperatures. It is to be expected that if the combination is adequately stable at the highest temperature reached the system may be suitable for commercial exploitation.

An additional complication arises when, as is usually the case, the available compressor requires lubrication. The importance of determining the stability and the lubricating effectiveness of working fluid/lubricant combinations has been discussed by Drakesmith (Ref.1). The equipment, analytical techniques and experimental procedures used by the present authors to determine the stability of working fluid/materials of construction/lubricant combinations under static test conditions have recently been presented elsewhere (Ref.2). It was there shown that these methods produced results which were compatible with earlier published work. The increased size of test sample made possible by this equipment has enabled the effect of time and temperature on the viscosity of the lubricant in such samples to be determined. Examples of such results are presented herein.

The above studies are, of necessity, time consuming. Many samples can be lost by failure of the pressure seals of the reaction vessels which are constructed like bomb calorimeters with gas sampling facilities added. A rapid method of rejecting unsuitable combinations is desirable. This is the purpose of the continuous unit here reported for the first time. It is hoped that the equipment will permit the time taken to achieve a given breakdown, when a heated stream of lubricant/working fluid impinges on a metal target heated to, say, 450°C, to be related to the time taken to obtain the same breakdown in the static test equipment at, say, 180°C. Some preliminary results from this equipment are reported.

Because of its thermodynamic advantage as a heat pump working fluid at temperatures up to 125°C, stability studies on R11 combinations are also reported.

2. FLOW METHOD FOR THE STUDY OF THE STABILITY OF HEAT PUMP WORKING FLUIDS

In order to overcome the inherent limitations of the static method and to increase the flexibility of the testing method an equipment has been designed which simulates as closely as possible the cyclic conditions existing in a vapour compression heat pump. The working fluid is alternately heated and cooled between two levels of temperature while impinging at one part of the flow cycle onto a target. This target is replaceable and can be made of a variety of materials. The target represents the hottest surface encountered in the heat pump cycle and can be varied in temperature up to about 450°C. The effects of temperature range, hot spot temperature, pressure, lubricant type and quantity, materials of construction, etc., on the stability of potential heat pump working fluids can be studied with this equipment.

The equipment is shown schematically in Fig.1. It can be seen that the apparatus mainly consists of two independent loops, one for the circulating working fluid and one for the lubricating oil. The streams can be passed independently or as a mixture over a heated target inside a stainless steel reaction vessel.

The circulation of the working fluid is achieved by an open compressor. The compressed working fluid passes through electrically heated tubes with bypass facility. The temperature of the working fluid can be varied in the range 100°C to 200°C within $\pm 2^{\circ}$C. The rate of flow of the working fluid can be measured by an orifice meter located at the vapour entrance into the reaction vessel.

The oil is circulated by means of a metering pump. At present no provision has been made to heat the oil before it enters the reaction vessel, but if needed in future, it is possible to provide electrical tape heating.

The reaction vessel consists of two parts. The top part of the reaction vessel houses the electrically heated replaceable targets. The temperature of the target can be varied to within $\pm$ 5°C up to 450°C. The bottom part of the reaction vessel acts as a collector for the mixture before it enters the cooler.

The cooler is made of copper tube with the hot vapours flowing inside and water on the shell side. Another similar cooler is connected in a line, provided for startup purposes, which bypasses the reaction vessel.

A liquid flow rotameter, connected between the cooler and the separator, was used for the calibration of the orifice meter. It may, however, be bypassed when running, particularly when inert gas is present in the system.

A separator is provided between the coolers and the compressor where the lubricating oil is separated from working fluid vapours before re-entering the compressor.

Sampling points have been provided before and after the reaction vessel to enable gaseous and liquid samples to be removed for analysis.

The following are the important features of the equipment.

   i) Continuous and independent circulation of working fluid and lubricating
       oil is possible.

  ii) It is possible to study the stability of different mixtures of working
       fluid and lubricating oil.

 iii) The temperature of the working fluid can be varied in the range 100°C
       to 200°C.

  iv) Heated targets of various metallic and non-metallic materials can be used
       to study their effect on breakdown rates of the working fluid and lubricant
       mixtures.

   v) The system can be operated with the total pressure and temperature
       independently varied by the inclusion of an inert gas. This should enable
       the effective reaction rate to be slowed at higher temperatures thus allow-
       ing greater control of these reactions.

  vi) The target temperature can be maintained at any desired level up to 450°C.

 vii) Continuous and intermittent sampling of both the gas and liquid phases
       is possible.

It is proposed to study the type and rate of degradation of various potential high temperature heat pump working fluids at different temperatures. This can be done with various concentrations of different lubricating oils in the presence of different materials. In the presence of an inert gas the stability of the lubricant alone can be studied. Gas, liquid and material samples will be analysed using suitable techniques which have been described elsewhere (Ref.2).

An attempt will be made to relate the extent of degradation of a combination of working fluid and lubricating oil to the appearance of any deposit formed on the target. If successful, this would provide an accelerated test to determine the suitability of that working fluid combination for use in high temperature heat pumps.

3. EXPERIMENTAL

3.1 Static decomposition of R11

The experimental equipment and techniques were expected to be similar to those previously described for the investigation of the decomposition of R12B1 under static conditions (Ref.2). From literature references, and after carrying out trial analyses with different gas chromatographic columns, a 6.3mm outside diameter by 3m long stainless steel column packed with a mixture of 6 per cent Bentone 34 and 6 per cent Silicone MS 550 on 80 to 100 mesh Chromosorb W (acid washed) was selected as being suitable for the analysis of R11 and its probable breakdown products.

Calibration of the detector with R11 posed some unexpected difficulties. Different volumes of vapour samples from an R11 sample cylinder were drawn into 'Pressure-Lok' syringes and injected into the chromatograph in the usual way. The results did not produce a reproducible calibration curve. This was found to be due to partial condensation of R11 inside the syringe since the ambient room temperature was lower than

23.7°C, the normal boiling point of R11. An unsuccessful attempt was made to avoid this problem by raising the temperature of the testroom.

The R11 sample cylinder was heated to a constant temperature in the range 40-60°C in a thermostatic water bath to give a vapour pressure higher than atmospheric and the sample syringe was heated by means of a hot air blower. The samples again failed to give a reproducible calibration curve. The reasons were found to be (i) the existence of a temperature differential between the syringe and the sample cylinder, and (ii) the presence of comparatively cooler regions somewhere in the syringe or needle due to uneven heating.

Calibration by injecting liquid samples was not attempted because (i) the range of electrical output signals that would be obtained from 1 to 25µl of liquid R11 would be much greater than those expected from gaseous samples of actual degradation product mixtures, and (ii) the volume of vapourised samples would be too large for the flame ionisation detector.

If a suitable solvent can be found which has a much longer retention time on the column of the chromatograph than a given working fluid or its decomposition products, it may be possible to calibrate for that fluid by using dilute solutions of known composition. Ethyl acetate was found to be such a solvent for R11 and produced reproducible calibration curves provided that the sample syringe was chilled. The method has the advantage that ethyl acetate may be used as an internal standard, which simplifies quantitative analysis of R11 in the gaseous decomposition products.

The technique is to take a gaseous sample from the reaction vessel, heated if necessary to a temperature above the critical temperature of the products to be detected, and pass this sample into a chilled sampling loop where it is condensed. A solution is made by mixing a known volume of this condensate with a predetermined volume of ethyl acetate. Aliquots of this solution are then injected into the gas chromatograph using a chilled microsyringe.

It was found that R21, one of the main decomposition products of R11, overlapped the chromatogram of R11. This difficulty was overcome by injecting the samples into the column operating at 35°C and then raising the temperature of the column to 100°C after the R21 had been eluted from the column.

4. RESULTS

The objective of the present work was to compare the stability of R11 with that of R12B1 when used as high temperature heat pump working fluids. It was previously shown (Ref.2) that at 150°C, in the absence of mineral oil, the decomposition of R12B1 was very slow even in the presence of copper, steel or both. It was also shown that at 150°C, in the absence of R12B1, the decomposition of mineral oil was negligable in the absence of copper and steel and only slight in the presence of either or both. There resulted, however, a solution containing a few parts per million of copper in the oil.

It is the aim of other work in progress in the Department of Chemical Engineering in the University of Salford to develop high temperature heat pumps in which certain parts may reach temperatures in excess of 200°C. As stated elsewhere, the static tests can be extremely time consuming so, apart from confirmatory tests, the reaction rates were accelerated by working at temperatures in this region, namely 200°C and 220°C.

Fig. 2 indicates the percentage decomposition of R11 at these temperatures for varying times in the presence of an equal weight of either a naphthenic mineral oil (38 per cent aromatics) or a special synthetic oil (mixed alkyl benzenes). It can be seen that, in the presence of the synthetic oil less than 20 per cent of the R11 has decomposed after 70 hours at 200°C with only a slight increase in decomposition to 20 per cent at 220°C. By contrast, in the presence of the mineral oil, 20 per cent decomposition was reached in only 18 hours at 200°C, while decomposition reached 40 per cent in that time at 220°C. Decomposition had reached 62 per cent in 30 hours at 220°C by which time the pressure in the reaction vessel had reached the safety limit so that decomposition at longer periods could not be determined. However, after 30 hours at 220°C, the mineral oil was found to be solid when the reaction vessel was opened at room temperature. The oil in this condition would not, therefore, be likely to be effective as a lubricant. After 70 hours at 220°C the synthetic oil was still fluid at room temperature but had darkened in colour. The presence of 0.1 per cent by weight of water added to the reaction mixture had a marked effect as can be seen from Table 1.

At 220°C the mineral oil was solid after 20 hours and the synthetic oil after 50 hours.
The presence of this small amount of water also greatly increased the fraction of R21
in the gas phase at these times.

Fig. 3 presents decomposition data at 200°C and 220°C for R12B1. It can be seen
that, in the presence of mineral oil, 45 per cent had decomposed in 18 hours at 200°C
and 55 per cent at 220°C in the same time. This is much greater than the decomposition
of R11 in the same time. Comparison with Fig. 2 shows that the same is true for the
synthetic oil. Thus, with either lubricating oil, R11 is more stable than R12B1 at
200°C and at 220°C. It is, therefore, likely to be more stable over the desired temperature range. This is probably because the bond energy is smaller between carbon and
bromine than between carbon and chlorine. It is noticeable that neither oil went completely solid in the presence of R12B1 even after 70 hours at 220°C.

For comparison, at 220°C for 70 hours in the absence of either oil, the breakdown
of R11 is only about 2 per cent while the breakdown of R12B1 is about 25 per cent.

5. ANALYSIS OF METAL PLATES

As was found when using R12B1 as the test material, the copper and steel plates in
contact with either mineral or synthetic lubricant were found to be coated after static
tests when using R11. The coatings from those parts of the plates immersed in the oil
and from those parts exposed only to the vapour were examined separately using the
scanning electron microscope. An analysis of the elements present in the coatings was
obtained using EDAX techniques. Typical micrographs and EDAX peaks are presented in
Figs. 4 to 11. Such results indicate that the crystalline material found in the coatings is similar whether in contact with the liquid or vapour except that the crystals
are better shaped when grown beneath the liquid surface. This is similar to the find-
ings with R12B1 except that the crystals here consist of copper chloride instead of
copper bromide.

Unlike R12B1 there was no evidence of spherical globules of halogenated hydrocarbons being deposited from the oil phase as the specimens cooled down. Also, unlike
R12B1, attack on the copper was not very pronounced when using R11 and did not result
in a serious weakening of the copper.

6. THE EFFECT OF TIME AND TEMPERATURE ON THE VISCOSITY OF OILS IN THE PRESENCE OF R11 AND R12B1

Although it is not the only criterion of the power to lubricate a bearing, the
viscosity of an oil is an important technical parameter. Most machines are designed to
operate at a particular running temperature with an oil of specified viscosity. The
desired viscosity is often specified at a temperature of 40°C although this may be remote from the operating temperature of the surfaces to be lubricated. The logarithm of
the absolute viscosity of a simple liquid component varies approximately linearly with
the reciprocal of the absolute temperature. Viscostatic oils are blended so that this
variation shall be small over a range of temperature. In other cases an estimate of
the temperature variation of viscosity may be obtained by reporting the values at two
or more temperatures and assuming the above relationship. The straight line obtained
may then be translated into the more usual exponential curve showing the variation of
viscosity with temperature, from which the assymptotic value of viscosity at high temperatures may be estimated.

In Figs. 12 to 15 viscosity data for the mineral oil and the synthetic oil are presented at 40°C and 60°C for specimens obtained after exposure to R11 or to R12B1 for
different times at 200°C or at 220°C. These results are combined in Fig. 16 which
shows the percentage change in viscosity with time under different test conditions.

7. RESULTS FROM THE CONTINUOUSLY RECYCLING TEST EQUIPMENT

As already stated, this equipment is designed to examine the effect of cyclic heating and cooling of a heat pump working fluid while being subjected to impingement on a
hot spot in the equipment. The unit has only recently been commissioned but the initial results using R11 as the working fluid may be of interest. Fig. 17 indicates the
trend of percentage decomposition with time for 4 kg of R11 cycled between 180°C and
140°C while impinging in each cycle upon a target of copper maintained at 220°C or
400°C. The curves bear a resemblance to those obtained in the static equipment.

It is intended to vary the temperature parameters of the equipment in an attempt to
obtain a match for the shape of the curves obtained under static conditions displaced
in time. This should allow the curves to be estimated in a shorter test period.

The equipment will shortly be used to establish the effect of circulating the
lubricant oil as well as the working fluid alone.

8. SUGGESTED MECHANISM FOR THE BREAKDOWN OF R11

The gas phase from the samples was subjected to analysis by gas chromatographic
and mass spectrographic techniques as described in detail elsewhere (Ref.2). On the
basis of the components detected by such analyses and the EDAX analyses, the following
mechanism is suggested for the reaction of a hydrocarbon oil in the presence of R11.

$$CCl_3F \quad + \quad R.CH_2.CH_2.R_H \longrightarrow CHCl_2F \quad +$$

$$R11 \qquad\qquad oil \qquad\qquad R21$$

$$R.CHCl.CH_2.R_H \xrightarrow{-HCl} R.CH:CH.R_H \longrightarrow brown\ oil \longrightarrow sludge$$

$$chlorinated\ oil \qquad unsaturated\ oil$$

The HCl can react with copper or steel to give the chloride crystals previously
described. R and R_H represent hydrocarbon radicals.

The mass spectrographic tests detected the presence of $CHCl_2F$ (R21), CCl_2F_2 (R12),
$C_3H_3F_3$ (R243). This together with the presence of chlorinated hydrocarbons in the
oil phase and the presence of copper and steel chlorides on the plates suggested the
breakdown mechanism for R11 proposed below.

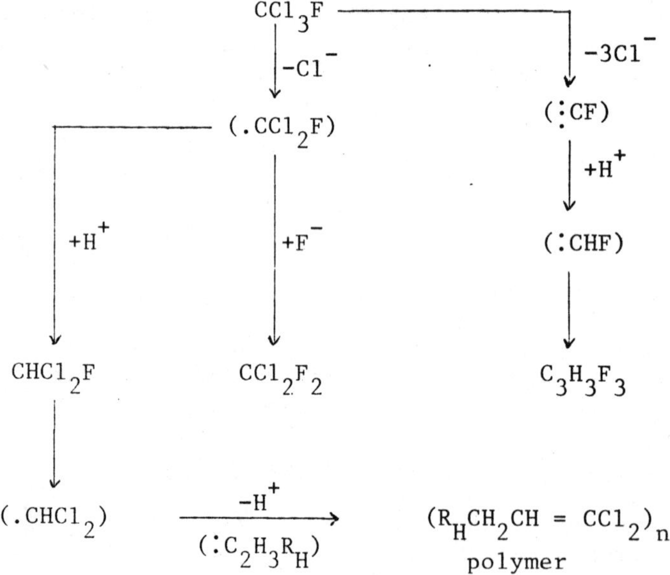

The hydrogen and chloride ions are assumed to be formed by an initial attack of the
R11 on the oil, the Cl^- being removed from the system as metal chlorides.

9. CONCLUSIONS

Thermodynamically, R11 appears to be a better high temperature working fluid than
R12B1 and is also lower in price. It appears, from the present work, that R11 is more
stable than R12B1 at temperatures up to $220^\circ C$ when in contact with copper and steel
whether or not a lubricating oil is used. It also appears that the presence of a lubricating oil greatly reduces the stability of R11 and R12B1 so that a given amount of
breakdown occurs in a shorter time. The increased breakdown rate is greater with a
naphthenic oil than with a synthetic alkylbenzene mixture. The latter is to be preferred until a dry compressor can be developed by which the maximum advantage will be
gained from the use of R11 as a high temperature working fluid.

10. <u>ACKNOWLEDGEMENT</u>

The authors wish to thank the Science and Engineering Research Council for its financial support, and the Departments of Pure and Applied Chemistry and of Pure and Applied Physics in the University of Salford, and the Electricity Council Research Centre, Capenhurst, for their invaluable help and assistance with the analysis of materials. In particular, special thanks are due to Dr. J. Clark, Mr. P.J. Diggory and Mr. S. Ghosh. Finally, grateful thanks are due to Professor F.A. Holland, Chairman of the Department of Chemical and Gas Engineering in the University of Salford, without whose initiative the work would not have been started and without whose active help and encouragement the work could not have continued.

11. <u>REFERENCES</u>

1. <u>Drakesmith, F.G.</u>: "Fluids for high temperature heat pumps", In: "Heat Pumps, energy savers for the process industries", North Western Branch Symposium Papers 1981 No. 3 (Salford, U.K.: April 7-8, 1981), Institution of Chemical Engineers, Paper 6, 11pp.

2. <u>Abbas, S.P., Srinivasan, P., Devotta, S. and Watson, F.A.</u>: "The stability of heat pump working fluids", In: "Heat pumps, energy savers for the process industries", North Western Branch Symposium Papers 1981 No. 3 (Salford, U.K.: April 7-8, 1981), Institution of Chemical Engineers, Paper 7, 13pp.

<u>TABLE 1.</u> EFFECT OF 0.1 PER CENT WATER ON THE DECOMPOSITION OF R11 AT 220°C

	R11/mineral oil/Cu/Steel/20h		R11/Synthetic oil/Cu/Steel/50h	
	per cent decomposition	area ratio* R21/R11	per cent decomposition	area ratio* R21/11
without water	41.3	0.05	15.2	0.07
with water	62.0	2.46	46.7	0.35

*area ratio calculated from the areas of R21 and R11 peaks in the chromatograms.

<u>NOTE</u>: The viscosities were determined on roughly two drops of clear samples of the oils, by means of the Ferranti-Shirley cone and plate viscometer. In order to ensure the absence of dissolved working fluid the bulk samples of oil were heated to about 100°C for ½ to 1 hour and any turbidity allowed to settle.

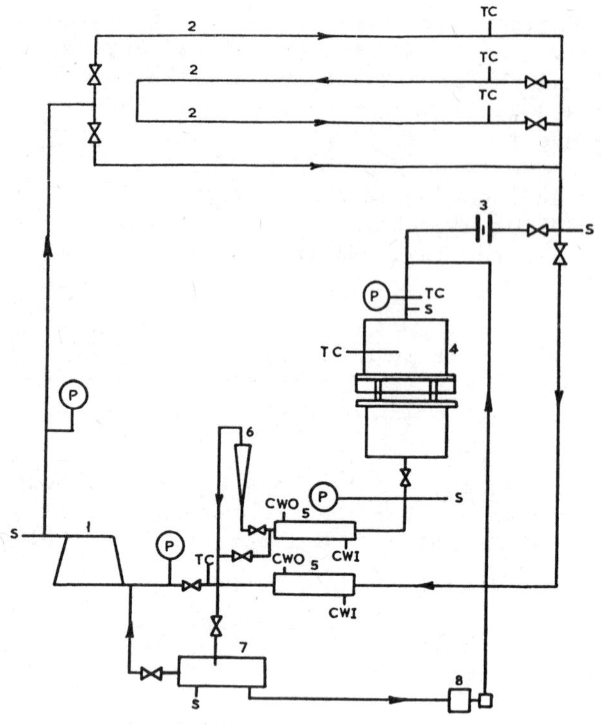

Fig. 1
Schematic diagram of equipment to
decompose heat pump working fluids
under recirculating conditions.

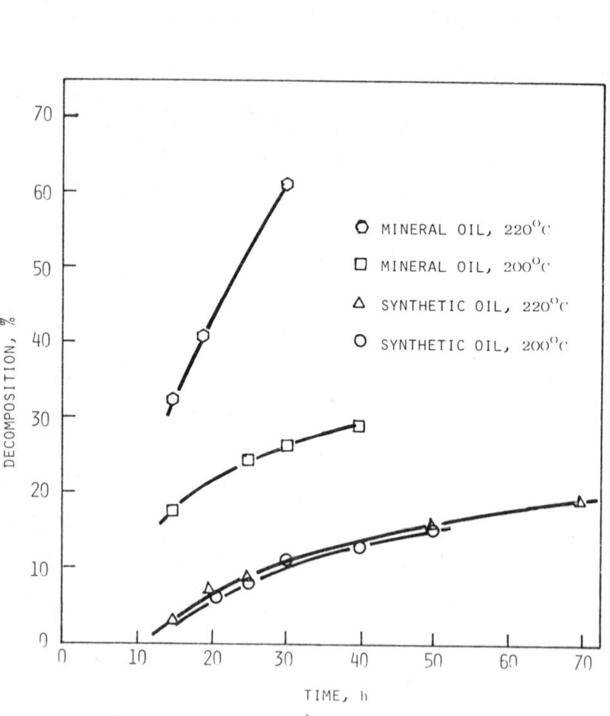

Fig. 2   Percentage decomposition of R11
with time in the presence of
copper, steel and lubricating
oil.

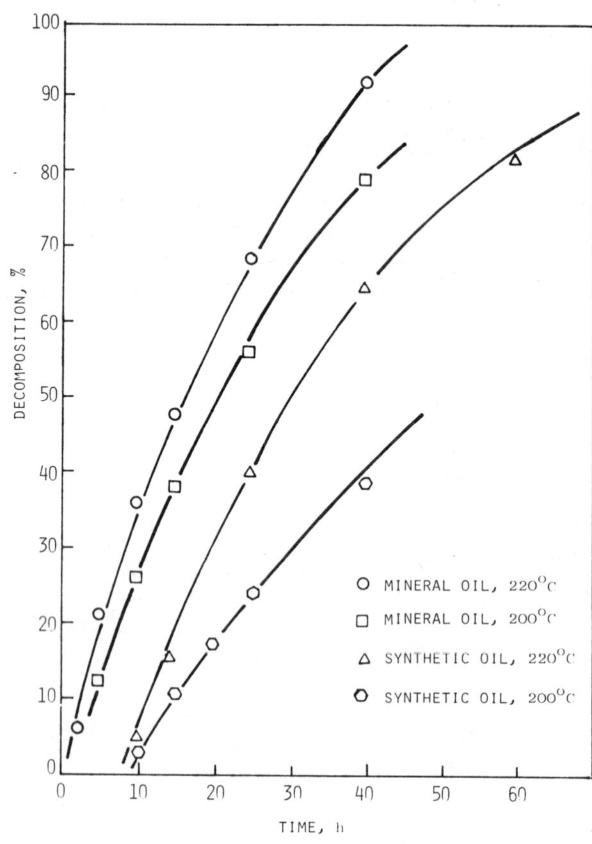

Fig. 3   Percentage decomposition of R12B1
with time in the presence of
copper, steel and lubricating oil.

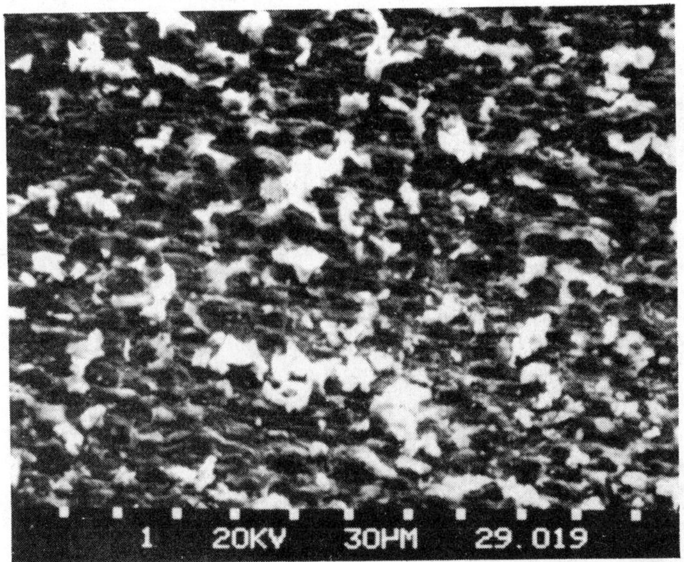

Fig. 4
Deposit on the surface of steel exposed to
the vapour of R11 at 150°C for 480h.

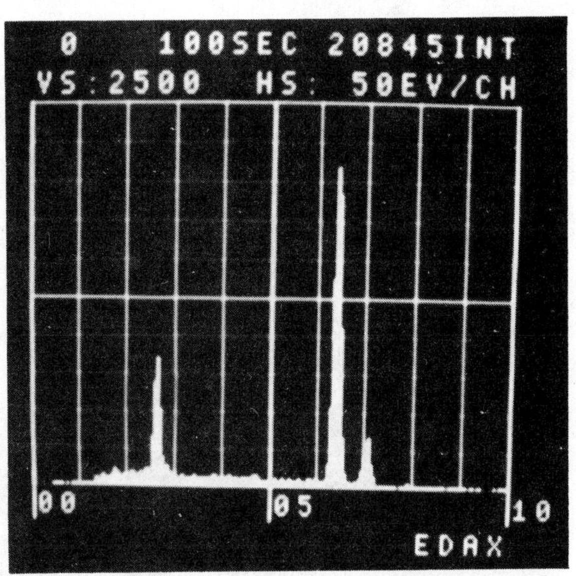

Fig. 5
EDAX of the surface shown in Fig. 4.

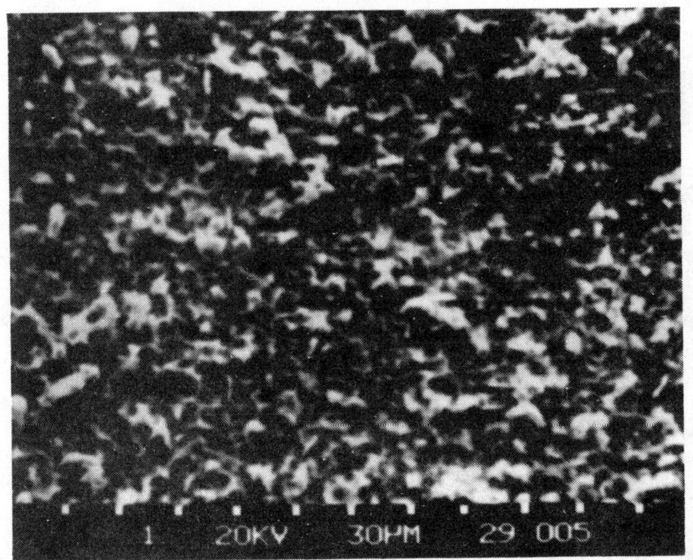

Fig. 6
Deposit on the surface of steel exposed to
the oil and R11 at 150°C for 480h.

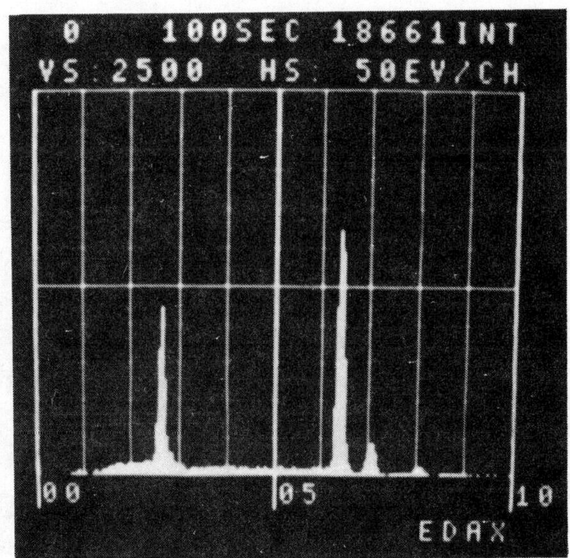

Fig. 7
EDAX of the surface shown in Fig. 6.

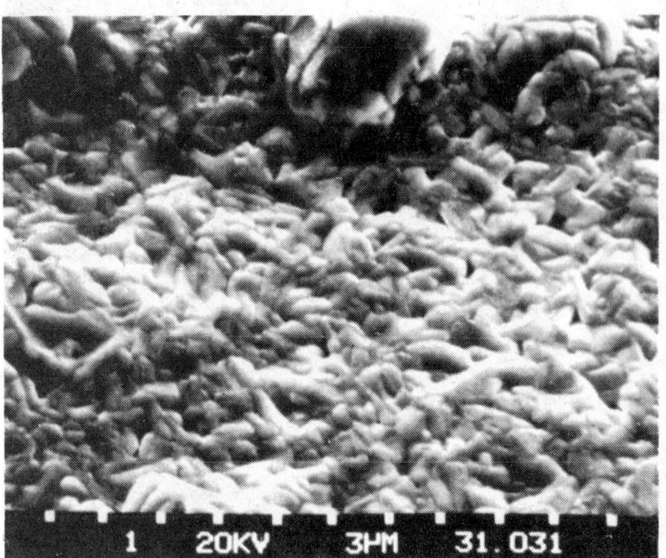

Fig. 8

Deposit on the surface of copper exposed to vapour of R11 at 150°C for 480h.

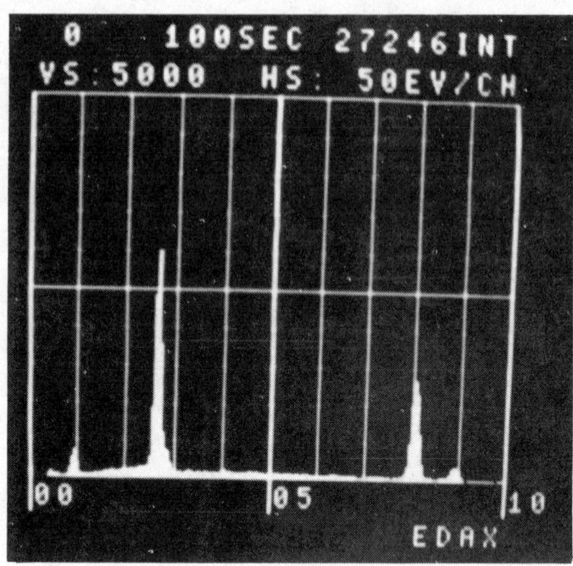

Fig. 9

EDAX of the surface shown in Fig. 8.

Fig. 10

Deposit on the surface of copper exposed to the oil and R11 at 150°C for 480h.

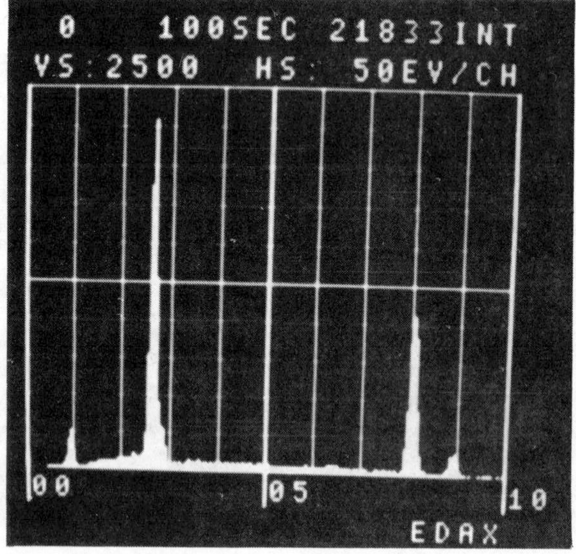

Fig. 11

EDAX of the surface shown in Fig. 10.

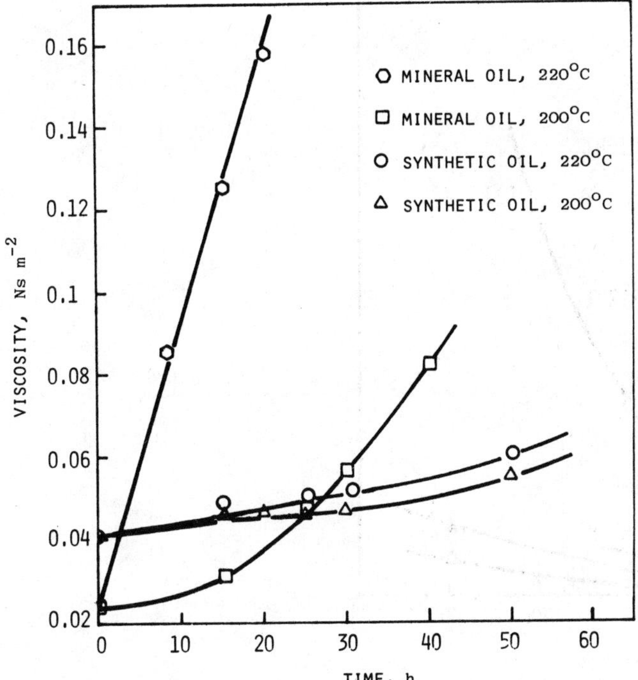

Fig. 12

Variation of the viscosity of lubricating oils as measured at $40^{\circ}C$ with reaction time in the presence of copper, steel and R11.

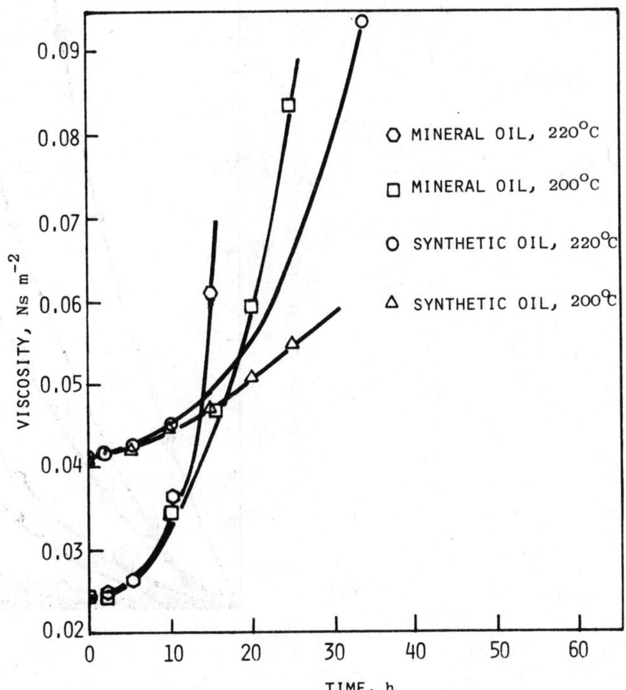

Fig. 13

Variation of the viscosity of lubricating oils as measured at $40^{\circ}C$ with reaction time in the presence of copper, steel and R12B1.

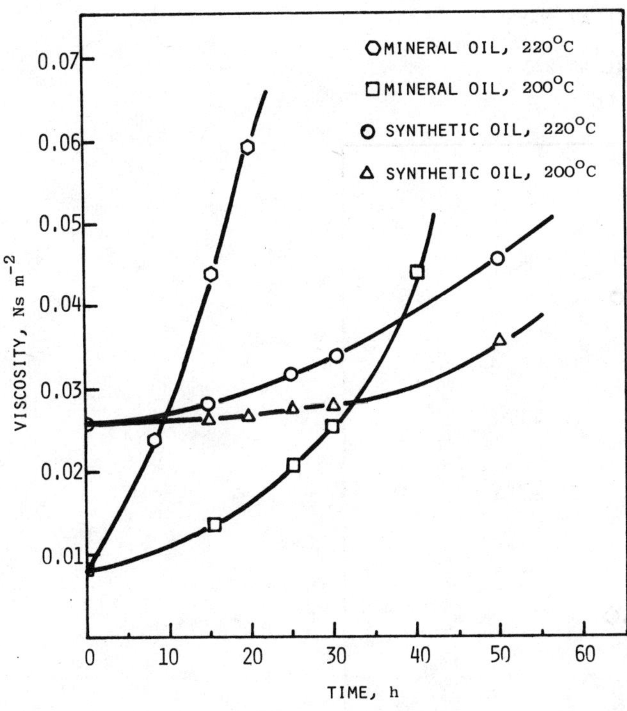

Fig. 14

Variation of the viscosity of lubricating oils as measured at $60^{\circ}C$ with reaction time in the presence of copper, steel and R11.

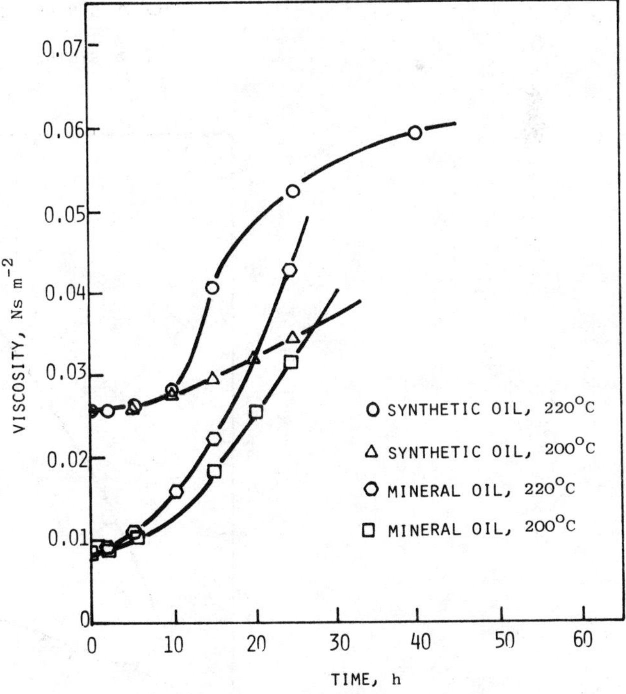

Fig. 15

Variation of the viscosity of lubricating oils as measured at $60^{\circ}C$ with reaction time in the presence of copper, steel and R12B1.

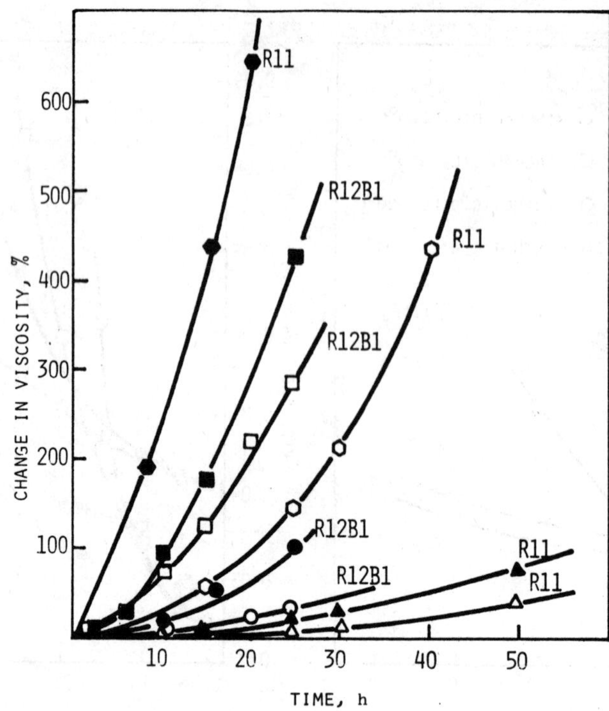

Fig. 16   Percentage change in the viscosity of lubricating oils as measured at 60°C with reaction conditions in the presence of copper and steel.

○ □ MINERAL OIL AT 200°C

● ■ MINERAL OIL AT 220°C

○ △ SYNTHETIC OIL AT 200°C

● ▲ SYNTHETIC OIL AT 220°C

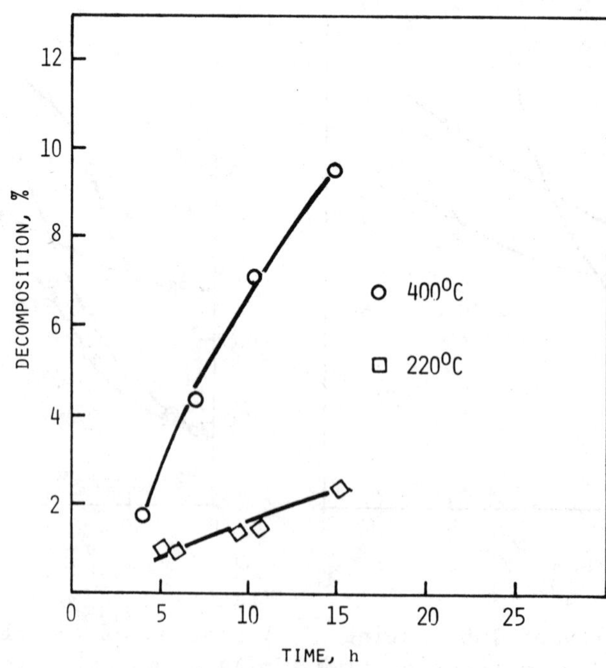

Fig. 17   Decompsition of R11 recirculating between 140°C and 180°C with a copper target at temperatures of 220°C and 400°C.

30

EXPERIMENTS WITH R113a
A NEW REFRIGERANT FOR HIGH TEMPERATURE HEAT PUMPS

Y. Almin, J.R. Feuga and L. Vuillame

Electricité de France, France

Summary

Using R114 as refrigerant in a heat pump allows condensing temperatures up to 120°C (250°F). However, the considerable superheat it necessitates, its low density and some difficulties in finding it, led us to look for another refrigerant which could replace it.

We finally chose R113a which we tested out for nearly 3000 hours on a 25 kW thermal power laboratory heat pump. Results of tests carried out (evaporating temperature between 40 and 70°C, i.e. 105 and 160°F, condensing temperature between 70 and 120°C, i.e. 160 and 250°F) showed this refrigerant might be a good substitute to R114. A high cost may however limit its diffusion.

Held at the University of Warwick, U.K.
Symposium organised and sponsored by
BHRA Fluid Engineering
©BHRA Fluid Engineering, Cranfield, Bedford MK43 0AJ, England.

1 - <u>INTRODUCTION</u>

Refrigerants usually used in refrigerating machines are not suitable for high-temperature heat pumps, for they do not allow a condensing temperature higher than 50°C. On the other hand, R.12 makes it possible to reach 75°C.

Beyond that, for a condensing temperature between 80 and 120°C, we have, until now, recommended R.114 because of its availability and good stability.

Nevertheless, the latter has two drawbacks (see Figure 1):

- a relatively low density during compressor suction (40°C < T < 80°C), whence the need to sweep along large volumes,

- the risk of isentropes' cutting the saturation curve at the time of compression, whence the need for a high superheat (additional exchanger) to avoid slugging.

Furthermore, its availability in the future is not guaranteed, its main outlet being its use as a propulsive gas in aerosol sprays.

That is why we have undertaken to investigate other fluids capable of being used as refrigerants in high temperature heat pumps. R.133a, denser than R.114 and requiring only a low superheat, has been given our attention.

Regarding it, we have some results of stability tests made at 160°C, in sealed test-tubes, in the presence of oil and the metals which go into the construction of a heat pump, as well as of a comparative study of the performances of R.133a and R.114 concluding that the refrigerating effect per unit of swept volume of the former is greater by 30 % (1).

We hence decided to try that fluid on a laboratory heat pump for the purpose of checking its stability and behaviour in operation with respect to the oil used, and of determining the performances to which it leads.

The results of that more than 3000 hour test are the subject of this report.

2 - <u>CHARACTERISTICS OF R.133a</u>

R.133a is an intermediary of synthesis in organic chemistry; it is not presently used as a refrigerant.

We are talking about chlorotrifluoroethane with the formula : $CF_3\ CH_2\ Cl$.

Its main characteristics (Rhône-Poulenc data) are :

- Molecular mass   : 118.486
- Critical temperature : 427.16°K
- Critical pressure   : $4.05\ 10^6\ N/m^2$.

Its pressure-temperature curve at saturation is very close to that of R.114; on the other hand, there is no chance of its isentropes' cutting the saturation curve during compression. (see figure 2).

3 - <u>TEST INSTALLATIONS</u>

The tests took place on a high-temperature water-water heat pump. It is equipped with
an open-type reciprocating compressor powered by an electric motor. It comprises, in
addition, a condenser, a liquid-receiver and an evaporator fed by a thermostatic
expansion valve. A low-flow water loop and two high-flow water loops ensure the heat
transfers between the evaporator and the condenser (see figure 3).

We shall describe briefly below the various components of the installation.

3.1 - <u>Refrigerating circuit</u>

3.1.1 - <u>Compressor</u>

COMEF belt-driven twin compressor
Swept volume .... 204 cm^3 per cylinder
Oil pump .... built in
Oil crankcase resistance 1100 W modulatable
LEROY SOMER 18 kW - 1500 RPM. electric motor.

3.1.2 - <u>Oil separator</u>

The installation comprises an oil separator placed on the discharge of the compressor
so as to prevent the condensed refrigerant returning into the crankcase. Oil is
allowed to return only after three minutes of heat pump operation.

3.1.3 - <u>Condenser</u>

Multitubular horizontal heat exchanger made up on the inside of smooth tubes and
of fins on the outside. The water circulates inside the tubes in two channels, the
refrigerant condenses on the outside.

3.1.4 - <u>Evaporator</u>

Horizontal, multitubular exchanger with baffles on the water side.

The refrigerant vaporizes inside the tubes in three channels. The tubes comprise ten
inner spiral fins at the rate of two turns per meter.

3.1.5 - <u>Expansion valve</u>

The evaporator is fed by a DANFOSS expansion valve equipped with a special power
system designed for R.114. The superheat control had to be re-set with every change
of the operating conditions.

3.1.6 - <u>Lubricants</u>

Two oils developed for high-temperature heat pumps were used:

- RHONE POULENC 801 E23 oil,
- MOBIL OIL SHC 234 oil.

3.2 - <u>Water loops</u>

A special hydraulic circuit provides for heat transfers between the evaporator and
the condenser.

3.2.1 - <u>High-flow loops</u>

The evaporator and the condenser are each connected to a loop comprising a 5 m^3/h
pump ensuring a sufficient flow on the exchanger, and a modulating mixer valve controlled by the temperature of the water upon entry into the exchanger.

3.2.2 - Low-flow loop

A loop with a 2 m^3/h pump provides for heat transfers between the two-flow loops. The temperature of this loop is maintained at a value intermediate between those of the high-flow loops by means of a 12 kW electric reheater or by means of a city-water fed exchanger, according to needs.

3.2.3 - Pressurized tank

This 0.3 m^3 tank has a double function: it acts as an exchanger to evacuate eventual excess heat from the low-flow loop, and as a heat accumulator, preventing short-cycling between the various loops.

Considering that condensing temperature is capable of reaching 120°C, this tank is pressurized with nitrogen at 4.10^5 N/m^2. Its level is maintained constant thanks to an electrode level control.

3.3 - Measuring

The installation comprises numerous means of measuring.

3.3.1 - Temperatures

Temperatures are measured and recorded with "100 ohms platinum" probes, calibrated and paired for measuring differences in temperature (thermal balance sheet on water).

- refrigerant circuit 11 measuring points
- water loops   10 measuring points.

3.3.2 - Pressures

Protais-type manometers, equipped with recopy potentiometer allow pressure measurements and monitoring of the behaviour of the installation.

3.3.3 - Flow rates

A volumetric meter, installed up-stream with respect to the expansion valve, enables the flow rate of the refrigerant to be measured.

The flow rate of the different water loops is measured with the help of float-equipped rotameters.

3.3.4 - Energy

An impulse-emitting electric meter measures the quantity of energy consumed by the compressor and a wattmeter constantly records the power drawn.

4 - TESTS CARRIED OUT AND RESULTS

4.1 - Measuring heat pump performances

To adapt the machine to operation with R.133a as a refrigerant, the compressor speed was set for 1,260 RPM for all the tests.

By varying the temperatures of the high flow-rate loops of the installation, performance of the machine has been measured for an evaporating temperature between 40°C and 70°C, and a condensing temperature between 70°C and 120°C.

Some 20 tests lasting 8 hours each were necessary.

To the degree such was possible, the superheat at the evaporator output was maintained at 5.5 degrees.

4.2 – Reliability test

To check the good behaviour of the machine with the oil-refrigerant mixture used, we also proceeded with a series of 225 start-stop cycles (evaporating temperature 70°C, condensing temperature 120°C) according to the sequence: one hour of operation, half an hour of rest. Voltage was applied to the electrical crankcase resistance during the time the compressor was off and enabled the oil to be maintained at around 70°C.

The conditions to which the machine was subjected for this test were the following:

- Upon starting the compressor, the low-pressure pressostat was bypassed for 30 seconds so as to enable the evaporator to be fed normally until the suction pressure went beyond that pressostat set value.

- Stoppage of the compressor by pump-down of the evaporator by cutting off its intake of refrigerant for 20 seconds. That delay is enough to make the pressure drop in the evaporator without causing any significant emulsion in the crankcase. Pump-down is necessary to limit the dilution of refrigerant in the oil during the shut-off period; it allows for satisfactory viscosity of the lubricant upon restarting.

Thanks to this procedure, the test went along without any problem, thus demonstrating the compatibility of the refrigerant and of the oil used with a high-temperature heat pump.

4.3 – Thermal power of the heat pump

The thermal balance sheet of the installation during stable operation was calculated on the basis of the different magnitudes measured on the circuit of the refrigerant and on the water loops.

For the different evaporating temperatures, figure 4 gives the evolution of the thermal effect per unit of swept volume in terms of the gap between the condensing and evaporating temperatures.

Figure 5 shows the volumetric efficiency of the compressor in terms of the compression ratio.

During our tests, sub-cooling was always nil.

4.4 – Coefficient of performance

Different coefficients of performance can be defined.

We have selected two of them:

$$COP_{Carnot} = \frac{T_2}{T_2 - T_1} \quad (*)$$

$$COP_{practical} = \frac{\text{Condenser effective power}}{\text{Motor electric power}}$$

(*) T_1 = cold reservoir temperature

   T_2 = hot reservoir temperature

During the tests, the motor was oversized, whence poor efficiency. We hence made the following correction:

- Calculation of the mechanical power produced on the shaft from the measured electrical power.

- Determination of an electrical power called corrected, by applying a motor efficiency of 0.88 to the mechanical power.

The evolution of the $COP_{practical}$ is represented on table 1.

The $\dfrac{COP_{practical}}{COP_{Carnot}}$ ratio is comprised between 0.4 and 0.5; the highest values corresponding to the lowest gaps between condensing and evaporating temperatures.

4.5 - <u>Overall performance of the installation</u>

The superheat on the discharge of the compressor never exceeded the condensing temperature by more than 15 K, whatever the compression ratio was.

With both oils used, the oil pressure was always at least two bars higher than the pressure in the crankcase. The return of the oil to the crankcase was very satisfactory. Lastly, we did not notice any emulsion upon compressor start-up, the oil having been maintained at 70°C during the rest periods.

5 - <u>CONCLUSION</u>

The tests carried out with R.133a on a high-temperature heat pump with an adapted lubricant have confirmed the hopes born of the previous studies with respect to the possibilities of using that fluid as a refrigerant in such machines. The thermal performances obtained compare fabourably with those of R.114. Lastly, the volumetric efficiency of the compressor is not affected by the use of this fluid and did not present any particular problem.

From the practical point of view then, it is an excellent refrigerant for high temperature heat pumps.

If the coefficients of exchange to which it leads do not penalize it, its price and its availability alone will determine whether it will compete with R.114.

REFERENCE

(1) P. GRAZIANI : "Comparative study of thermodynamical performances of R.133 and
    R.114". (Etude comparative des performances thermodynamiques du R133 et du R114).
    EDF report P33/4300/80-16 (In French).

Table 1 - Value of the practical C.O.P. in terms of the evaporating temperature and of
the gap between condensing and evaporating temperatures.

EVAPORATING TEMPERATURE	GAP BETWEEN CONDENSING AND EVAPORATING TEMPERATURES				
	30 K	40 K	50 K	60 K	70 K
40°C	5,35	4,10	3,35	2,50	
50°C	5,45	4,35	3,50	2,70	2,20
60°C	5,20	4,30	3,35	2,55	2,15
70°C		3,95	3,25		

PRACTICAL C.O.P. VALUE

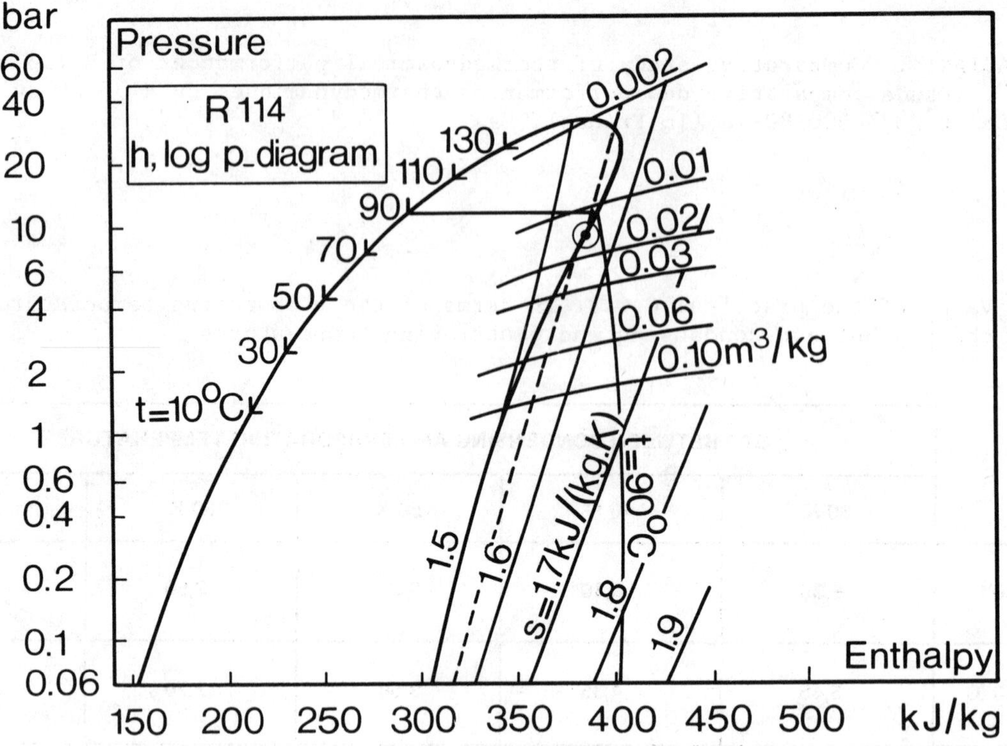

Figure 1 - R.114 h, log p - diagram.

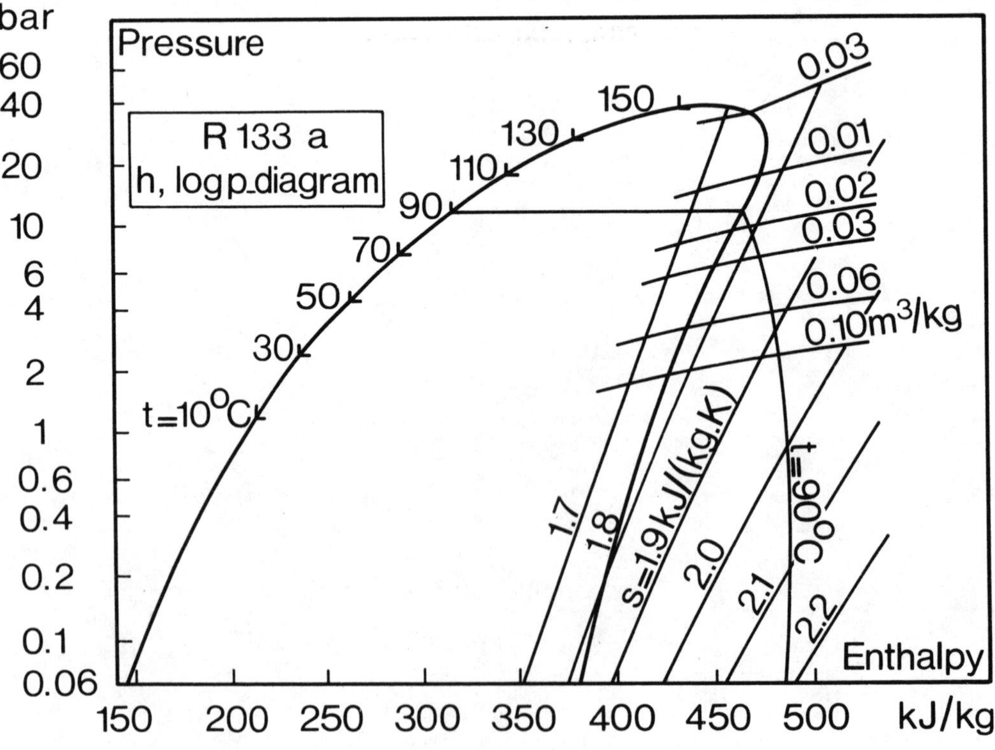

Figure 2 - R.133a h, log p - diagram.

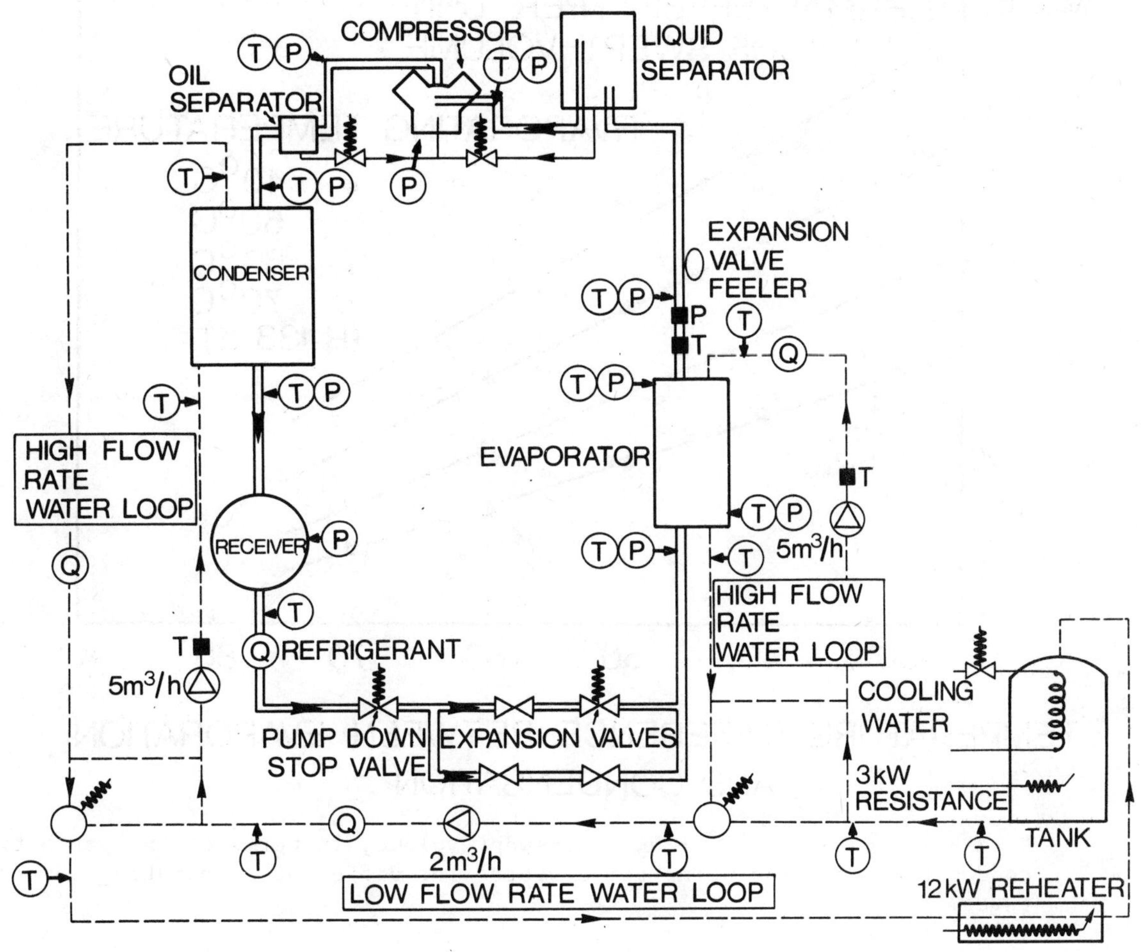

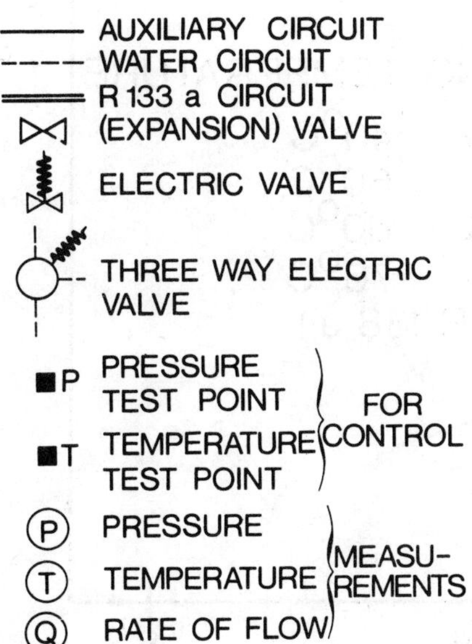

Figure 3 - Test facility circuits.

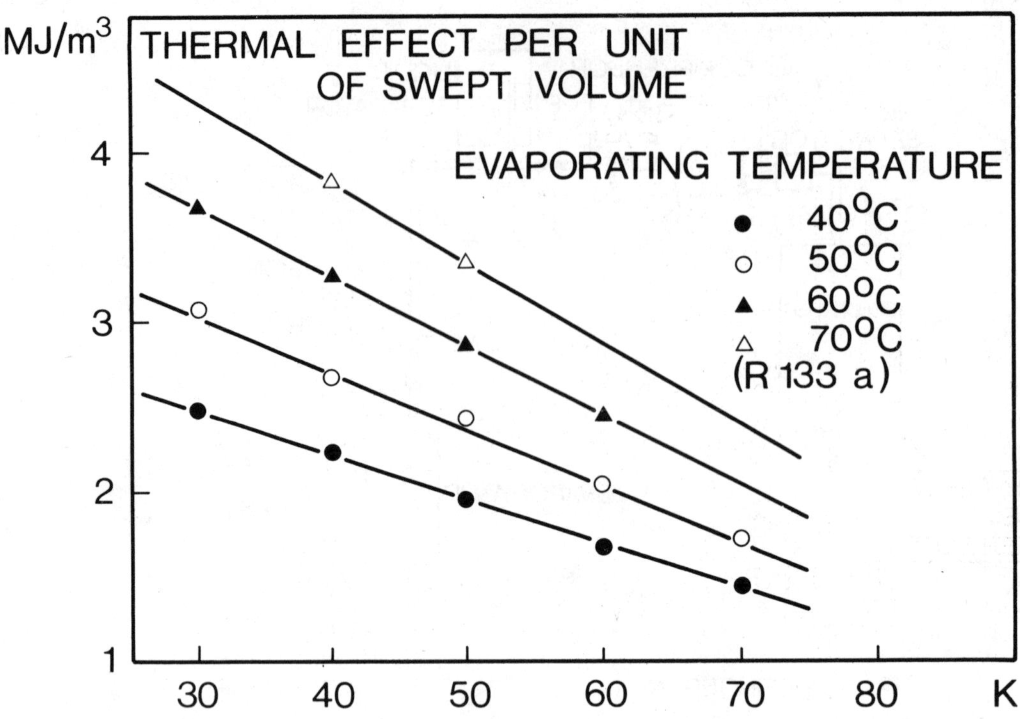

Figure 4 - R.133a thermal effect per unit of swept volume, in terms of the gap between condensing and evaporating temperatures for different evaporating temperatures.

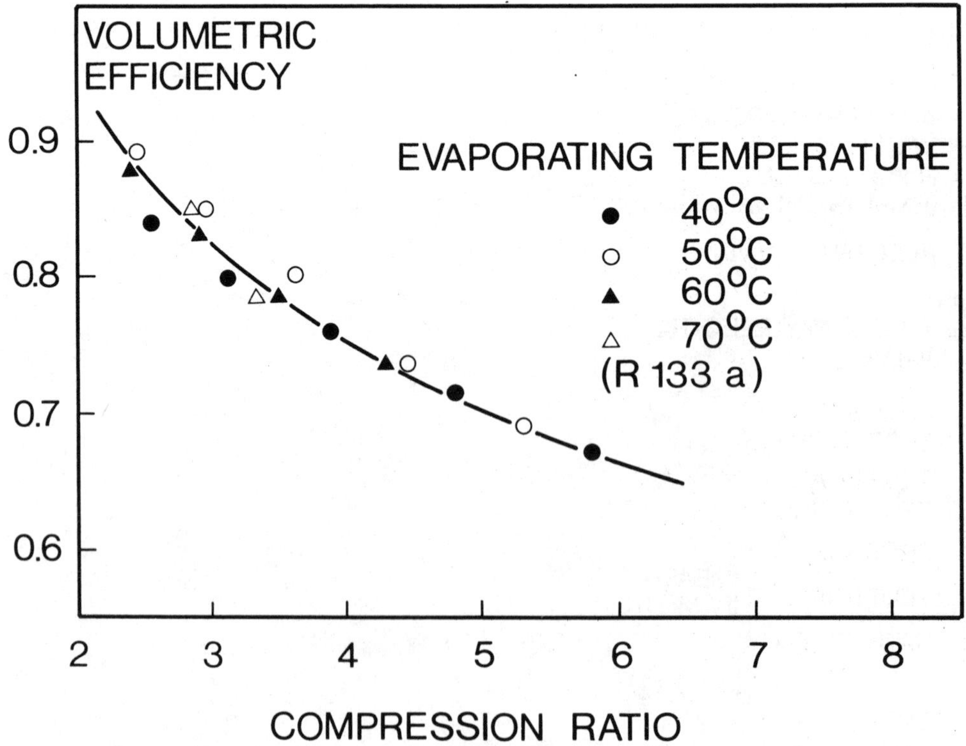

Figure 5 - Volumetric efficiency with R.133a, in terms of the compression ratio, for different evaporating temperatures.

PERFORMANCE OF SELECTED SYNTHETIC LUBRICANTS IN INDUSTRIAL HEAT PUMPS

G. Daniel

Mobil Oil Française, France

M. J. Anderson

Mobil Oil Co. Ltd., U.K.

W. Schmid

Mobil Oil A.G., Germany

M. Tokumitsu

Mobil Sekiyu Kabushiki Kaisha, Japan

Summary

As a result of the continuing efforts towards energy savings, development of industrial heat pumps will increase in the future. The progress made in improving the reliability of this equipment now allows their extension to higher temperature applications such as drying processes and other examples.

This paper reviews the technical lubrication requirements specific to industrial heat pumps operating at relatively high condensing temperatures. This study indicates the potential limitations of mineral oil based lubricants in these applications. Based on laboratory tests, selected synthetic lubricants, particularly synthetic hydrocarbon fluids of the polyalphaolefin type (SHF) and selected Polyglycols, appear very well adapted for these severe applications. A review is made of the comprehensive physical data developed on mixtures of these selected synthetic lubricants with halogenated refrigerants, e.g. viscosity, vapour pressure, stability. The capability of synthetic hydrocarbon fluids to lubricate industrial heat pumps was then demonstrated at several heat pump builders. Experience gained from actual industrial heat pump applications has confirmed definitely their outstanding performance. In fact, these lubricants are necessary for the satisfactory unit operation under these severe conditions.

Held at the University of Warwick, U.K.
Symposium organised and sponsored by
BHRA Fluid Engineering
©BHRA Fluid Engineering, Cranfield, Bedford MK43 0AJ, England.

1. INTRODUCTION

The lubricant plays a fundamental role in the operation and efficiency of a heat pump plant. The lubricant, diluted with refrigerant, should primarily retain a sufficient viscosity level at the highest temperature in the system. Friction and wear should be minimized in reciprocating compressors whilst a satisfactory sealing effect is required in oil-injected rotary compressors. Furthermore, the lubricant should have a good thermal and chemical stability at high temperature but it should also have satisfactory fluidity performance at the lowest temperature in the system.

This paper surveys the evaluation of suitable lubricants for modern industrial high temperature heat pumps. The resulting lubrication requirements are best met by selected synthetic lubricants. Comprehensive laboratory evaluations indicate that synthetic hydrocarbon fluids (SHF) (Ref. 1) and selected Polyglycols are potential candidates.

2. TRENDS IN LUBRICATION REQUIREMENTS OF INDUSTRIAL HEAT PUMPS

Whilst domestic heat pumps can be used both for heating and cooling buildings, most of the industrial heat pumps are non-reversible and are used only for heat upgrading. They recover heat from a relatively low temperature source and deliver it at a relatively high temperature where it can be used directly.

Some industrial applications require heat at high temperature, for example at 90-100°C, possibly leading to condensing temperatures close to 120-130°C. The operability of such a high temperature heat pump has been previously demonstrated at a pilot plant stage (Ref. 2). We understand that the development in technology now allows trouble-free operation of high temperature heat pumps in the field, leading to their possible extension in the future.

Various refrigerants have been considered for industrial heat pumps. The following table lists the most common refrigerants currently found or tested in such applications.

Refrigerant	Boiling Point at 1 bar, °C	Pressure at 30°C in condenser, bar	Indicative Temperature at condenser, °C
R 22	− 40.7	11.03	Up to 55
R 500	− 33.3	7.82	Up to 65
R 12	− 29.7	6.43	Up to 70
R 114	+ 3.5	1.52	Up to 125

R 114 is of particular interest for high temperature applications. In practice, the risks of condensation of R 114 during the compression are alleviated by over-heating the vapour. This operation is carried out in a liquid/gas heat exchanger, by recovering heat from the condensed refrigerant which is further cooled before the evaporator to improve efficiency (Ref. 2).

The correct selection of the lubricant is as critical as that of the refrigerant for the heat pump manufacturer. The severe operating conditions now encountered in service impose more severe lubrication requirements than in traditional units. Table 1 summarizes the typical properties of selected commercial synthetic lubricants, SHF A, B, C and Polyglycols D, E, F in comparison with a conventional mineral oil. The extremely high viscosity and naturally high viscosity index offered by the low pour point SHF and selected Polyglycols represent a considerable advantage over conventional mineral oils, especially in the case of high dilution with refrigerants. Another major advantage results from the excellent stability of these synthetic lubricants in the presence of halocarbon refrigerants.

3. PROPERTIES OF HALOCARBON REFRIGERANTS/SHF AND POLYGLYCOL MIXTURES

The lubrication technology of heat pump systems is complicated by the numerous refrigerants available and by their complex behaviour with lubricants.

The properties of mixtures of typical halocarbons with SHF and Polyglycols are surveyed; whenever possible, a comparison with current mineral oils has been made. Low temperature fluidity properties, as well as low temperature miscibility data, are not covered in this paper since these performance aspects are generally of secondary importance to high temperature heat pump applications. The high temperature performance of synthetic lubricants has also been assessed, based on laboratory determinations.

3.1. Dissolution of Halocarbon Gas by Lubricants

Lubricating oils absorb gases with which they come into contact. According to Henry's law, the amount of gas dissolved is proportional to its partial pressure. For each condition of pressure and temperature, if sufficient gas is present, the oil reaches its saturation point. When the pressure is decreased, the gas will come out of solution until a new equilibrium is reached.

The amount of halocarbon gas absorbed depends upon the lubricant as well as the halocarbon. Figure 1 gives a comparison of the absorption of refrigerant R 22 in current naphthenic mineral oils, SHF and Polyglycols at two temperatures and at various vapour pressures. R 22 is less soluble in SHF than in mineral oils whilst it is much more soluble in Polyglycols. On the other hand, R 12 is less soluble in Polyglycols than in SHF and mineral oils.

These data are best used by representing the vapour pressure curves of solutions of refrigerant in the lubricant, at increased dosage, as a function of the temperature (Figure 2, lower part). Such measurements were made in a vapour pressure cell with membrane contact switches which enable an accurate pressure compensation with nitrogen (Ref 3).

3.2 Effect on Dissolved Refrigerant Gas on Lubricant Viscosity

The dissolution of refrigerant gas in the lubricant will lower the viscosity.

The viscosity of the mixtures were determined in a Höppler type falling ball viscosimeter (DIN 53015) especially designed to withstand pressure. The dynamic viscosities were converted to the kinematic viscosities using the measured densities of the mixtures.

Examples of kinematic viscosity data are given for the following lubricant/refrigerant mixtures: naphthenic oil/R 22, SHF A/R 22, Polyglycol F/R 12 (Figures 2 to 4). Comprehensive data have been determined on current refrigerants with SHF A to C and Polyglycols D and E and are available.

Instructions how to use these data are given in Figure 2. In summary, the oil system pressure/temperature chart is combined with the viscosity/ temperature chart. Isobar curves of the resulting viscosity/pressure/ temperature relationship may be drawn. Of specific interest is the aspect of the isobar curve which shows a maximum called inversion point. This is a function of the pressure and temperature. As the temperature increases, the curve asymptotically approaches the 100% wt oil line viscosity. On the other hand, when the temperature decreases below that of the inversion point, a slight temperature decrease results in a rapid drop of the viscosity, resulting from an important refrigerant absorption.

The inversion point is significantly higher for synthetic lubricants than for mineral oils. For example, at a vapour pressure of 10 bars, SHF A will ensure an acceptable viscosity of 12 cSt whilst the same ISO viscosity mineral oil will have a borderline viscosity of 6.7 cSt. Similarly, selected Polyglycols will still offer a higher viscosity than an equivalent ISO viscosity SHF in the specific case of R 12 under high pressure and moderate temperature.

These viscosity/pressure/temperature curves have been determined in the laboratory once the refrigerant absorption by the lubricant has reached its equilibrium. This represents therefore a more severe case than the current situation in the compressor where the absorption will take place to a lesser extent. The resulting viscosities should be considered as guidelines to enable the heat pump manufacturer to select lubricants. The laboratory viscosity data show that both SHF and Polyglycols offer significant advantages over current mineral oils in terms of a resultant higher viscosity under the severe service conditions which lead to an important oil dilution by the refrigerant.

3.3 Stability of Lubricants with Halocarbon Refrigerants in Heat Pump Systems

The lubricant/refrigerant mixture is exposed, in the compressor and the condenser, to high pressure and temperature. These conditions combined with the presence of active metals in the system can lead to the degradation of the lubricant.

The evaluation of oil stability in the presence of halocarbons can be made using the Philipp test. The oil is heated at 250°C and the vapours come into contact with the refrigerant at 40°C in a sealed glass U tube. Every 24 hours the tube is inspected for reaction products such as condensate droplets of halogenated acid. The test is stopped when no reaction is evidenced after 96 hours. High quality mineral oils and the synthetic lubricants under review pass the Philipp test.

A more severe test has been developed by Linde in Germany to evaluate the superior stability performance of the synthetic oils. It consists of exposing a mixture of 90% oil and 10% refrigerant at a temperature of 100°C for one month. Strips of steel, copper and aluminium are used as catalysts. After completion of the test, the surface of the metal catalyst is inspected and the oil is thoroughly analysed after removal of the refrigerant. Under these severe test conditions, both SHF and Polyglycols are extremely stable in the presence of the halocarbons currently used in heat pump systems.

The excellent stability of the synthetic oils under review should guarantee clean heat pump systems allowing the maximum efficiency in heat transfer to be maintained. Furthermore, should some air remain trapped in the system, the above lubricants will offer a good resistance to oxidation and against sludge and deposit formation.

4. BUILDER EVALUATION OF SHF/POLYGLYCOLS IN HEAT PUMPS

Due to the inherent advantages found in laboratory tests, synthetic lubricants such as SHF and Polyglycols were presented for evaluation to builders and organizations involved in the development of industrial heat pumps.

Numerous tests were conducted in several countries in various equipments involving reciprocating or screw compressors as well as different refrigerants. Some tests were conducted under extremely severe test conditions which mineral oils could not tolerate. Table 2 gives a summary of the builder tests. Some of them are detailed hereunder as examples.

EDF (Electricité de France) have studied extensively the development of high temperature heat pumps operating at condensing temperatures up to 120°C, because of their potential interest for various industrial applications (Ref 2). Studies have covered the heat pump technology applied to high temperature environment, as well as the selection of the refrigerant in accordance with the lubricant (Table 2, Case 1).

The experimental heat pump run at EDF was operated by a 50 kW 6 cylinder reciprocating compressor, using R 114. Tests were conducted at 120°C condensing temperature. Preliminary tests with a conventional lubricant gave oil related problems. Examples of wear and deposit formation were experienced due to respectively poor lubrication performance, caused by the dilution of the lubricant by the refrigerant, and its unsatisfactory chemical stability. A lubricant offering higher performance at high temperature was required.

SHF C, the highest viscosity grade available, was selected for this application. The calculated viscosity at the temperature of 130°C prevailing at the compressor outlet, is of the same order as that of a current naphthenic oil at 80°C. We have seen that SHF C, in comparison with other lubricants, offers a high viscosity level in presence of refrigerant dilution. Furthermore in the presence of R 114, the stability of SHF C is excellent, based on laboratory stability tests as well as excellent performance developed in severe modern high temperature ammonia refrigeration systems.

In practice, SHF C gave overall satisfactory operation for two successive periods of 600 and 500 hours. COP values of about 4 were measured in this experimental unit for an evaporating temperature of 70-80°C at a constant condensing temperature of 120°C. The used oil analysis at the end of the test confirmed the satisfactory lubrication performance of SHF C. Further tests at EDF confirmed the suitability of SHF C to lubricate high temperature heat pumps.

Similarly, ECRC have developed heat pumps condensing at 120°C using R 114 and have found that successful results were obtained when using selected synthetic Hydrocarbon Fluids due to their high chemical and thermal stability and their favourable viscosity properties (Table 2, Case 4).

Another case of interest concerns heat pumps driven by rotary screw compressors. A major Japanese builder Mayekawa has conducted comparative tests between mineral oils and SHF in the presence of R 12 and R 114 (Table 2, Case 2) on large screw compressors (220 kW). SHF offered excellent thermal/chemical stability under high oil injected temperature conditions (up to 120°C) which are too severe for mineral oils. The shift from mineral to synthetic lubricants allowed the oil injected temperature to be significantly increased, thus improving the compressor efficiency.

A similar experience has been gained at a Swedish builder (Table 2, Case 7), using R 12. This refrigerant has a high solubility in mineral oils and SHF. In a screw compressor, the presence of refrigerant dissolved in the oil is a disadvantage with respect to efficiency. It is therefore desirable to increase the oil injected temperature to reduce the absorption of refrigerant, provided the oil viscosity remains at an acceptable level and its thermal stability is satisfactory. Results of tests conducted by this builder indicated that changing from a mineral to a synthetic oil of the SHF type, allowed operation at a higher oil injected temperature which led to a significant reduction of the power input to the screw compressor.

In order to keep a relatively high oil viscosity in the presence of R 12, a German builder prefered Polyglycol E rather than SHF (Table 2, Case 3). Under the operating conditions prevailing in the compressor (17 bars, 70°C), Polyglycols absorb significantly less R 12 than SHF or mineral oils. Polyglycol E has given trouble-free operation.

The Zimmern type single screw compressor is claimed less critical, with respect to lubricant viscosity requirement, than the current twin screw compressor (Ref. 4). Tests of the lowest viscosity grade SHF A in R 114 systems have given satisfaction for condensing temperatures up to 120°C (Table 2, Case 6).

The builder tests under review have demonstrated that mineral oils prove unsatisfactory as soon as the severity of the heat pump system is increased. SHF A, B, C, as well as Polyglycols D, E, F in specific cases, offer better lubrication at high temperature in the presence of refrigerants; in addition, they reduce the risk of carbonizing effects due to their better thermal/chemical stability. Based upon the satisfactory results obtained, several builders are now commercializing high temperature industrial heat pumps operating with these lubricants.

5. FIELD EXPERIENCE WITH SHF/POLYGLYCOLS IN INDUSTRIAL HEAT PUMPS

Although the technology is now available, the utilization of industrial heat pumps has not found so far large extension for high temperature applications. A selection of industrial heat pumps representing, to our knowledge, the most severe field applications, is given in Table 3.

The following example, referenced as Case 1 in Table 3, illustrates the utilization of Polyglycols in R 12 systems. At Erlangen (Germany), a large heat pump is used to heat a group of 80 flats occupying an overall area of 6,400 m^2 (Ref. 5). This unit, designed by Linde A.G. Cologne, operates with two screw compressors driven by two gas motors. The evaporator temperature varies from −25 to + 10°C depending on the outside air temperature. The condensing temperature reaches a maximum of 65°C. The conditions of temperature (80°C) and pressure (17 bars) encountered in the compressor lead to a significant dilution of the lubricant with R 12. Nevertheless, Polyglycol E ensured satisfactory operation of the plant since its start-up at the end of 1980. In particular no oil loss is evidenced with this synthetic lubricant indicating that the oil carried over to the evaporator (−25°C) normally returns to the compressor. It is considered that a mineral oil of the same 220 ISO viscosity would not have a satisfactory low temperature fluidity for this application. After a 1500 hour operation, analysis of Polyglycol E indicates no changes in physico-chemical characteristics and therefore no significant degradation.

Also of interest is the heat pump used for plaster tile drying at Premaco (Table 3, Case 3) (Ref. 6). In brief, three heat pumps mounted independantly and driven by reciprocating compressors operate at a condensing temperature of 65°C and an evaporating temperature of 25°C.   The conditions of pressure and temperature encountered at the compressor, namely 17 bars and 90°C, lead to a significant dilution of the lubricant by the refrigerant R 12.

Preliminary tests conducted with a conventional mineral oil gave crankshaft seizing, due to excessive dilution with R 12. As a result, SHF B was recommended for this application. Since February, 1980 the heat pump plant operates satisfactorily. The oil was drained in summer 1981 after a 10,000 hour operation although the used oil analysis conducted at 9,700 hours showed no significant degradation and contained only very low amounts of wear metals. In conclusion, the use of an SHF lubricant has allowed the satisfactory lubrication of the plant.

More recently an industrial high temperature heat pump operating at 100°C condensing temperature has been installed at a dairy work in France (Table 3, Case 5). This heat pump is combined with a refrigerator plant. The condenser of the refrigerator plant exchanges heat with the evaporator of the heat pump. The experience gained at EDF has been implemented in this application and the same refrigerant R 114 was used. Similarly, SHF C was selected as offering the best guarantee of satisfactory performance under the high compressor outlet temperature of 115°C. No oil related problem has been experienced since the start up of the plant.

We understand that other builders are commercializing high temperature heat pumps, driven by screw compressors, operating at condensing temperature up to 120°C. These units are currently lubricated by SHF exclusively.

6. CONCLUSION

The development in technology has now made possible the commercialization of high temperature heat pumps. Such units require high performance lubricants to ensure satisfactory operation. Comprehensive laboratory data developed in the presence of halogen refrigerants under simulated field conditions show that synthetic products such as SHF A, B, C and Polyglycol D, E, F are best suited. The satisfactory performance of these selected SHF and Polyglycol lubricants has been confirmed during builder tests conducted on reciprocating and oil-injected screw compressors, in the presence of various refrigerants and under severe operating conditions. In the field, industrial heat pumps operating at elevated condensing temperatures have confirmed the potential of the synthetic lubricant. Thus it is believed that the use of the synthetic lubricants described makes a significant contribution to the successful operation of the high temperature industrial heat pump.

ACKNOWLEDGMENT

The authors wish to thank the heat pump builders and users for having provided helpful information and particularly Electricité de France, Electricity Research Council, International Research and Development, Hall Thermotank Products, Linde and AEE (Agence pour les Economies d'Energie).

7. <u>REFERENCES</u>

1. <u>Manley L. W. and Jublot R. M.</u> : "New Developments in Synthetic Lubricants". The Tenth World Petroleum Congress. Bucharest, Roumania 9-14th September, 1979. Vol. 4, Heyden and Son Ltd, London, Philadelphia, Rheine 1980.

2. <u>Laroche G. and Degueurce B</u>: "High Temperature Heat Pumps" (Les Pompes à Chaleur Haute Température); Revue Générale du Froid, 3rd March, 1979, pp 139-152 (In French).

3. <u>Mall K.</u> : "Measuring Equipment for the Determination of the Vapour Pressure of Oil Refrigerant Mixtures" (Messeinrichtung Zur Ermittlung Des Dampfdruckes von Oel-Kaeltemittel-Gemischen).Kaeltetechnik-Klimatisierung, 1970 Heft 8, 22 Jahrgang, p 257. (In German).

4. <u>Chan C.Y., Haselden G.G. and Hundy G.</u> : "The Hall Screw Compressor for Refrigeration and Heat Pump Duties" International Journal of Refrigeration, 5th September, 1981, Vol. 4, N° 5, pp 275-280.

5. <u>Meissner, Ebert, Bub</u>: "Gas Motor Driven Air-Water Heat Pumps for Monovalent Operation" (Gas Motorisch Betriebene Luft-Wasser Waermepumpe Fuer Monovalenten Betrieb). Bundesministerium fuer Forschung und Technologie - Rationelle Energie Verwendung - Status Bericht 1980. (In German).

6. "Heat Pump Application for Plaster Tile Drying" (Séchage des Carreaux de Plâtre par Pompe à Chaleur). Booklet issued by the "Agence pour les Economies d'Energie (A.E.E.).

COMPARISON OF TYPICAL PROPERTIES OF NAPHTHENIC MINERAL OILS, SYNTHETIC HYDROCARBONS AND POLYGLYCOLS

	Naphthenic Mineral Oil	Synthetic Hydrocarbon Fluids (SHF)			Polyglycols		
		A	B	C	D	E	F
Density at 15°C, g/ml	0.911	0.833	0.843	0.852	1.012	1.006	1.05
Flash Point, °C	208	245	266	285	280	280	280
Pour Point, °C	−35	−55	−45	−43	−31	−30	−30
Kinematic Viscosity, cSt							
20°C	220	155	680	1300	460	700	1000
40°C	55	62	205	400	161	225	470
100°C	6.5	10.1	22.1	42.3	23.1	30.9	79.5
Viscosity Index	50	148	148	150	173	181	200
Thermal/ Chemical Stability with Harlocarbon Fluids	Fair	------- Excellent --------			--- Excellent---		

SUMMARY OF BUILDER EVALUATION OF
SHF/POLYGLYCOLS IN HEAT PUMPS

Country/Builder	Compressor Type (Power, kW)	Outlet Temp. °C	Condenser/Evaporator Temperature °C	Refrigerant	Lubricant
France					
1. E.D.F	Recip.	130	120/70	R 114	SHF C
Japan					
2. Mayekawa	Screw (220)	80-125	80-115/30-60	R 114	SHF C
	Screw (220)	70-120	30-75/-20-30	R 12	SHF B
Germany					
3. Linde	Screw (130)	70	Up to 65/ down to -25	R 12	Polyglycol E
UK					
4. E.C.R.C.	–	–	120	R 114	SHF A
5. International Research and Devt.	Recip. (75)	130	120/65	R 114	SHF C
6. Hall Thermotank Products	Single screw (250)	–	120	R 114	SHF A
Sweden					
7. Major builder	Screw	45-100	–	R 12	SHF A

Comments

1. SHF assessed suitable for high temperature applications. Other tests conducted at other builders led to the approval of SHF and Polyglycols for R 114 and R 12 systems respectively.
2. SHF offered excellent thermal/chemical stability.
3. Polyglycol E gave satisfactory lubrication performance in presence of R 12.
4. SHF gave good results due to their high chemical and thermal stability and their favourable viscosity properties..
5. No oil related problems were encountered.
6. Performance of SHF A assessed satisfactory.
7. The use of SHF permitted operation at higher oil injected temperature.

FIELD EXPERIENCE WITH POLYGLYCOLS AND SHF IN
INDUSTRIAL HEAT PUMPS

Country/ End use	Service Conditions, Hour/year	Refrigerant	Lubricant	Compressor		Condenser/ Evaporator Temp., °C
				Type (Power, kW)	Outlet Temp., °C	
Germany						
1. Building Heating	–	R 12	Polyglycol E	2 Screw (2 x 130)	80	Up to 65/ –25 to +10
France						
2. Building Heating	–	R 12	Polyglycol D	3 Screw	–	–
3. Plaster Tile Drying	8000	R 12	SHF B	2 Recip. (30, 45)	90	65/25
4. Plaster Tile Drying	7500	R 12	SHF B	2 Recip.	90	–/32
5. Dairy Work	6000	R 114	SHF C	Recip. (38)	115	100/43

Comments

1. Since start up of the plant in September, 1980, no oil related problems have been encountered. The lubricant carried over to the evaporator normally returns to the compressor at the lowest evaporator temperature. Oil analysis after 1500 hours shows no sign of degradation.

2. No oil related problems since unit start-up in 1978.

3. Under conditions giving important dilution of the lubricant by R 12, mineral oils gave crankshaft seizing. Since February 1980, the heat pump plant operates satisfactorily with SHF B. The oil drained after 9700 hours service was found in good condition.

4. No oil related problems since unit start up in March, 1980. SHF B remained in excellent condition after 7600 hour service.

5. No oil related problems since unit start up in March 1980, using SHF C.

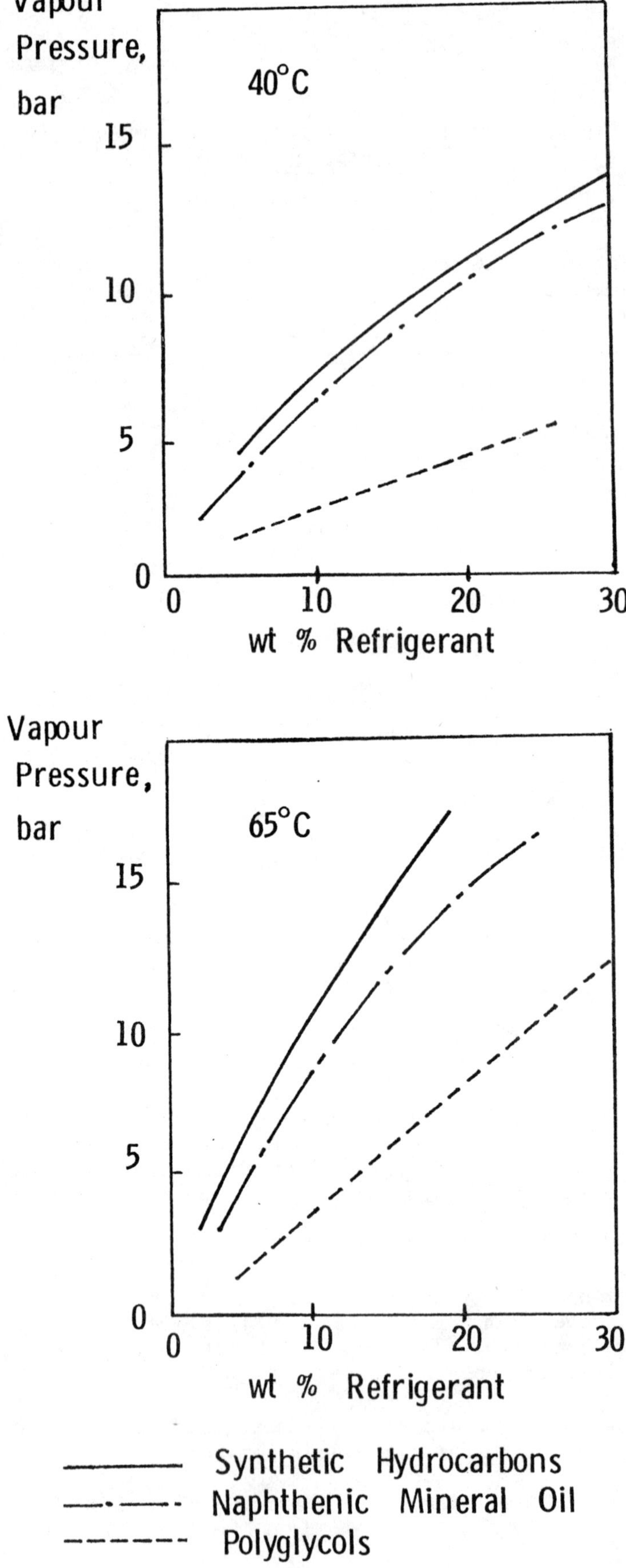

Fig. 1 Comparative Solubility Data of Refrigerant 22 in Mineral Oils, Synthetic Hydrocarbons and Polyglycols

<u>Example of Determination of Kinematic</u>
<u>Viscosity of Oil - Refrigerant Mixtures:</u>

For technical purposes, use of the oil
system pressure (approximately equivalent
to the refrigerant vapour pressure) in
 bar and the oil temperature in   °C
to describe the lubricant system of a
refrigerating plant is permissible, the
system pressure and the oil temperature being mesured at the same point.
The two values (pressure and temperature) give the point of intersection 1
on a refrigerant-wt. %-line (existing or
to be added) in Diagram   1
A vertical line plotted from the point
of intersection 1 to the corresponding
oil-wt. %-line (existing or to be added)
in Diagram 2 gives the point of
intersection 2. A horizontal line drawn
from the point of intersection 2 to the
left-hand scale gives the kinematic
viscosity of the oil-refrigerant mixture.
Vertical lines drawn from the points
of intersection 1;1' and 1'' in Diagram
  1 give the corresponding points of
intersection 2; 2' and 2'' in Diagram
  2 , which can be joined to obtain
the isobaric curve for 5 bar.
The right-hand part of the isobaric curve
asymptotically approaches the 100 wt.%-
oil-line. The left-hand part of the
isobaric curve is of critical importance
for bearing design, since a relatively
small decrease in temperature results
in a very large fall in viscosity.

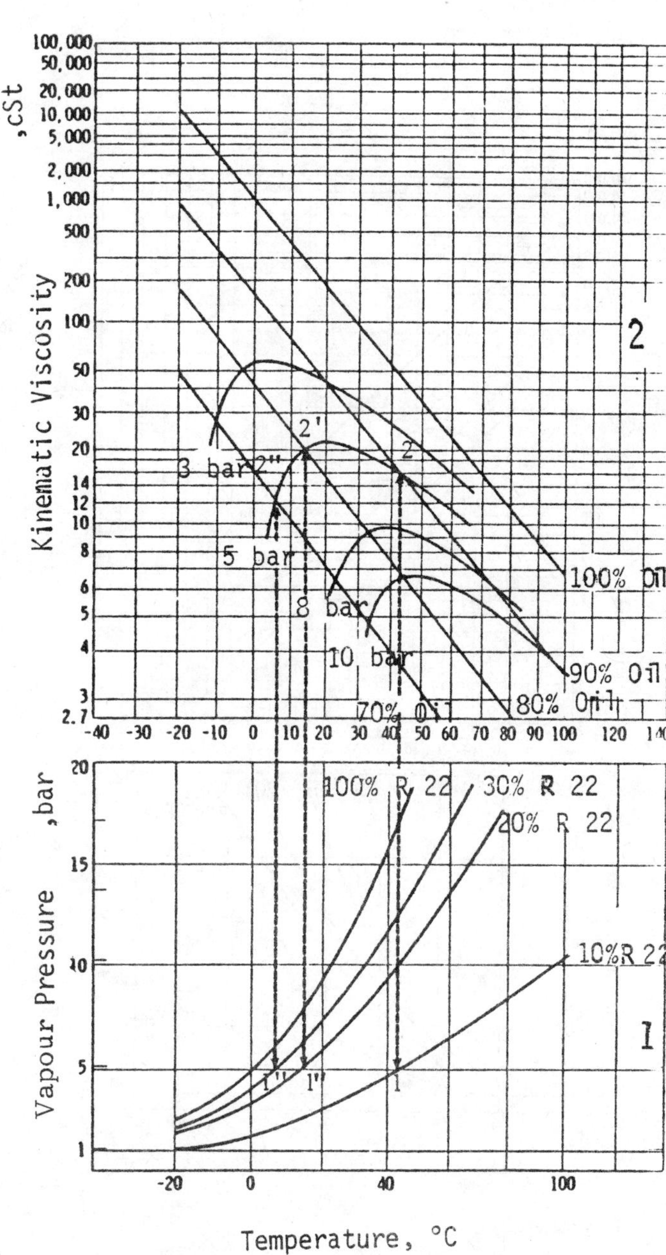

Fig. 2 Viscosity/Pressure/Temperature Chart ISO 50 Naphthenic Oil with R 22.

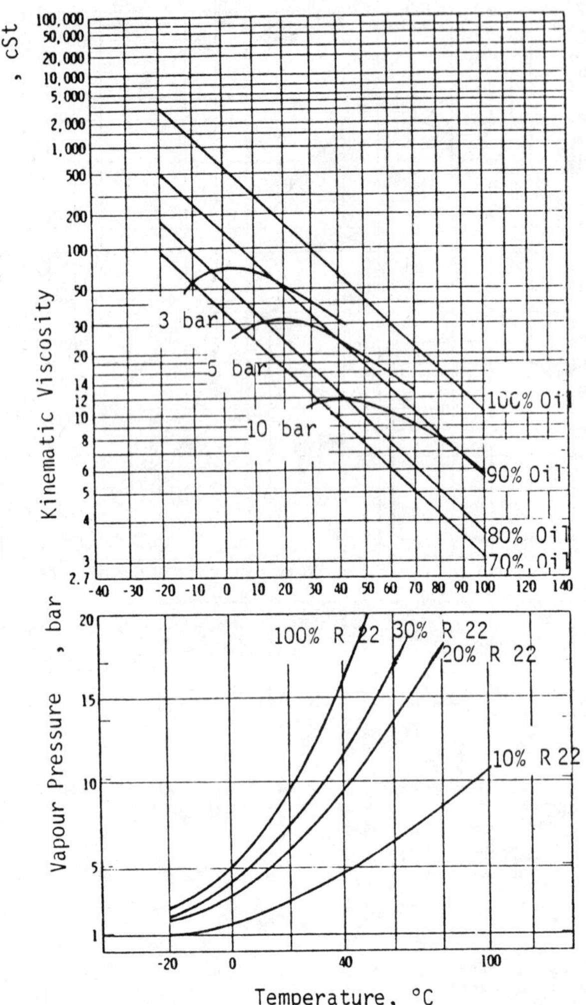

Fig. 3
Viscosity/Pressure/Temperature Chart
SHF A with R 22.

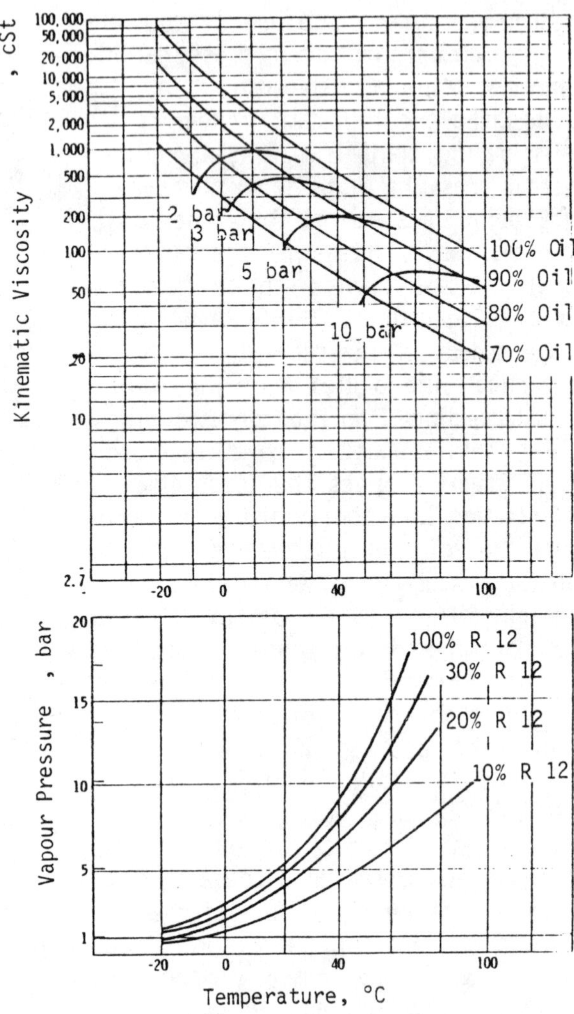

Fig. 4
Viscosity/Pressure/Temperature Chart
Polyglycol F with R 12

A SYNTHETIC LUBRICANT FOR HIGH TEMPERATURE HEAT PUMPS

F. Giolito and M. de Moncuit

Rhone-Poulenc, France

Summary

The thermal stability and main characteristics of a new synthetic lubricant fitted to heat pumps using R 114 as working fluid had been established, which allowed experimental service tests without any trouble, up to 120°C as condenser temperature. Complementary results on thermal stability with R 113 - R 133 a - are given.

Held at the University of Warwick, U.K.
Symposium organised and sponsored by
BHRA Fluid Engineering

1. <u>INTRODUCTION</u>

This work was initiated in 76-77, when ELECTRICITE DE FRANCE began to pro-
mote high temperature heat pumps using R 114 as working fluid in order to obtain condenser temperature of 120°C. The question was to have a lubricant viscous enough when
diluted by R 114, chemically and thermally stable at the maximum temperature (at the
exhaust of the compressor, temperature of the gas reach 140-150°). First attempts with
heavy distillates of branched alkylbenzene, with or without special light viscosity
improvers, were unsuccessful in service on compressors.

So, special alkylterphenyls were synthesised [3] whose viscosity and other
properties can be adjusted by the alkylation ratio. One of these products -801 E 23-
had the viscosity requirements (25 cS at 100°C) and had been therefore widely experimen-
ted with on a lab. scale, and on prototype heat pumps run by EDF.

The job of RP's labs concerned stability tests - in the presence of the three
kinds of components : fluids - lubricants - metals ; the viscosity - pressure - temperature - concentration charts for pratical use were then established from experimental
measurements (2). The main characteristics of this lubricant are presented in table 1
and viscosity characteristics with R 114 in diagrams 2a and 2b.

The first results on stability were good enough to encourage EDF to start
running prototype heat pumps, while RP's lab. was going on very long stability tests.
Several experimental batches of some hundreds of kg of lubricant were manufactured from
RP's pilot facilities.

2. <u>EXPERIMENTAL PROCEDURES FOR STABILITY TEST</u>

Tests had been adapted from our usual practice on organic heat transfer fluids studies.

Samples of products are pyrolysed in small pyrex flasks (fig.3) of about 12 ml internal volume, which can stand pressure up to 50 bars without exploding. Typical quantities of products are :

                         Fluid   : 6 to 8 grams
                         Lubricant   : 0.7 to 0.9 gram
ratio lubricant to fluid is about 15 %.

Metals are generally introduced in the flasks as small wires through the capillary tube ; some special flasks had been prepared including rectangular steel sheets for corrosion tests. Geometric characteristics of metallic samples are given on table (4).

Lubricant is introduced through a capillary fitted to a syringe.
Fluid is introduced through a capillary directly fitted to the storage bottle, while the flask is placed for a short moment in liquid nitrogen. The flask is then fitted to a vacuum device and sealed (pressure : 1-0.1 millibar).

The flasks are then placed in a furnace (fig.5) which is an aluminium cylinder (Ø 200 mm H 200 mm) drilled with 21 holes - electrically heated and quite well insulated.

Temperature homogeneity around each flask is about 1°C, and the reference temperature of the block remains in the range of 1°C during the very long tests (up to 8000 hours).

Concerning R 12 - because of its very high vapor pressure at the test temperature - the pyrex flasks were pyrolysed inside a stainless high pressure autoclave, with a nitrogen counter pressure of 60 bars, to mininize Δp between the inside of the flask and outside.

After pyrolysis, flasks are broken in a special device (fig.6) and the fluid mixed with 15,6 g isopropanol for acidity measurements by a potentiometric method (reactant : tetra N butyl ammonium hydroxide 0.1 N).Four or five peaks can be observed ; the first peak comes from HCl, and the second from HF. Others come from weaker acids. Acidity value is reported for 1g of working fluid.

Steel sheets devoted to corrosion tests had been weighed and observed under optical microscope or scanning (MEB).

3. <u>RESULTS AND CONCLUSIONS</u>

Preliminary tests included a general scope on R 114, R 113, R 133a, R12, R 142 b, R 12 B1.

They confirmed that R 12 B1 was very bad at 160 and 175°.
R 113 was rather fair without lubricant up to 160°/2000 hours.
R 114 - R 133 a - R 12 - R 142 b were in the same range of stability rather good at 160° and limit at 175°.

So the long term program was focused on R 114, with or without lubricant, with or without different metals diversely combined (Fe, Cu, Al).

Results are reported on diagrams and the main conclusions are :

. the global rate of decomposition of R 114 ($CF_2Cl - CF_2Cl$) and R 133a (CF_3CH_2Cl) at 160°C keeps low enough in the presence of metals and lubricant 801 E 23 after 8000 hours. After the beginning period, decomposition seems to be rather linear.

. There is no important problem of corrosion with dry fluids (20 ppm water max.). Generalised corrosion of steel keeps very low.
However corrosion was not studied thoroughly and care must be taken when installations are opened.

. The lubricant 801 E 23 has been tested by EDF on heat pumps working with R 114, between 70°C (evaporator or temperature) and 120° (condenser temperature) a COMEF compressor 2 CC 68 driven by a 18 kW electrical motor (flow 35.5 m^3/h at 1450 rpm).

After more than 500 h for instance, no mechanical problem had occured. The lubricant remained clear and its acidity below the threshold of the detector ($<$ 3 mgKOH/g).

The different characteristics of a typical run are given on table 10

These results encourage the running of heat pumps with R 114 up to 120°C, in the condenser.

4. <u>REFERENCES</u>

4.1. <u>LAMY - BRUN</u>   RHONE-POULENC Report 22 Mai 1980

Etude du système R 114/801 E 23

VISCOSITY VS PRESSURE - TEMPERATURE

4.2. <u>LAROCHE - VUILLAME</u>   EDF REPORT HE 142 W 1526 14.11.79

Essai d'un nouveau lubrifiant de synthèse RHONE-POULENC 801 E 23
sur pompe à chaleur haute température.

4.3. <u>DESBOIS Michel</u>   Unpublished work

5. <u>ACKNOWLEDGEMENTS</u>

This work has been initiated and financed on a large part by EDF.

Table 1

| Specific gravity : | g/cm³ | at | 20°C | 0.899 |
| | | at | 100°C | 0.851 |

| Viscosity (typical) | cS | at | 40°C | 282.5 |
| | | | 100°C | 24.8 |

VIE   (viscosity index extended)   112

Pour point	°C	-18
Fire point	°C	300
Aniline point	°C	95

1 - 1500	cP	5 - 400	cP
2 - 1000	cP	6 - 300	cP
3 - 700	cP	7 - 200	cP
4 - 500	cP	8 - 150	cP
9 - 100	cP	13 - 30	cP
10 - 70	cP	14 - 20	cP
11 - 50	cP	15 - 15	cP
12 - 40	cP	16 - 10	cP

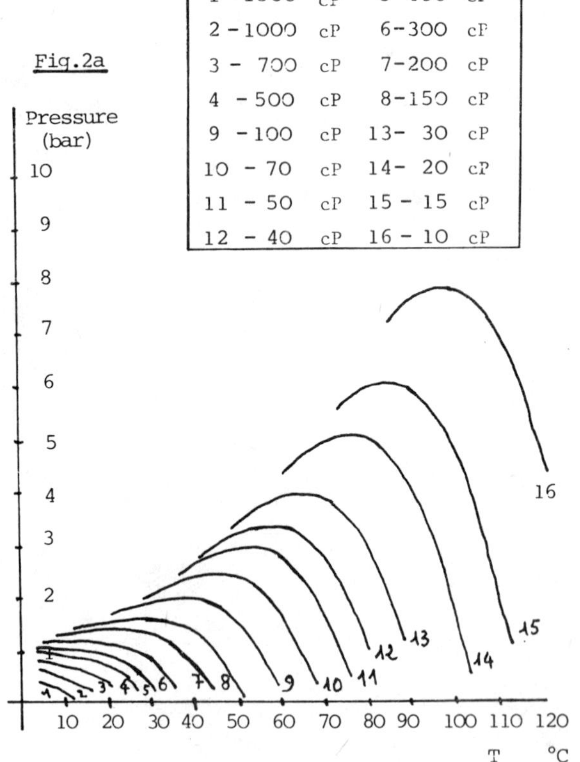

Iso viscosity curves
Mixtures of R 114 and 801 E 23

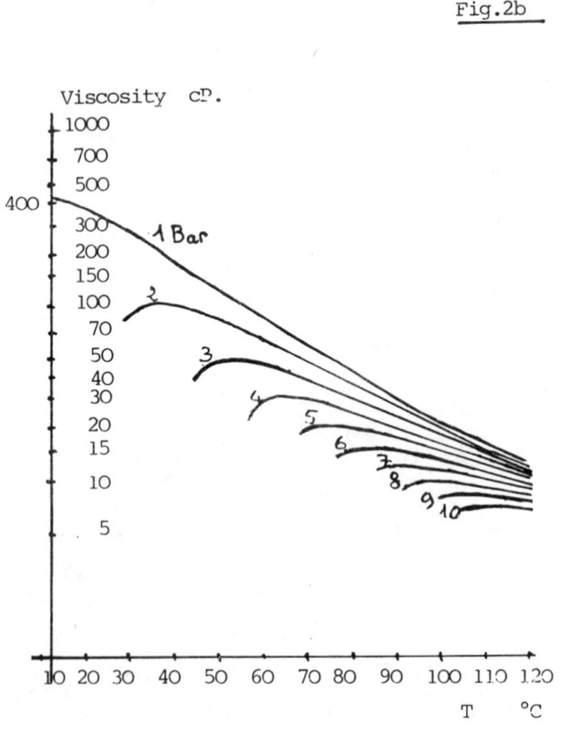

Iso bar curves

PRESSURE - VISCOSITY DIAGRAMS

Fig.3

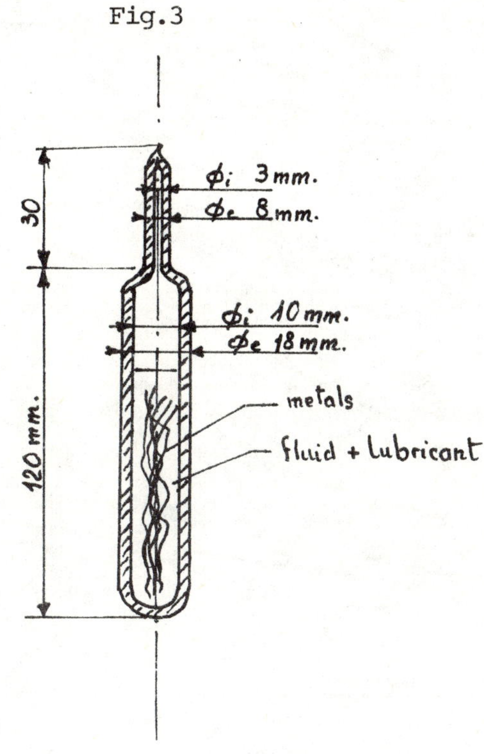

Table 4 - metallic samples

Steel   Ø 0.02 mm

Copper   Ø 0.1 mm

Al   Flat ribbon 0.15 mm

Corrosion steel samples

80 x 7 x 3 mm

Fig.5

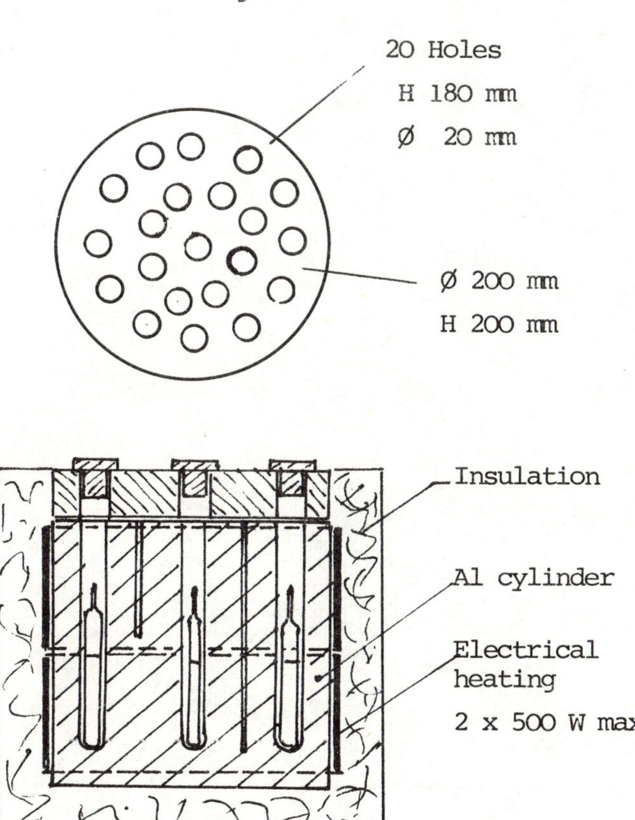

Fig.6

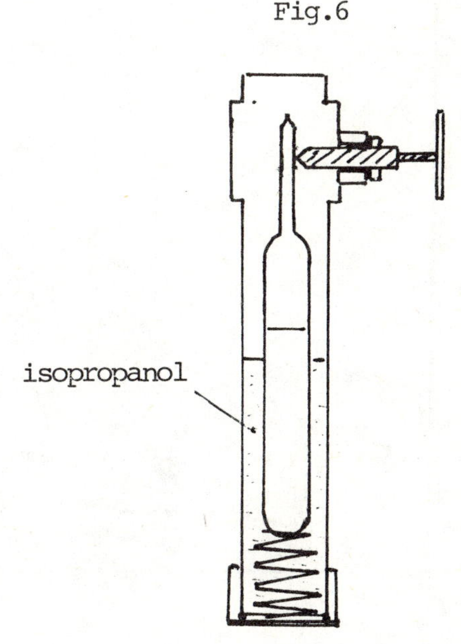

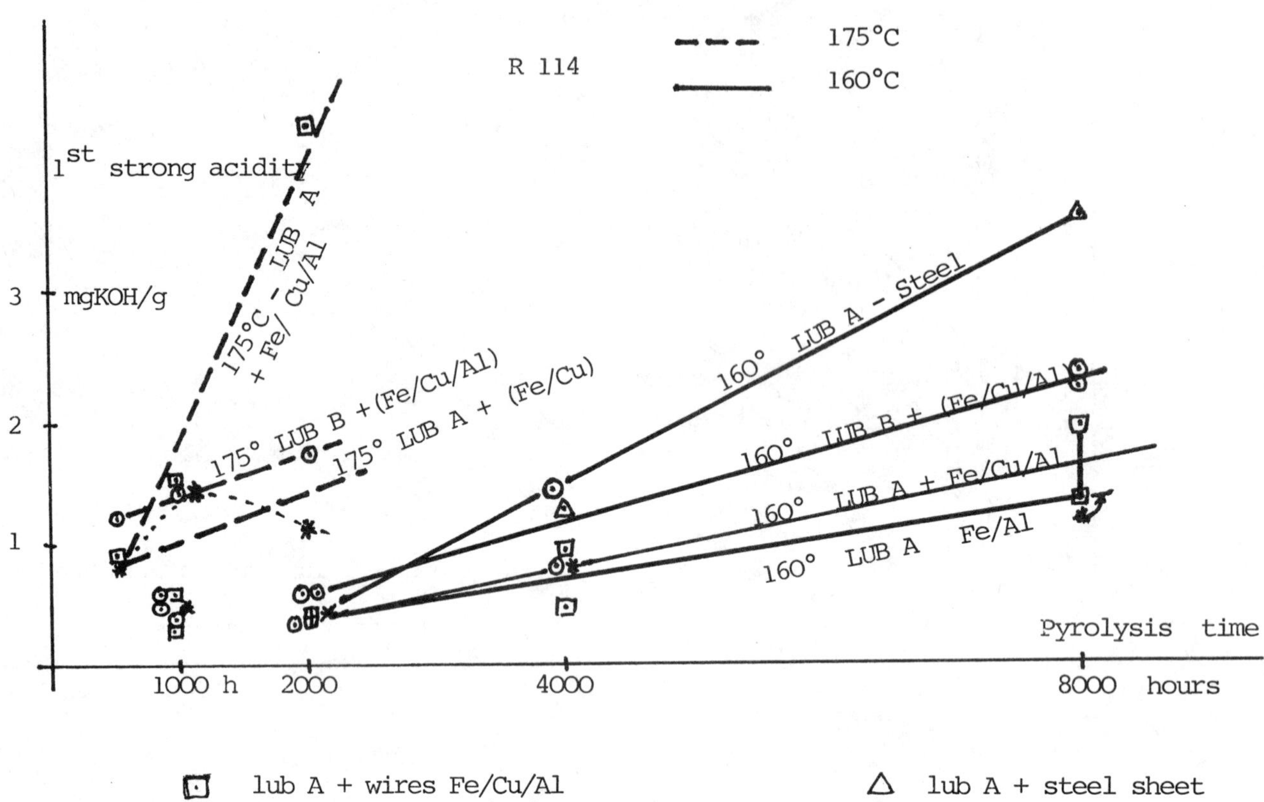

Fig 7
R 114
175°C
160°C
1st strong acidity
3 mgKOH/g
2
1
175°C - LUB A + Fe/Cu/Al
175° LUB B +(Fe/Cu/Al)
175° LUB A + (Fe/Cu)
160° LUB A - Steel
160° LUB B + (Fe/Cu/Al)
160° LUB A + Fe/Cu/Al
160° LUB A Fe/Al
Pyrolysis time
1000 h   2000   4000   8000 hours
lub A + wires Fe/Cu/Al
lub B + wires Fe/Cu/Al
lub A + steel sheet
lub A + wires Fe + Cu

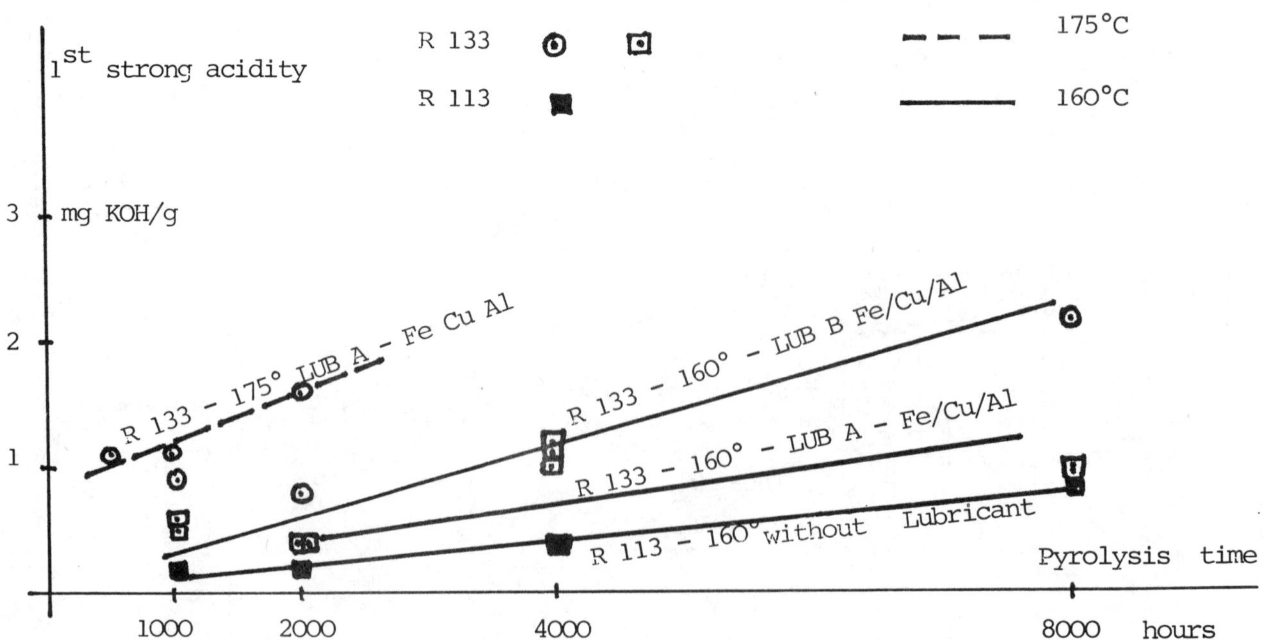

Fig.8
1st strong acidity
R 133
R 113
175°C
160°C
3 mg KOH/g
2
1
R 133 - 175° LUB A - Fe Cu Al
R 133 - 160° - LUB B Fe/Cu/Al
R 133 - 160° - LUB A - Fe/Cu/Al
R 113 - 160° without Lubricant
Pyrolysis time
1000   2000   4000   8000   hours

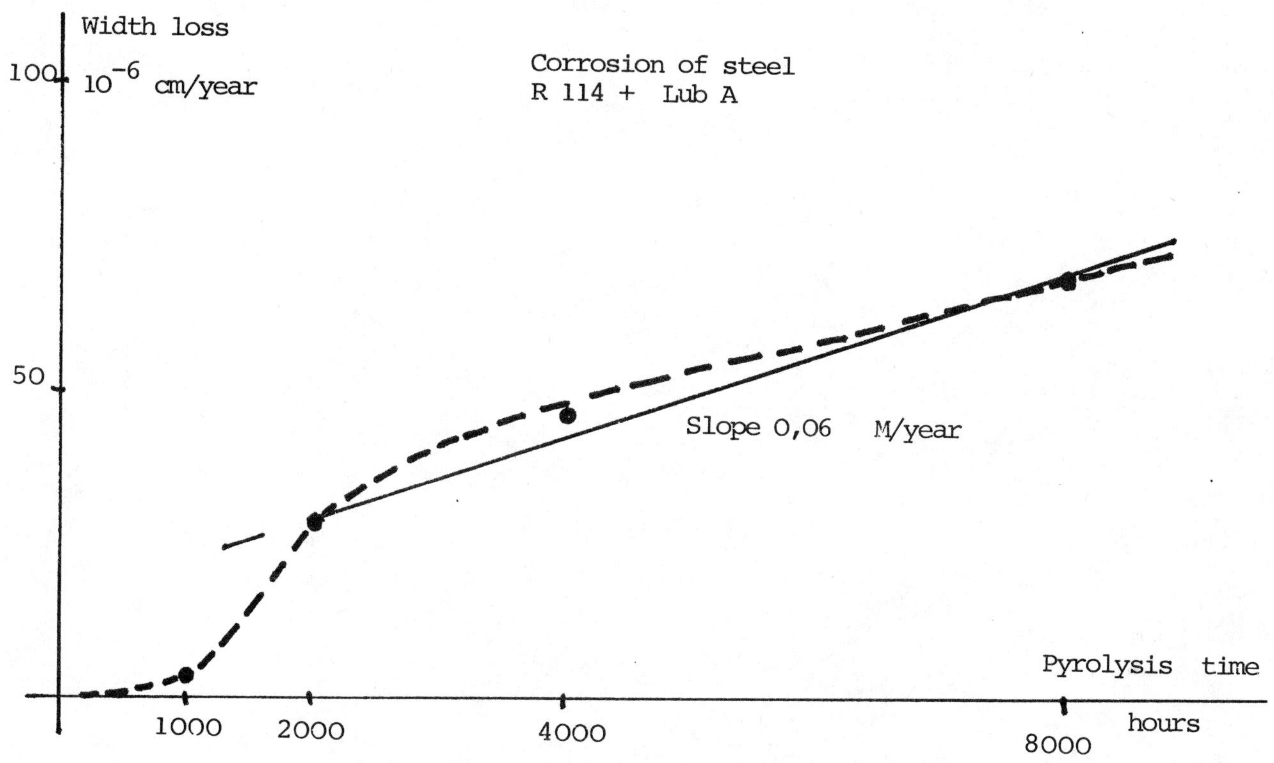

Table 10

A TYPICAL RUN OF HEAT PUMPS WITH R 114

Inlet temperature	°C	90.1
Inlet pressure	°C	7.4
Inlet saturation temperature at aspiration	°C	70
Super heating R 114	°C	20.1
Lubricant temperature in cankrase	°C	105
Lubricant super heating	°C	35
Lubricant viscosity	cS	21.5
R 114 concentration mass %		15.5
Viscosity of pure lub.	cS	22

INDUSTRIAL HEAT PUMP DEMONSTRATIONS

G. J. Newbert

ETSU (Energy Technology Support Unit, Harwell), U.K.

Summary

This paper describes some recent industrial heat pump installations in the UK. Three examples are taken from a list of heat pump installations, supported through the Department of Energy's Energy Conservation Demonstration Projects Scheme. The three examples illustrate the economic application to the electric, the engine and the steam turbine driven heat pumps in three different applications in the food industry. This section of industry is now appearing as the first market for heat pumps in the UK. In the longer term as the heat pump is adopted by other industrial sectors the energy savings from heat pumps in industrial processes could built up to be worth £40-50M a year.

Held at the University of Warwick, U.K.
Symposium organised and sponsored by
BHRA Fluid Engineering
©BHRA Fluid Engineering, Cranfield, Bedford MK43 0AJ, England.

1. INTRODUCTION

Since 1974, with the rise in energy costs, energy saving measures have been increasingly introduced into British industry. Good housekeeping, improved insulation, better controls, condensate return and direct heat recovery are clearly among the first things to be done, but they can only do so much. The heat pump is unique in that it can recover heat from a low temperature and give out that heat at a higher and useful temperature. For many industrial processes the heat pump is the only effective means of further reducing energy costs.

Over the past twelve months there has been an increasing number of heat pumps going into industry. Some of these are supported under the Department of Energy's Energy Conservation Demonstration Projects Scheme, to show the heat pump's technical and economic effectiveness in working situations to other potential users.

This paper outlines three recent heat pump installations. The three show the variety of systems that are available to industry and show the conditions that need to exist for a successful heat pump application.

2. HEAT PUMP APPLICATIONS

The heat pump is not a panacea for energy saving. Other forms of energy saving should always be considered first and the heat pump should be considered only when these have been exhausted. Then, for the heat pump to be an economic proposition there are a number of basic conditions that must be satisfied :-

(i)   the waste heat is available at a temperature not much lower than that required; this is because the performance of a heat pump improves as its temperature range or lift is reduced.

(ii)   there must be no high grade waste heat available to satisfy demand; if there is then it will always be cheaper to use conventional heat exchangers rather than heat pumps.

(iii) the heat must be required at a relatively low temperature; present equipment is limited in the temperature at which it can deliver heat; this effectively limits operation to less than 120^0C.

(iv)   the heat pump must have a high duty cycle; this is because it is a costly device and it will only pay if its utilisation is high.

In short, because of their capital cost and technical limitations heat pumps are attractive only where they can operate over long periods, within a temperature range of $30-120^0$C, and where simpler and cheaper energy management and heat recovery techniques cannot be applied. However within these conditions there are a large number of heat pump applications to be found in industrial processes.

2.1   Waste Heat Recovery.

It is common practice in many parts of industry to heat up water for process use
and then discharge warm effluent to waste or reject the heat from a cooling tower.
The heat pump can usefully recover this low grade waste heat for preheating feedwater.
The heat pump upgrades the waste heat, feeds it back into the process and reduces the
overall energy requirement of the process.

2.2   Combined Heating and Cooling.

In this application the heat pump does two jobs. It can remove heat from one
stream and give out heat to another stream. Hence in a process that requires a
product to be repeatedly heated and cooled as in the pasteurisation of milk, the heat
pump can perform both functions. A very efficient system can be achieved when the
required temperature changes are well matched to the heat pump.

2.3   Drying.

Drying is a necessary process in many parts of the food, paper and chemicals
industry. The heat pump is well matched to the drying process as it can recover the
heat in the air that is discharged from the drying chamber for heating the air going
into the chamber. A high efficiency can be achieved when the temperature in the
drying chamber doesn't need to be too high.

3.   HEAT PUMP TYPES

For each application a range of heat pumps can be used. The main types that are
available to Industry are based on the vapour compression or Rankine cycle. This
technology has become well known from refrigeration practice. Other cycles can be
used for heat pumps, such as the Brayton or Joule cycle, but they require more
development before they become commercially available. Of the possible drives for
heat pumps there are three main sorts that can be used. These are the :-

- Electric drive

- Internal combustion engine

- Steam turbine

3.1   The Electrically Driven Heat Pump.

The electric heat pump is the simpler system. The electric motor is clean,
highly efficient and reliable. At relatively low temperature lifts, of the order of
30^0C the electrically driven heat pump can deliver more than five times the
electrical energy it uses as heat.

3.2   The Engine Driven Heat Pump

With the engine driven heat pump heat taken from the engine cooling water and
the exhaust is used. This forms a large part of the heat delivered by the heat pump,
which can be at a fairly high temperature. This increases the overall output
temperatures and the temperature lifts that can be delivered economically by the heat
pump.

3.3   The Steam Turbine Driven Heat Pump.

The steam turbine is commonly used in industry to drive machines such as
electrical generators and pumps. The heat pump is another piece of equipment that
can take advantage of this well tried drive. They have a very attractive feature
for industrial sites that generate steam at high pressure and use some steam at
low pressure. The pressure reducing valve can be replaced with a steam turbine to
provide a heat pump that has virtually no running costs.

4.   INDUSTRIAL HEAT PUMP DEMONSTRATIONS

With the heat pump coming in a variety of configurations the choice of system depends on the available sources and possible uses of heat and their temperature levels. The heat pump demonstration projects aim to illustrate a range of applications that could be attractive to Industry. Fig.1 is a matrix which shows the industrial heat pump opportunities according to the type of drive, size and type of application. The boxes which are filled indicate the sector in which a demonstration project is being mounted or is in the pipeline.

These projects are listed in Fig.2. All of these projects have target energy savings associated with them, based on an assessment of the market that can use similar systems. The overall target to date is 133,000 tonnes of coal equivalent a year.

The demonstration projects are an opportunity for UK industry to see at first hand a heat pump working in an application that can be of direct benefit to them. The demonstrations show how a heat pump can be used, what the benefits are and as operating experience is gained how they have worked out in practice.

Three examples are taken to illustrate progress so far.

4.1   Heat Recovery By Heat Pumps in a Dairy.

4.1.1 Operation.

The first installation in the UK of a packaged electrically driven heat pump for waste heat recovery was in a Unigate Dairy at Walsall. The heat pumps are capable of providing 975 kW of heat from 181 kW of electrical energy - a coefficient of performance of over five.

The Unigate Dairy produces sterilised milk. In this process the milk is heat treated and filled into plastic bottles. These bottles are then passed to a Hydrolock steriliser, before being labelled and packaged. Here the bottles of milk are heated and then cooled and the steriliser discharges large quantities of warmed water. Normally around 200 pints of milk a minute are processed; when the factory is operating on double shifts, however, it produces nearly three-quarters of a million bottles a week.

A schematic of the operation of the heat pumps at Unigate is shown in Fig.3. The heat pumps recover the heat from the discharge of the steriliser and use it for process and space heating. They are also used to chill the discharge from the steriliser so that it can be re-used for cooling in the steriliser. In earlier, similar plants the cooling water was poured down the drain.

The process water is then used for pre-heating both incoming milk in the dairy processes and boiler feedwater, and providing washing-down water and factory heating. To allow for some diversity in the demand for hot water and to enable the heat pumps to operate for reasonable periods of time more or less continually, the hot wells provide some storage.

4.1.2 Economics.

The financial benefits from the use of heat pumps are twofold. First, there are energy cost savings and secondly water cost savings. Together they give a simple payback period of 1½ years on the cost of the heat pumps when the dairy operates on double shift. If the energy savings only are taken into account and the heat pumps save 50% of the energy that would have been used for the same purposes, the simple payback period is three years.

4.1.3 Performance.

The performance of the installation at Unigate is being monitored. The monitoring began in earnest in the summer of 1981 when the heat pumps began working full-time. Early results show that the heat pumps are providing COPs of over 5. Earlier problems had arisen at the start up of the sterilising process, when too high temperatures were fed in as the source of the heat pumps, but these were overcome when bypasses were installed.

4.2   The Gas Engine Driven Heat Pump at ABM, Louth.

The first heat pump to be installed in the maltings industry in the UK was put
in by ABM at Louth.

4.2.1 Operation.

The heat pump operates in conjunction with a simple run round coil as shown in
Fig.4. The combined system provides 3 MW of heating. The run round coil takes heat
for the initial preheating of the kiln inlet air. The heat pump circuit takes some
more heat from the exhaust, which is raised in temperature for further heating the
inlet air. Heat is also recovered from the engine, which drives the heat pump for
heating the air into the kiln.

4.2.2 Economics.

The bulk of energy use in maltings is in drying and the greatest scope for energy
saving is in waste heat recovery. With the discharge temperatures of 30^0C
conventional heat recovery can be done but not all the available waste heat can be
used in that way. Expected energy savings at Louth are 47% compared with kilns
operating without heat recovery. Also the kilns normally operate all the year round
and so give the system at Louth a 3.5 year payback period.

An additional advantage of the heat pump is that it provides indirect heating -
in contrast to traditional direct firing the air stream itself. The benefits of this
include a shorter kilning programme and longer kiln life.

4.2.3 Performance.

The heat pump was started in March 1981. During the course of 1981 there has
been a steady reduction in fuel usage corresponding to the learning curve for
operating the new machinery. In July 1981 ABM's target energy consumption figure of
20 therms/tonne was reached, and since then the fuel usage has been within the range
of 18-20 therms/tonne corresponding to an overall saving of 47% in fuel gas usage.
In cash terms the reduced gas usage, at 30 pence/therm, equates to an annual saving
of £148,800 but after allowing for maintenance and electrical running costs this is
reduced to £135,200.

Of particular concern to the other potential users of heat pumps is the long
term reliability of the gas engines, the control of the plant under varying
operating conditions and the behaviour and optimisation of the refrigerant circuit.
The engines have, after 4,800 hours of service up to the end of October, given
trouble-free operation, with routine maintenance being carried out at 1,000 hour
intervals. However, under abnormally low load conditions there has been a
tendency for the operating compressor(s) to trip out on low oil level. This is not
causing ABM any undue operating problems. For future installations this difficulty
may be avoided by increasing the size of the oil rectification system.

4.3   Heat Recovery By a Steam Turbine Driven Heat Pump.

The first steam turbine driven heat pump in the UK is going to be installed by
the Marfleet Refining Company in Hull. The processing of edible oils involves the
use of large quantities of steam, a substantial amount of which is used in steam
ejectros for vacuum raising. Heat from the steam ejectors is rejected to cooling
water. Due to contamination the condensate cannot be used directly for boiler
feedwater.

The heat pump is the only means of recovering the heat from the cooling tower
circuit for boiler feedwater preheating to 80^0C. Of the three options that were
considered for the job by the Marfleet Refining Company including electric and
engine driven systems, the steam turbine drive was chosen since it provided the best
financial returns. The steam turbine took advantage of the circumstances on site at
Marfleet that steam was generated at medium pressure but could be used at low
pressure hence providing a very economic drive.

4.3.1 Operation.

A schematic of the operation of the steam turbine driven heat pump is shown in Fig.5. Medium pressure steam is produced by a central boiler at 17 bar. Part of this is used in the steam turbine to drive the heat pump and produce low pressure steam at 2 bar. The heat pump recovers heat from the cooling circuit and provides 1067 kW of heat for boiler feedwater preheating.

4.3.2 Economics.

The financial benefit from the heat pump is by replacing the oil to heat the boiler feedwater. The heat pump is expected to operate for 6,000 hours a year and the saving is anticipated at 630 tonnes of fuel a year to give a simple payback period of $2\frac{1}{2}$ years.

4.3.3 Performance.

The heat pump should be installed and operating by mid-1982. The performance of the installation to be monitored and the results will be published in due course.

5.   OTHER INDUSTRIAL APPLICATIONS

In ETSU we have made an assessment of the potential for heat pumps in industry at large. The results are shown in Fig.6. The proportion of total industrial energy use is quite small, but this is still equivalent to 600,000 tonnes of coal equivalent per annum, worth perhaps £40-50M per annum. The food, chemicals and textiles industries account for two-thirds of the potential.

6.   CONCLUSION

The food industry is appearing as the first market for heat pumps in the UK and is providing a very good grounding for them. There are short term prospects for further installations similar to those at Unigate, ABM (Louth) and Marfleet, saving 83,000 tonnes of coal equivalent a year. In the longer term in the rest of the market there are much larger prospects.

We need more demonstration projects to fill the gaps in Fig.1 and we need widespread replication of all these projects so as to achieve a worthwhile part of the very large potential that is available.

The heat pump represents a great opportunity for Industry to save on its energy costs. The annual saving is reckoned to be worth £40-50M a year. In addition this represents a great opportunity for the equipment supply industry for electric, engine and steam turbine systems. This could be worth at least £150M in the UK over the next 10 years.

Figure 1 Principal relevant sectors for heat pump demonstrations

Industrial heat pump applications	Electrically driven		Gas engine driven		Other drives
	Large	Small	Large	Small	
Drying		Paper	Malting		
Heating & cooling		Dairy	Food	Food	
Process heat recovery	Food			Textiles	Steam turbine

Figure 2 Heat pump demonstration projects

Heat pump type	Project	Energy saving target tce/a
large electric	Heat recovery from a steriliser in a dairy	6,000
	Heat recovery from refrigeration plant	22,000
large gas engine	Gas engine driven heat pump in maltings (Three projects)	60,000
small gas engine	Heating and cooling in a cheese making plant	12,000
	heat recovery in a textile finishing works	16,000
other drives	Heat recovery by steam turbine driven Heat pump	17,000
all	Eight projects	133,000

Figure 3 Schematic of the heat pumps at Unigate

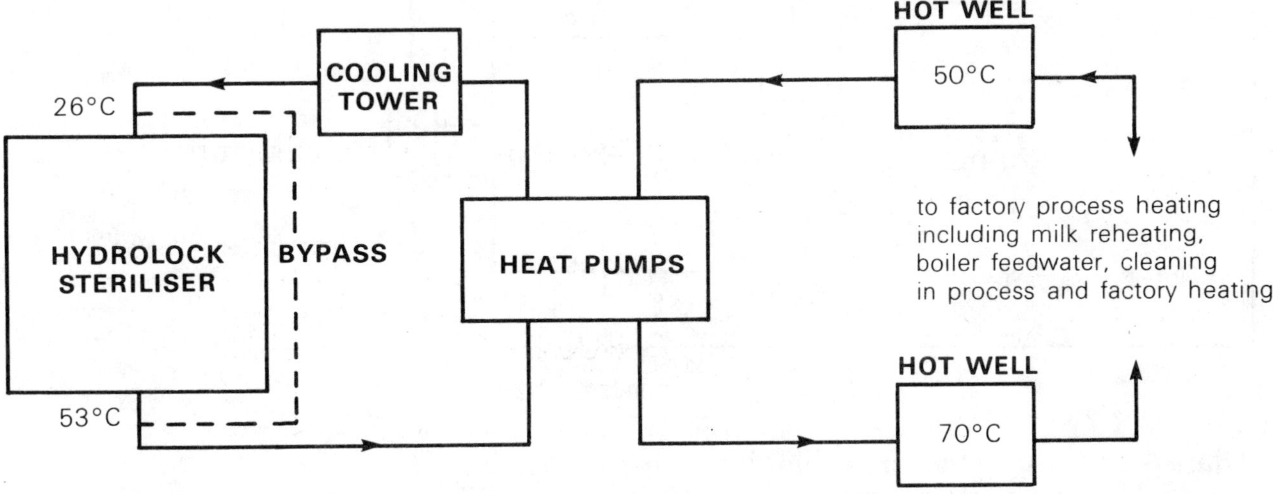

Figure 4 Schematic of a gas engine driven heat pump for malt kilning

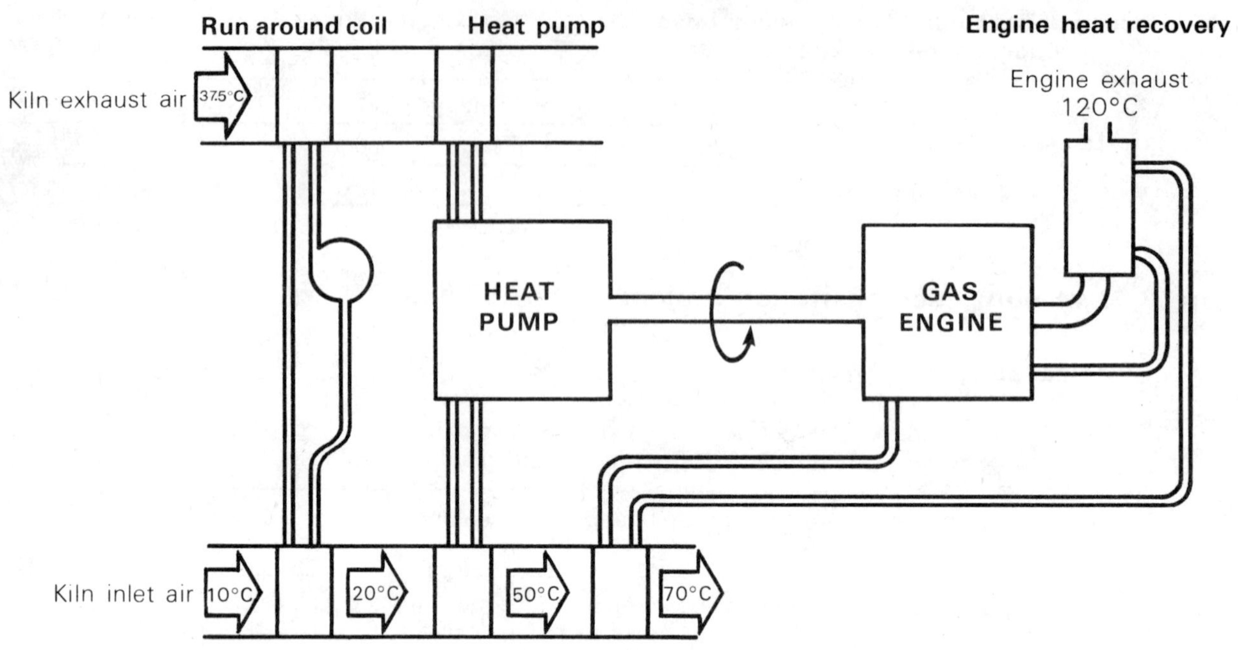

Figure 5   Schematic of the heat pump at Marfleet

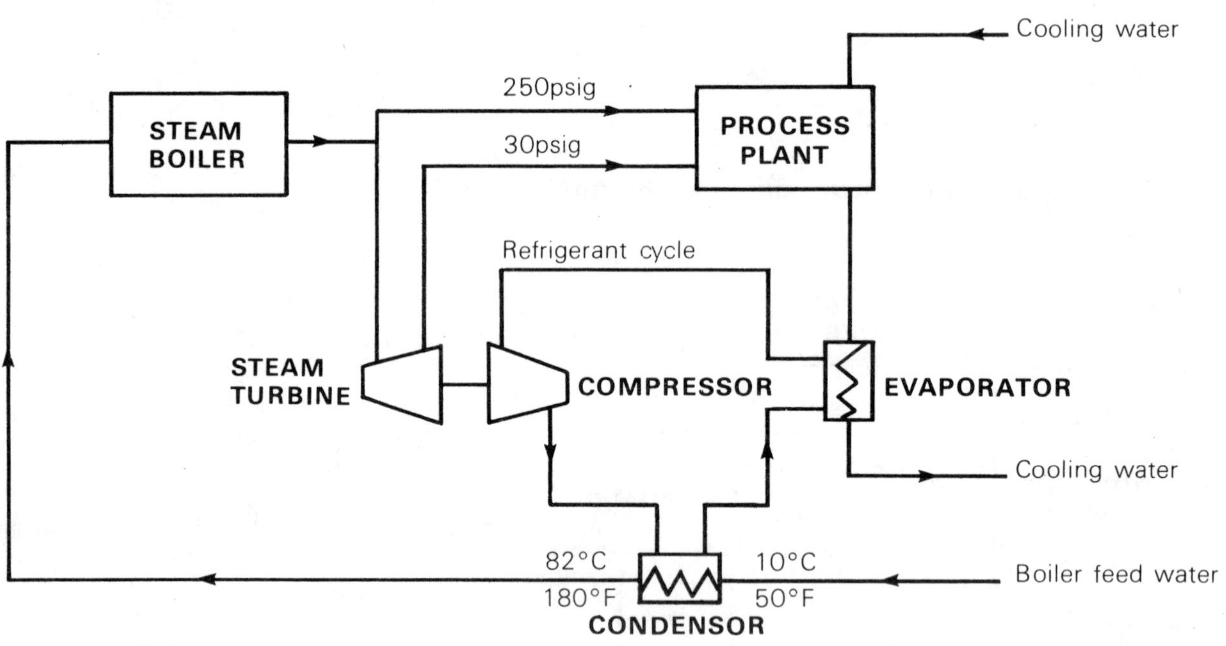

Figure 6 Estimate of heat pump potential energy saving (TCE)

Industrial sector	Potential energy saving
Engineering	50,000
Food	200,000
Chemicals	100,000
Textiles	100,000
Paper	50,000
Building	50,000
Other	50,000
Total	**600,000**

PROCESS DRYING WITH A DEHUMIDIFYING HEAT PUMP

T.N. Oliver
Westair Limited, U.K.

Summary

This paper outlines the energy saving potential of the heat pump dehumidifier when applied to the process drying industry. Specific areas of application are identified and estimates are made of the heat pump efficiency in comparison with convection type heat and vent dryers. Several applications are presented in detail showing substantial energy use and cost reductions over the conventional methods, and recent and future developments of the system are briefly discussed.

Held at the University of Warwick, U.K.
Symposium organised and sponsored by
BHRA Fluid Engineering
©BHRA Fluid Engineering, Cranfield, Bedford MK43 0AJ, England.

NOMENCLATURE

C.O.P.	=	Coeffiecient of performance
Q_h	=	Heat rejected at condenser
T_1	=	Ambient temperature of air into a convection dryer
T_2	=	Temperature of air after passing over heating coil
T_3	=	Temperature of exhaust air
V	=	Ratio of recirculated air to total air flow within dryer
W	=	Thermal equivalent of energy input to compressor
η_t	=	Thermal efficiency

1.1   Energy Use in Process Drying

Substantial amounts of energy are used within industry in a variety of processes and by a variety of methods but with one aim, namely to physically remove water from the product. It has been estimated (Ref 1) that of the total energy consumed by British Industry, other than in the Iron and Steel sector, approximately 6% is used simply to vaporise water. In 1979 the total energy used by these 'other industries' has been given (Ref 2) as 1920×10^6GJ and therefore about 115×10^6GJ of energy is being used solely in drying. This figure is possibly conservative as the estimate assumes a thermal efficiency in drying of 50%.

Although drying processes are energy intensive few users have any knowledge of their dryer's efficiency or drying costs. Most systems have little or no monitoring or instrumentation equipment and any cost analysis is usually considered to be too difficult to achieve. As a result most conventional dryers which may have evolved over many years are very often neglected and rarely considered as a section of the total process where substantial savings in energy usage and cost can be made.

1.2   The Drying Process

Products are normally dried by either one of or a combination of convection, conduction, radiation and vacuum. It is within the area of convection dryers where much work has been done to increase drying efficiency particularly by the application of heat pump dehumidifiers. In a convection drying process the product is dried by the passage of pre-heated air across its surface, the rate of drying being related to air velocity, air temperature and humidity. There are basically two types of convection dryers, batch or tray and continuous. In the batch type a single product load is dried for a pre-determined length of time with the chamber requiring recharging for subsequent product loads. In the continuous process, the product is fed onto a moving belt or track into a tunnel or multi-pass stove emerging after a time determined by the speed of the track or belt.

A batch or tray type convection is illustrated in Fig 1. An important feature of the conventional convection dryer is that once water has been removed from the product, it can only be removed from the chamber by being exhausted to atmosphere in its vapour form. As air is exhausted or vented, more ambient air must be drawn into the chamber and heated. This process is more often termed 'heat and vent' drying.

Drying efficiency is a term which has many variants but in its most useful form it is a measure of the quantity of energy expended in removing unit mass of water from a product. Normally this is measured in terms of kJ/kg although when considering systems incorporating electrically driven heat pumps units of kWh/kg are more readily useful. In the conventional heat and vent dryer the thermal efficiency can be increased by reducing the ventilation rate. The efficiency can be written by the approximation (Ref 3).

$$\eta_t = \frac{T_2 - T_3}{(T_2 - T_3) + (1 - V)(T_2 - T_3)}$$

If the rate of ventilation is reduced, the ratio 'V' of recirculated air to total air flow will increase and therefore the efficiency will increase. This, however, will tend to increase drying times as the process moisture removal rate will fall. Although in theory it is possible for the thermal efficiency of a convection dryer to approach 100% by reducing the ventilation rate to zero, in practice there must be a compromise where an economically viable drying time is achieved for a reduction in thermal efficiency.

A useful yardstick when comparing drying processes is the latent heat of vaporisation of water at ambient temperature. This figure is approximately 2300 kJ/kg and serves as an indication of the efficiency of a convection dryer operating at 100% efficiency. In addition to high ventilation rates there are many other factors which will reduce efficiency including:

conduction losses through the chamber structure
poor seals to doors
low burner efficiency
energy to drive primary air circulation equipment

In practice actual efficiencies can be very low with levels below 10%, 23000 kJ/kg, not unusual. Such low efficiencies are more likely where old steam raising plant is involved. Modern drying chambers which incorporate direct fuel firing and some form of heat recuperator have helped increase thermal efficiencies to around 60%, or 3800 kJ/kg.

2.   THE HEAT PUMP DEHUMIDIFIER

2.1   Method of operation

The dehumidifier is an air to air heat pump which functions in a similar manner to the domestic mechanical refrigerator. It comprises a vapour recompression refrigeration circuit with a refrigerant fluid operating on the reversed Rankine cycle. It is not intended here to detail the thermo-dynamic principles on which the dehumidifier operates, but more to explain its usefulness as an efficient moisture removal device.

A typical heat pump dehumidifier is shown in Fig. 2 and in schematic format in Fig 3. Its major components are a cold heat exchanger (evaporator), compressor, hot heat exchanger (condenser), expansion device and a fan to provide air movement. In common with all heat pumps heat is absorbed at the evaporator and rejected at the condenser with the compressor increasing the pressure of the refrigerant fluid and therefore the temperature at which heat is rejected. The quantity of heat delivered at the condensor, 'Q_h' will comprise the heat absorbed at the evaporator plus the thermal equivalent of the energy input 'W' required to bring about the compression, less any small losses that may occur. The efficiency of the heat pump, its coefficient of performance, can then be written:

$$C.O.P. = \frac{Q_h}{W}$$

In a heat pump dehumidifier the source for the heat absorbed at the evaporator is the humid air drawn into the dehumidifier from the dying process. As this air passes through the evaporator it is rapidly chilled to a temperature below its dew point resulting in water condensing out and being drained from the process. The heat recovered, approximately 2300 kJ for every kg of water condensed is rejected at the condenser at a much higher temperature than the air incident on the hot coil. Air is therefore returned to the process dryer and at a higher temperature than when it entered the dehumidifier. Should additional heating be required an auxillary heater bank is usually fitted within the dehumidifier.

In the dehumidification process water is removed from the chamber without ventilation, the system is totally recirculatory leading to a thermal efficiency approaching 100%. By removing water in its liquid rather than vapour state only a small amount of sensible heat is discharged and no latent heat is lost.

It is usual to determine the efficiency of a heat pump from its coefficient of performance as described above, however with a dehumidifier a more useful measure is the quantity of water condensed per unit of electricity consumed. This is termed the specific moisture extraction rate or SMER and is usually in units of   kg/kWh Additionally when designing a drying system incorporating dehumidifiers water extraction per unit time at any given air condition is necessary to enable the dehumidifier requirement to be matched to the system moisture removal rate. Performance charts for a typical dehumidifier are shown in Fig 4. The specific moisutre extraction rate for most well designed dehumidifiers lies between 1 to 4 kg/kWh with an average of about 2.5 kg/kWh. It is useful to compare this figure with that of the latent heat of vaporisation of water of 2300 kJ/kg or 1.56 kg/kWh. The dehumidifier therefore has the ability to remove water from a process at specific moisture extraction rates above that necessary for 100% efficiency of evaporation.

2.2   System Efficiencies

The figure of most importance to an Energy Manager when considering drying efficiencies is the total energy input per unit water output. Total system efficiency is lower than that of the dehumidifier alone because other components within the system also require energy. Fig. 5 shows a dehumidifier drying chamber which includes fans for primary air circulation but is totally enclosed. During the drying process in addition to electrical energy to drive the compressor it is also needed to:

> pre-heat the product and chamber structure
> produce the primary airflow
> replace conduction and air leakage losses.

The heat pump dehumidifier system is more suited to the batch than continuous system as the former allows total recirculation with a very low air leakage rate thus giving high thermal efficiencies. Only recently are continuous dryers becoming available which include effective air sealing of the product in feed and outfeed. Much of the recent work on dehumidifier application (Ref. 4) has also involved improved drying chamber design with particular attention to a high degree of thermal insulation and a gas tight seal to the chamber superstructure. Fans for primary air circulation have been chosen to give required air velocities over product surfaces using drive motors that will withstand the drying conditions of high temperature and high humidity. In this way residual heat from the motors is absorbed within the drying chamber instead of simply being lost to atmosphere.

The average system efficiency of a well designed heat pump dehumidifier is approximately 2300 kJ/kg and, as will be shown in section 3, in many cases the efficiency can be much higher resulting in specific moisture extraction as low as 1650 kJ/kg.

It has been shown (Ref. 1) that taking into account the efficiency of generation and distribution of electricity at about 30% a typical heat pump dehumidifier drying process will use less primary energy than a steam heated conventional dryer with a thermal efficiency of 75% or a direct gas fuel dryer of 58% efficiency. As in practice thermal efficiencies of dryers without total recirculation are substantially less than those figures, there is considerable scope for a reduction in energy consumed in process drying.

2.3   Areas of Application

Fig. 4 shows that the specific moisture extraction rate of a heat pump dehumidifier decreases as the relative humidity level of the incident air decreases. The dehumidifier is therefore more efficient when applied to processes which have high average relative humidities, typically above 50% RH. Additionally temperature constraints are placed upon a dehumidifier due to the physical characteristics of refrigerant fluids. Until recently the maximum permissable temperature of air drawn into a dehumidifier was approximately $50^{\circ}C$ as the maximum condensing temperature of the refrigerant under normal operating conditions was approximately $55^{\circ}C$. Ideal areas of application for these units are those which require drying temperatures of no more than $50^{\circ}C$ with an average relative humidity during drying of well above 30% RH. Processes which require close or schedule control of drying conditions are well suited to the totally enclosed heat pump system.

One of the most successful applications of dehumidifiers is in the seasoning of timber, where the wood is 'kiln dried' by following a pre-determined drying schedule in which the severity of the air condition is gradually increased as the wood moisture content decreases. Careful control is required to prevent or reduce timber degrade to within acceptable limits. A second and more recent area of application is within the clay products industry where a wide range of products have to be dried before a high temperature firing process. Drying must take place at temperatures below $50^{\circ}C$ and at high humidities to prevent too rapid drying and possible degrade. In Fig. 6 the similarity between a timber drying schedule and the prevailing conditions in a pre-drying process in the clay products industry can clearly be seen, with the main difference being the overall drying time.

Many other industrial sectors have recently become more energy concious and the $50^{\circ}C$ heat pump dehumidifier is now used in a large variety of industries including:

     textiles both natural and synthetic
     leather
     gypsum
     carpet
     confectionery
     refractories

It has also become apparent that there are many areas requiring heat pumps capable of operating at much higher temperatures and the first step in increasing operating temperature has now successfully been made.

2.4   Recent Developments

The problems of increasing the operating temperature range of dehumidifiers required considerable research and development work and was carried out in collaboration with the Electricity Council Research Centre, Capenhurst, Chester, UK (Ref4). Careful consideration had to be given to the choice of refrigerant fluid and lubricant to ensure compatibility at the pressures and temperatures existing in a system operating at 80°C. An extensive test programme was undertaken in which the heat pump and its components were tested under differing load conditions with the result that the final working unit was built from standard components.

This unit has now been available commercially since November 1980 and has already found application again within the timber seasoning industry. It differs from the earlier lower temperature heat pump dehumidifiers in that it stands outside the drying environment with air ducted to and from the process. This configuration reduces the corrosion and deterioration that would occur in a high temperature/high humidity environment and makes for ease of servicing.

In Fig 7 it can be seen that with this ducted arrangement of the 80°C unit, a proportion of the incident humid air bypasses the evaporator. This is to achieve a greater air flow rate over the condenser thus reducing condensing temperature of the refrigerant. A reduction of the difference in refrigerant temperature between evaporator and condenser increases the coefficient of performance of the unit and hence the drying efficiency of the system.

A typical drying system layout incorporating the high temperature dehumidifier is shown in Fig 8.

3   CASE HISTORIES

3.1   General

The case histories presented are intended to show the high drying efficiencies achievable with a heat pump dehumidifier with the most important figure to the industrialist being the on site fuel consumption or drying cost reduction. Comparisons are made of the site energy consumption and where information is available comparisons are also made of primary energy input. This is not always possible as a detailed breakdown of fuel types and consumption of each is required and has not always been recorded on older systems. For primary energy estimates the production efficiencies of the fuels discussed have been taken as 78% for oil and natural gas, and 27% for electricity. The final case history outlines a project undertaken with the objective of comparing scientifically the cost of drying timber in a conventional oil fired heat and vent chamber and an 80°C heat pump dehumidifier system.

3.2   Textiles - Wool Drying

This company produces woollen berets and with escalating costs of fuel oil together with highly inefficient drying chambers they decided to change completely to heat pump dehumidifier drying. The hats together with yarn in the form of cheeses, hanks and ribbons all need to be dried following the dye-house process. Previously these were dried in two oil fired fan assisted steam heated ovens at a temperature not exceeding 50°C. Batch drying times were in the order of 5 3/4 hours which necessitated the periodic use of a third oven to relieve bottlenecks in the drying process. Trials were conducted which proved that the heat pump dehumidifier could improve drying times and substantially reduce energy usage and drying costs. As a result the two main ovens were upgraded by increasing insulation to give a wall panel 'U' value of 0.37 W/m2oC, equivalent to 100mm of thermal insulation material. The system was totally enclosed and pnuematic seals fitted to doors to ensure a gas tight structure.

The before and after situation for both drying chambers is summarised in Table 1. In 1980 the company compared the energy usage and running cost situation of the dehumidifiers with that of the previous heat and vent dryers. As can be seen from the table there have been reductions in on site energy use of approximately 90%, in primary energy use of an average 80% and in drying cost of 70% to 75%. Shorter drying time has also removed the bottleneck situation and the product quality, on a subjective basis, is said to be improved.

Fig 9, gives an interior view of one of the drying chambers drying cheeses of material. The 7.5kW package heat pump dehumidifier unit can be seen to the right of the product load. Drying temperature is 50°C and over the course of the drying run the relative humidity level falls from 70% to 30% RH.

Based upon the total cost of converting the chambers including additional insulation, the payback period on capital invested has been estimated by the company as 2.4 years at 1980 price levels.

3.3   Clay Products - gypsum mould drying

Ceramic tableware is often produced by pouring a liquid clay slip onto or into a gypsum mould. Excess slip is removed from the mould by a simple pouring action or by a centrifuge and water from the remaining clay is rapidly absorbed by the gypsum mould. After a short period of time the clay becomes 'leatherhard' when sufficient water has been taken up and the ware is separated from the mould. Both the ware and the mould now need drying although in most cases the plaster mould can be used several times. Moulds have a limited life and are continually replaced and new moulds have to be dried prior to use. Typically moulds are dried from 30% to less than 1% moisture content at temperatures less than 50°C, as higher temperatures can give rise to 'mould burn' which reduce their lifetime. Fig 10 shows a typical mould drying chamber drying two stillages of 800kg of wet plaster. In this case the customer had a considerable number of direct gas fired chambers but needed an increase in drying capacity. Trials were undertaken with a small 3kW package heat pump dehumidifier which proved that the system could give high product quality at a much reduced energy consumption.

Detailed fuel consumption figures were not available for the gas fired dryers but estimates made have put the specific energy consumption at approximately 4600kJ/kg which includes a factor for electricity used of about 400kJ/kg. The dehumidifier chambers, two were installed, were fitted with electricity meters and a number of detailed trials have been undertaken, the results of which are given in Table 2. Specific energy consumption for the heat pump system has been measured as 1676kJ/kg representing a 64% reduction in energy usage on site and a 12% reduction in primary energy.

3.4   Timber Drying

The first installation for the new 80°C heat pump dehumidifier described in section 2.4 was made at a sawmill in Boroughbridge, North Yorkshire, UK. There were previously three other dehumidifier systems on the site all operating at air temperatures up to 50°C, but when the sawmill operator decided to increase capacity the high temperature system was chosen for it's low energy costs and high throughput rate.

In Fig 11 the layout of the equipment plant room clearly shows the two 7.5kW heat pumps with ductwork connection to the drying chamber. The other equipment shown are the ancillaries required for control and humidification. The 80°C upper limit allows standard drying schedules (Ref.5) working up to these high temperatures to be used with a resultant reduction in drying time. Being the first installation of its kind, the local electricity board closely monitored the power consumption of the drying kiln compared it with the 50° systems on the same site. The results of the trials are given in Table 3 in which two similar loads of timber compared. It is, however, extremely difficult to achieve two identical loads and the results should serve as an indication only of the efficiency of both heat pump systems.

The 80°C system has an energy usage almost 50% below the 50°C system. Although a reduction is expected a substantial amount of the reduction is due to shorter drying times and therefore a smaller energy component from primary air circulation fans. The reduction in drying time is due to two factors, the higher operating temperatures and the thinner section timber in the high temperature load.

In section 2.1 the average specific moisture extraction rate of a dehumidifier was given as 2.5 kg/kWh or 1440 kJ/kg. This compares to a specific moisture
extraction rate for a system of 3520 kJ/kg as shown in Table 3.   The major factors
contributing to this increase in energy are:

> energy to produce air circulation
> conduction losses through walls and floors
> incidental air leakage
> sensible heat energy expelled with dry product

With products requiring slow moisture removal and long drying times the most
significant of these components will be energy required for air circulation.

As part of the research programme to develop a high temperature heat pump
dehumidifier the Electricity Council Research Centre undertook with the help of a timber
kilner a trial to dry identical loads of timber both by dehumidifier and in an oil fired
heat and vent kiln (Ref. 6). A load of beech was dried at the same time and over the
same schedule at the kilner's site and in the test kiln at Capenhurst with energy
consumption being closely monitored. The results in Table 4 show that for these two
loads over the same moisture content reduction the heat pump dehumidifier system used
only 29% of the specific energy consumed by the heat and vent kiln per unit volume of
timber. Translated into energy costs, this equates to a saving of 39% based upon fuel
costs applying at the time of the experiment and to a reduction in primary energy usage
of 18%

4.   FUTURE DEVELOPMENTS

Work is currently underway to further increase the operating temperature
range of heat pump dehumidifiers and this will create many new areas of application.
Considerable work remains to be done to develop the application within other areas of
industry. At present dehumidifiers are used primarily on batch type systems but modern
continuous dryers which are being developed with very low air leakage rates will further
widen the scope for the heat pump dehumidifier.

Much emphasis is also being placed on further reducing the energy consumption
of dehumidifiers although as yet techniques employed have not proven economically viable
with increased performance being outweighed by increased capital cost. The development
of a direct fossil fuel fired heat pump has been highlighted (Ref. 7) as a future
requirement in the timber industry.

5.   DISCUSSION

A substantial reduction in energy use in convection drying can be achieved
by converting to a heat pump dehumidifier drying process, ideal applications being those
requiring air temperatures of below 80°C. Care must be taken with regard to drying
chamber structure with full advantage of the dehumidifier only being made in a total
recirculatory system. To date the batch type dryer has proven most suitable but modern
continuous dryers with very low air leakage rates will introduce another area of
application.

In comparing the heat pump process with conventional heat and vent dryers
the main obstacle is one of poor instrumentation. Rarely is information available on
fuel use or running cost, although it has been shown that where data has been recorded
the savings in both energy usage and drying cost of a dehumidifier system over a conventional system are considerable. The degree of saving depends upon the quality of the
system being replaced, but energy reductions on site are usually greater than 50% and
cost savings can be as high as 75% resulting in short capital recovery periods.

Many new areas of application will become apparent in other industry sectors
as operating temperatures increase and as more managers become energy conscious. The
heat pump dehumidifier is now proving itself as a highly efficient energy reduction
device throughout the process drying industry.

6.   REFERENCES

1.   Hodgett, D.L. : "Efficient drying using heat pumps". The Chemical
          Engineer, July/August 1976

2.   Department of Energy : Digest of United Kingdom Energy Statistics 1980"
     London, HMSO, 1980

3.   Ministry of Technology : "The efficient use of fuel". London, HMSO, 1958

4.   Driscoll, J.L. et al : "The Capenhurst Westair high temperature heat pump
     dryer"

5.   Pratt, G.H. : "Timber drying manual". London, HMSO, 1974

6.   "High Temperature Kilning on Trial" : Timber Trades Journal, pp 28-29,
     8th August 1981.

7.   Department of Industry : "Energy use in the timber furniture and related
     industries". IETS report 15, London, May 1980

TABLE 1 - CASE HISTORY - TEXTILE DRYING

		Chamber One	Chamber Two
General	Product	Woollen berets	Hanks, cheeses, ribbons
	Chamber dimensions	9.74m x 2.75m x 2.05m	9.1m x 2.4m x 2.05m
	Construction	Galvanised steel panels with high thermal bridging, poor insulation and door seals.	
	Modifications to construction	Insulation improved to U value of 0.37W/m² $^{\circ}$C, all door seals improved.	
Oil Fired System	Oil usage	16.64 litres/hr or 599000 kJ/hr	13.54 litres/hr or 487700 kJ/hr
	Electrical power consumption	2.29 kW or 8244 kJ/hr	5.59 kW or 20124 kJ/hr
	Cycle time	5.75 hrs	5.75 hrs
	Water removal	154 kg	54 kg
	Specific energy consumption		
	on site	24080 kJ/kg	54074 kJ/kg
	primary	31748 kJ/kg	74515 kJ/kg
Heat Pump Dehumidifier System	Unit capacity (nominal)	15 kW	7.5 kW
	Total power consumption (average)	18.37 kW or 66132 kJ/hr	9.88 kW or 35550 kJ/hr
	Cycle time	4.5 hrs	4.7 hrs
	Water removal	145 kg	54 kg
	Specific energy consumption		
	on site	2052 kJ/kg	3095 kJ/kg
	primary	7600 kJ/kg	11463 kJ/kg
Comparison	Energy usage reduction		
	on site	91%	90%
	primary	76%	85%
	Energy cost reduction (customer's estimate)	70%	75%
	Payback period (including cost of total conversion).	2.4 years at 1980 fuel price levels	

TABLE 2 - CASE HISTORY - GYPSUM MOULD DRYING

General	Product	Gypsum moulds
	Chamber dimensions	3m x 2m x 2.5m
	Construction	Standard cold store panels with an insulation value of 0.25W/m² °C.
Gas Fired System	Specific energy consumption (esimated) on site primary	 4600 kJ/kg 7050 kJ/kg
Heat Pump Dehumidifier System	Heat pump capacity (nominal)	3 kW
	Product load (wet)	800 kg
	Water to remove	232 kg
	Cycle time	18 hrs
	Electricity consumed	108 kWh or 388800 kJ
	Specific energy consumption on site primary	 1676 kJ/kg 6207 kJ/kg
Comparison	Energy usage reduction on site primary	 64% 12%

TABLE 3 - CASE HISTORY - TIMBER DRYING UP TO 80°C

Low Temperature 50°C, Heat Pump System	Specie	Ash
	Thickness	25 to 50mm
	Kiln load	24m³
	Moisture content reduction	30% to 13%
	Water to remove	103.7 kg/m³
	Total energy consumption	4330 kWh
	Specific energy consumption	6300 kJ/kg
High Temperature 80°C, Heat Pump System	Specie	Ash
	Thickness	25mm
	Kiln load	24m³
	Moisture content reduction	32% to 13%
	Water to remove	115.9 kg/m³
	Total energy consumption	2720 kWh
	Specific energy consumption	3520 kJ/kg

Drying of 32mm thick W/E Beech from 27% to 12.3% mc

Direct Oil Fired System	Kiln load	21m³
	Oil used	27.5 litres/m³ or 101 x 10⁶ kJ/m³
	Electricity used	29.3 kWh/m³ or 0.11 x 10⁶ kJ/m³
	Water removed	91.1 kg/m³

	Specific energy consumption
	  on site
	  primary
		12,294 kJ/kg
		15,794 kJ/kg

High Temperature Heat Pump Dehumidifier	Kiln load	11.5m³
	Electricity used	88 kWh/m³ or 0.32 x 10⁶ kJ/kg
	Water removed	91.1 kg/m³

	Specific energy consumption
	  on site
	  primary
		3,477 kJ/kg
		12,878 kJ/kg

Comparison	Energy usage reduction
	  on site
	  primary
		71%
		18%

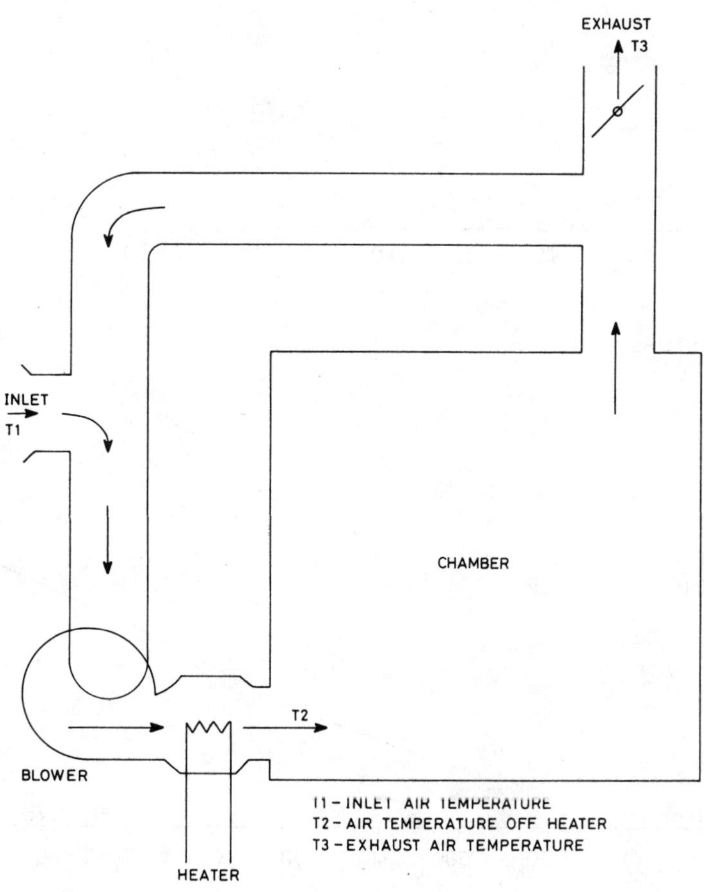

FIG.1 CONVECTION DRYER

FIG. 2 HEAT PUMP DEHUMIDIFIER

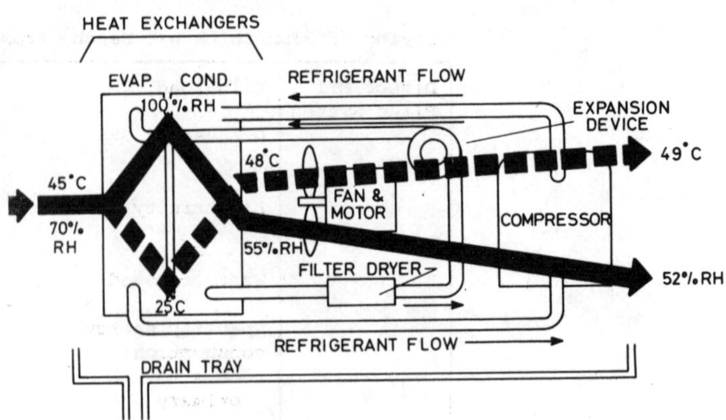

FIG. 3 HEAT PUMP DEHUMIDIFIER (50°C)
-SCHEMATIC

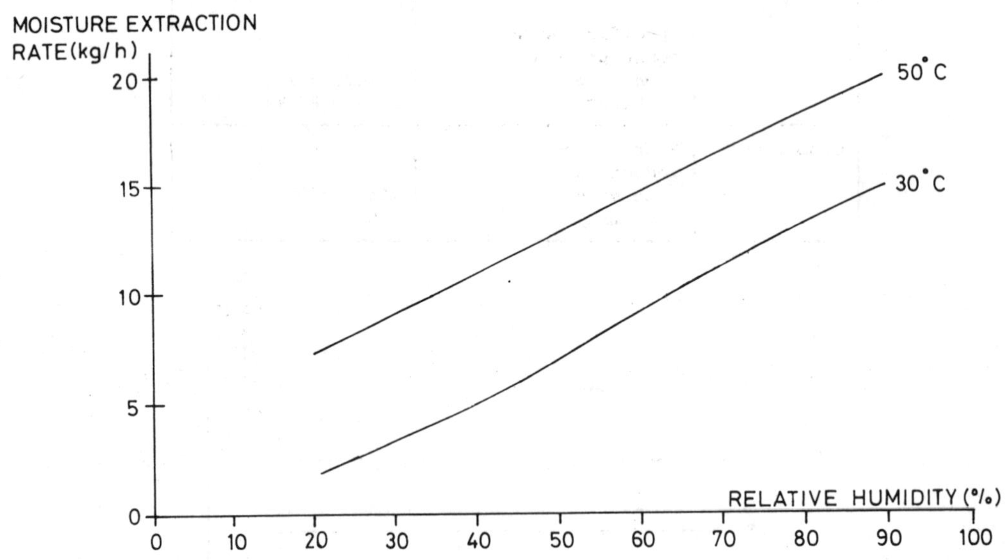

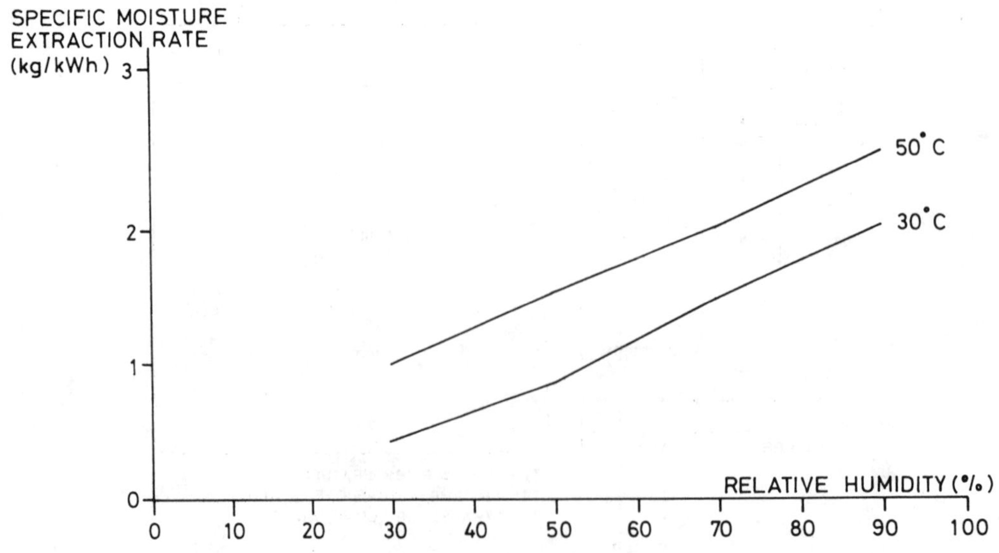

FIG. 4. PERFORMANCE CURVES NOMINAL 7·5 kW UNIT

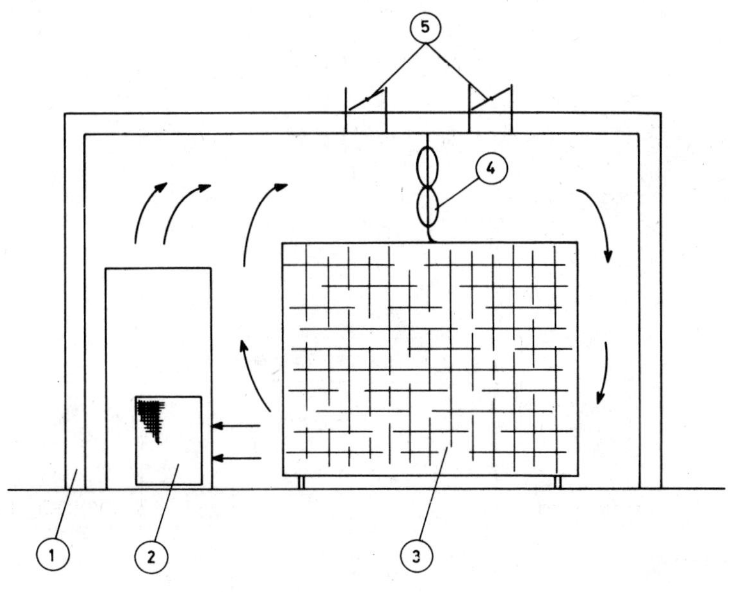

FIG. 5 DEHUMIDIFICATION DRYING SYSTEM – SCHEMATIC

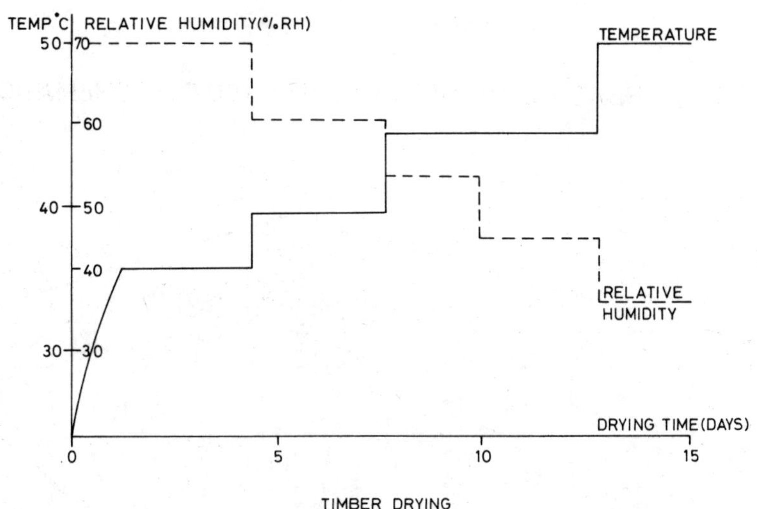

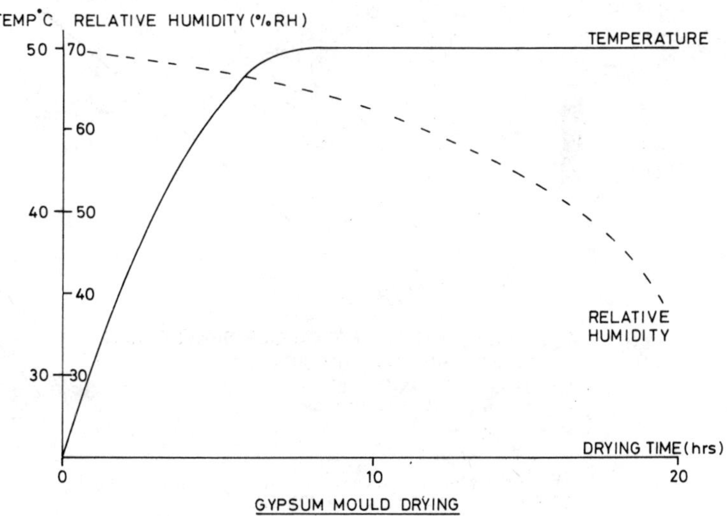

FIG. 6 DRYING CONDITIONS

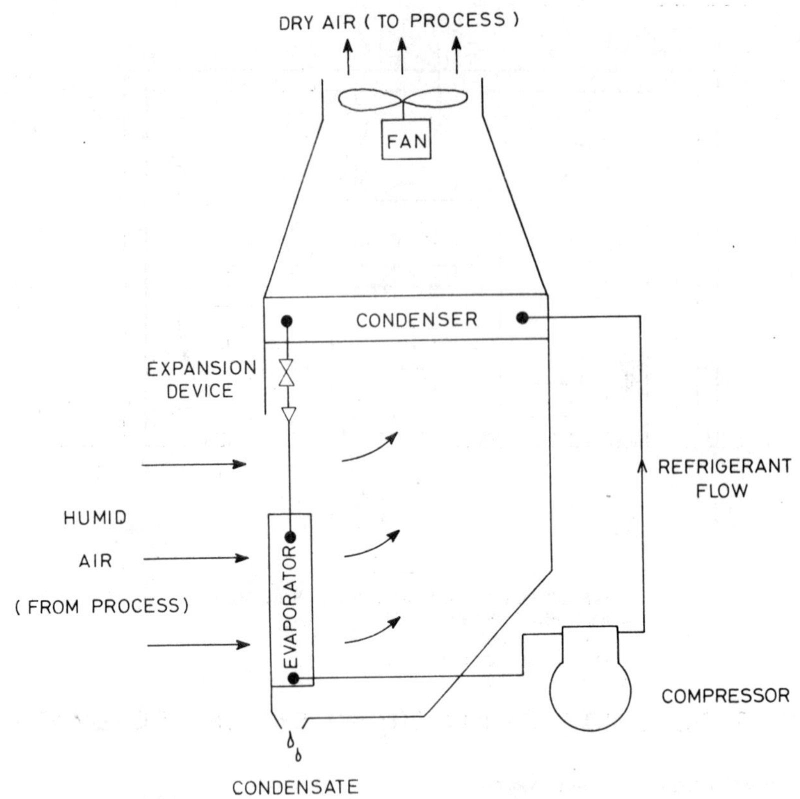

FIG.7 HEAT PUMP DEHUMIDIFIER (80°C)-SCHEMATIC

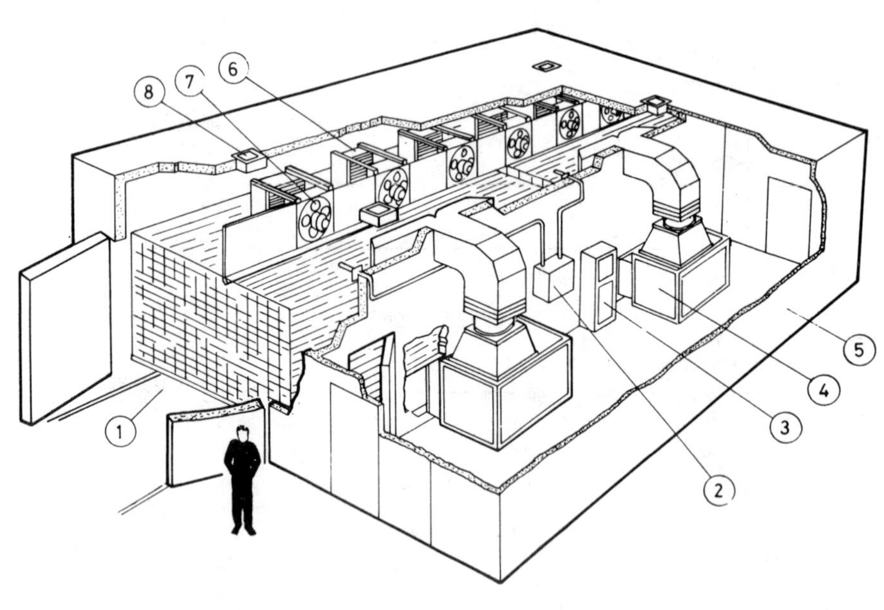

ITEM LIST

1. THERMALLY INSULATED VAPOUR SEALED KILN
2. ELECTRODE BOILER HUMIDIFIER
3. CENTRAL CONTROL SYSTEM
4. HEAT PUMP DEHUMIDIFIER
5. PLANT ROOM
6. HEATER BANKS
7. PRIMARY AIR CIRCULATION
8. TEMPERATURE BALANCE SYSTEM

FIG.8. TYPICAL HEAT PUMP DEHUMIDIFIER TIMBER DRYING SYSTEM

FIG. 9. TEXTILE DRYING WITH A 7·5 kW HEAT PUMP DEHUMIDIFIER

FIG. 10. GYPSUM MOULD DRYING WITH A 3 kW HEAT PUMP DEHUMIDIFIER

FIG. 11. TIMBER DRYING WITH TWO 7·5 kW HIGH
TEMPERATURE HEAT PUMP DEHUMIDIFIER

LARGE GAS ENGINE DRIVEN HEAT PUMPS

C.D.R. North

GEA Airexchangers Ltd., U.K.

Summary

This paper describes the application of heat pump technology to an industrial process in order to achieve a dramatic savings in energy.

Initial experience with a 3.0 MW gas engine driven system is reported. In addition, the development of a larger 8 MW system, incorporating a total energy concept to halve energy consumption, is described. (The system will become operational in Mid-1982).

The breakthrough for large industrial heat pumps has occurred in the Malting Industry. Some of the factors which have contributed to the breakthrough are discussed to help identify future applications.

Held at the University of Warwick, U.K.
Symposium organised and sponsored by
BHRA Fluid Engineering

NOMENCLATURE

T_M = Delivered air temperature

T_A = Ambient air temperature

T_E = Exhaust air temperature

η = Temperature efficiency %

<u>1). TYPICAL MALT PRODUCTION</u>

The production of malt for the brewing and distilling industries involves several stages:-

(a) Steeping the barley in water and allowing it to germinate.

(b) Drying the resulting product with hot air from 45% moisture content down to 12% moisture content, when it is said to be "hand dry".

(c) Curing the "hand dry" malt by driving off moisture, until the water content is about 2%, in which condition the malt is marketable. Kilning consists of heating air, usually by the direct combustion of oil or gas, and then propelling it through the bed of malt.

In Stage 2, the moisture is free and evaporates rather easily; with hot air entering the bed at about 70 Deg. C. the air above the bed is saturated at about 350 Deg. C. when exhausted to atmosphere.

Many kilns recirculate air after "hand dry" to reduce firing requirements.

In other systems, malt beds are operated in series either by locating beds one above the other or by means of transfer ducts. On the most efficient plants over-heating of the exhaust is avoided and saturated exhaust conditions are maintained for most of the operating cycle.

The kilning process usually takes 14 to 20 hours and down-time to the loading of the next batch is kept to a minimum.

Typically, energy usage is in excess of 8000 hours per year.

Being large energy users sensitive to the increasing price of prime energy, European maltsters began installing heat recovery systems about five years ago. Two basic types developed.

<u>2). CROSSFLOW HEAT RECOVERY</u>

The first method utilised to reduce firing requirements involves crossing the kiln inlet and exhaust through a tubular exchanger. In the case of an existing malting, this may require a large transfer duct from the exhaust to the inlet, with consequent major building modifications.

As the heat transfer is air to air, bare tubes in either stainless steel or glass are required to resist acid corrosion on the exhaust side. The performance is limited by the exhaust temperature despite its large heat content (often 100% R.H.). Most systems are designed for a temperature efficiency of up to 85%.

Translating this into total energy saving leads to an upper limit of about 35%, overall saving in the most advantageous situation is more typically 25%-30% (as in the example in Fig. 1).

The Crossflow Heat Recovery System is at its most competitive when the kiln is designed around the heat recovery system with inlet and exhaust flows adjacent as large transfer ducts are expensive and disruptive to kiln operation during construction.

The Crossflow unit is a viable heat recovery system but its performance is limited by temperature efficiency. To obtain the energy savings associated with heat pumps would require extensive modifications and additional exchangers.

3). THE INTERMEDIATE FLUID SYSTEM

In this system, heat is recovered from the wet sulphurous kiln exhaust using corrosion resistant exhaust coils and transported to air preheater bundles, by means of a water/glycol intermediate fluid (see Fig. 2).

Achievable performance is similar to the Crossflow system. However, the connection between the inlet and exhaust is small bore piping compared with large ducts and thus installation costs tend to be lower, particularly on existing plants. Indeed, it has proved possible to install such systems without interrupting the round-the-clock operation of the kiln.

The intermediate fluid circuit is lightly pressurised and is a sealed system. Expansion is allowed   by means of a diaphragm expansion tank.

The concentration of glycol is selected so as to ensure that the system does not freeze at minimum ambient temperature. Temperature efficiency is almost dependent of ambient temperature.

In addition, intermediate fluid heaters can be added to make the system indirect which has several advantages from a product quality point of view (Fig. 3).

Typical inlet and exhaust exchangers are shown in the photographs 1 and 2. The exhaust units have to resist the aggressive exhaust environment, but a simpler construction is possible for the inlets which, in photograph 2, are mounted in the wall of the kiln.

Considerable flexibility exists in the shape and positioning of the coils to suit a particular plant.

Because of the limitation on tempeature efficiency it soon becomes apparent that much more energy could be saved in a Malting equipped with an intermediate fluid heat recovery system, by the addition of a heat pump to extract further heat from the saturated kiln exhaust and to deliver it to the inlet air at a temperature above the kiln exhaust temperature.

4). THE GAS ENGINE DRIVEN HEAT PUMP

The heat pump is a very efficient means of raising the temperature level or large quantities of waste heat such as that found in the saturated exhaust of a malt kiln.

The system shown in Fig. 4 maintains compatability with the basic intermediate fluid system utilising the same inlet and exhaust bundles but more low grade heat is extracted from the exhaust and pumped to a useful temperature. The heat pump is a compact unit driven by reciprocating gas or diesel engines with water glycol circuits carrying the heat to the various exchangers.

In fact, a heat pump system could be designed with air preheater as a condenser and the exhaust recovery unit as an evaporator; higher coefficients of performance would be possible but large quantities of refrigerants would need to be piped around the kilns and compatability would be lost with the simple Ecoflow system.

The system described in Fig. 4 achieves an energy saving of about 55% in the period before the 'break' and a nett saving in the range 45%-50% over the complete kilning cycle with a system C.O.P. of approximately 4.

The system requires some additional heat at start-up but this can be indirect in a similar way to that described in Fig. 3 if total indirect heating is required. The large energy savings with this system demonstrates the exciting potential of the heat pump in the Malting industry. The indirect heat from the engine cooling circuit is utilised to 'top up' the system contributing to the overall efficiency.

The first such system was commissioned at the Louth Malting of Associated British Maltsters in the Spring of 1981., (Ref. 1) and supporting patents taken out (Ref. 2).

Engine room and Freon packages are shown in photographs 3 and 4.

The packages consists of a two engine driven compressor connected to a common Freon package. Exhaust economisers and recovery of jacket, gearbox and radiated heat leads to a thermal efficiency in excess of 90% on the engine. This is important when compared with the electric heat pump where the efficiency of primary generation is usually less than 37%.

Use of local generation and electric drives would still require generator and motor efficiencies to be deducted from system efficiency.

Thus the Direct Engine Driven Heat Pump is a thermally efficient system when, as in the Malting industry, the waste heat from the driver can be directly applied to the process.

5). THE TOTAL ENERGY SYSTEM

From the original system installed at Louth, Lincolnshire, it becomes evident that several improvements could be made from an energy saving and system performance point of view.

    (a)   The system could be made totally indirect by installing small 'top up' burners in the engine exhaust prior to the economiser system.

    (b)   When integrated with a microprocessor logic which nominated the priority kiln, the complete system could be controlled using the heat pump compressor.

    (c)   With a thermal efficiency of about 90%, electricity generation is both cheap and attractive.

Taking the three points above into consideration, the new system being installed at Wallingford, near Oxford, incorporates a combined afterburner/economiser capable of maintaining heat output with one engine of the twin engine system down.

The use of screw type compressors will improve the controlability of the system and enable integration with the plant microprocessor system. Each 750 KW engine drives its 500 KW compressor through a 550 KW generator enabling the system to satisfy the plant electricity deman as a second priority with a considerable improvement in the economics of the process, particularly when the compressor is at part load.

The heat pump is connected to the engine drive train by an electromagnetic clutch system which enables one compressor to be disconnected at part load conditions and the system to operate on a single compressor.

The system is due to become operational in July 1982. The installation has been supported by the Energy Technology Support Unit, Harwell, and detailed performance data will be available from a performance monitoring package.

6). FUTURE APPLICATIONS OF LARGE INDUSTRIAL HEAT PUMPS

There have been several reports on possible industrial heat pump applications (Ref. 3).

An initial breakthrough for large industrial heat pumps has been made in the Malting industry. A review of the characteristics which make the Malting industry attractive to heat pumps is instructive with regard to possible future applications.

(a)   Malting is a batch process, but kilnings are so close together as
        to give energy usage in excess of 8000 hours per year.

        In general the payback period is crucial to the viability of any
        installation. In the Malting industry paybacks are in the range
        of 3-4 years (before any grant support) at current energy costs.

(b)   In the Malting industry, prime energy forms a major part of the
        operating cost of the process. Thus all Maltsters are receptive
        to systems which can improve their competitive position. In this
        context, engine driven heat pumps have to compete with cheaper
        fuel sources such as heavy fuel oil and coal.

(c)   There is a ready source of heat in this case (the saturated kiln
        exhaust).

(d)   The required process temperature is one which can be achieved by
        heat pumps:-

        viz: Refrigerant 22 approximately 55 Deg. C.
              Refrigerant 12 approximately 75 Deg. C.
              Refrigerant 114 approximately 110 Deg. C.

        Above these levels, new technology such as the "steam" heat pump
        will be required.

(e)   The package is self-contained with the recovered heat used again
        in the process to achieve economy.

(f)   The packages installed have demonstrated the advantages of the
        engine driven system with its high thermal efficiency. They have
        also demonstrated the "compact Freon package" concept with trans-
        port of heat around the process by means of an intermediate fluid
        on the grounds of practicality rather than C.O.P. considerations.

Very many industrial processes satisfy sufficient of the above criteria to
suggest that the engine driven heat pump/total energy system has a major role in the
future in all countries exposed to the full World energy cost levels.

The two installations described above were built as 'packaged process plant' on
site in existing rooms in the Maltings.

The Author's current development work is directed towards developing factory-
assembled "Engine Driven Total Energy Systems" which produce:-

(1)   Low grade heat.
(2)   Higher grade heat.
(3)   Electricity.

The heat discharge circuits and the heat source terminate in shell and tube
exchangers which connect to intermediate fluid circuits in the specific process.

This modularisation should reduce installation costs and improve further the
validity of the system in different industrial applications.

1).   Parsons, N.E. and Marsh, J.B.: "Waste heat recovery and Internal Combustion Engines in Malt Kilns". Paper 69, 18th International Congress, European Brewery Convention, Copenhagen, 1981.

2).   Voges, W. and Chiappetta, L. and Wittech, U.: "Air Temperature Control Apparatus". U.K. Patent Application GB 2 052 704 A.

3).   Currie, W.M.; Jonas, P.M.; Newbest, G.J. and Rowe D.N.E.: "Heat Pumps in the United Kingdom". ETSU Note N-2/8- (Unclassified). Energy Technology Support Unit, C/O Building 156, A.E.R.E., Harwell, OX11 0RA.

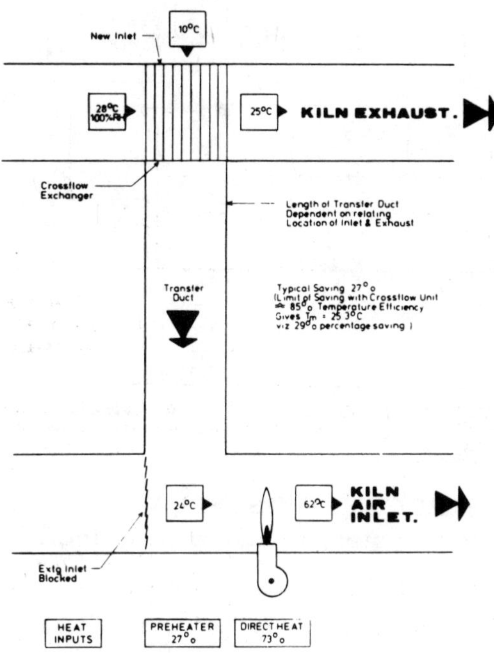

Fig 1.   HEAT RECOVERY SYSTEM
Incorporating Cross Flow Exchanger. (Glass or Stainless Steel)

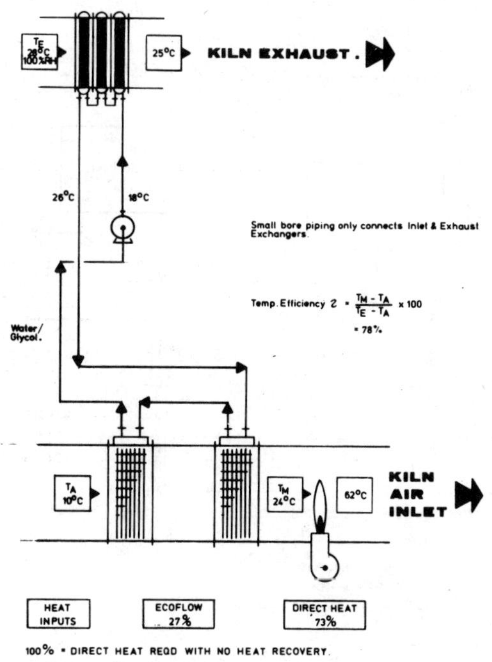

Fig 2.   ECOFLOW HEAT RECOVERY SYSTEM

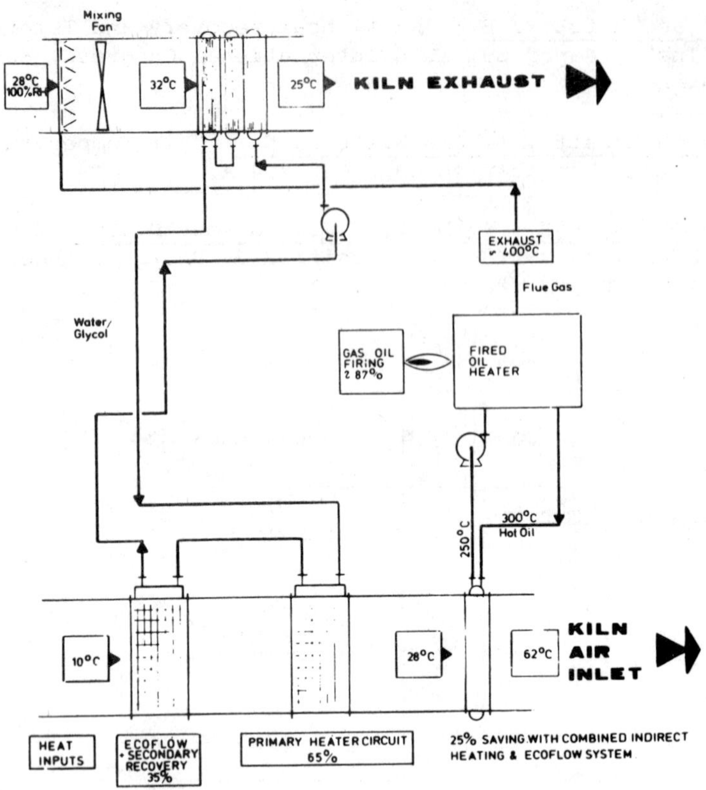

Fig 3.   HEAT RECOVERY SYSTEM
Ecoflow System incorporating Indirect Heating.

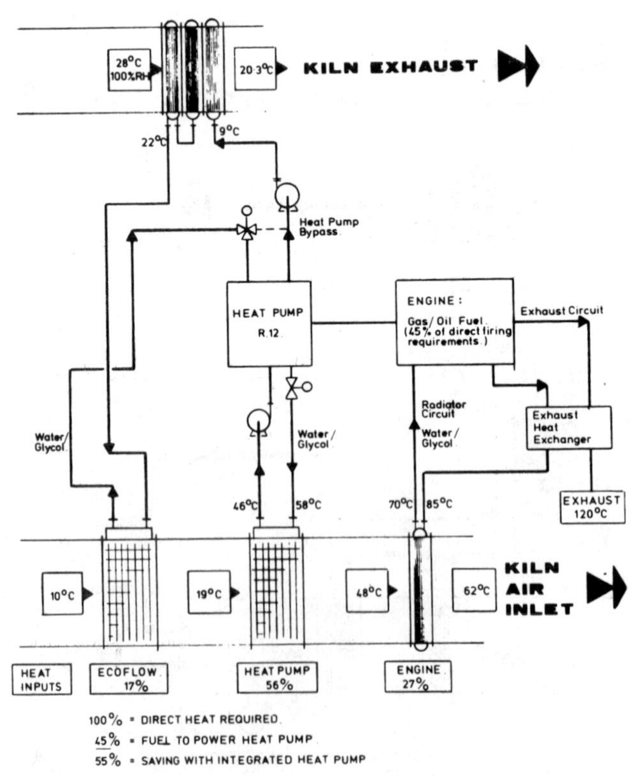

Fig 4.   TYPICAL HEAT PUMP SYSTEM (before hand dry - no recirculation.)
Evaporator in series with Ecoflow system.

(1)   Typical Inlet Coils

(2)   Heat Source - Kiln exhaust

(3)   Gas Engine Driven Compressors

(4)   Freon Package.

GAS MOTORS DRIVING HEAT PUMPS
IN THE MALTING INDUSTRY

O. Curis and J.D. Laine

CEM Systems, France

Summary

After reviewing the malt drying process (desiccation followed by rest of diastase reactions by raising to 80°C), the authors discuss the drying cycle implemented by CEM at MAUCLAIRE malting on two twin malt kilns.

This installation with two compression heat pumps in series driven by gas motors and placed on exhaust air and fresh air flows, with thorough sub-cooling, serves to guarantee all needs (maximum throughput 110.000 kg/h of dry air, maximum temperature 80°C).

The authors present the reasons for the different alternatives selected (drying cycle, heat pump cycles, compressor drives, sizing of aeraulic circuits) as well as the energy balances.

Held at the University of Warwick, U.K.
Symposium organised and sponsored by
BHRA Fluid Engineering
©BHRA Fluid Engineering, Cranfield, Bedford MK43 0AJ, England.

<u>INTRODUCTION</u>

Energy used in malting process representing a huge cost, many maltsters are faced with the problem of reducing its consumption. Moreover, systems used to conserve energy must comply with classic criteria in terms of return on investment.

In certain case, where a high coefficient of performance can be obtained for a low price of the installed thermic kW, the heat pump is the answer to both the above conditions.

For this purpose, MAUCLAIRE malting in ARCIS-sur-AUBE ordered a system of heat pumps driven by gas motors to equip their malt kilns, from CEM - New Techniques for Energy department.

Continuous exploitation of the malting, the high incidence of power costs on the price of the end product and the offered technical solutions make this installation particularly interesting.

1 - <u>MALTING</u>

Malting is a semi-agricolous activity which produces barley-malt, used for
the brewing and distilling industries.

During the malting process, diastasis contained in malt serve to change grain
starch to fermentible sugars.

The industrial process consists in making malt from barley by means of bioche-
mical reactions. Its main stages consist of cleaning the barley, steeping, germination and kilning.

After it is rid of impurities and foreign grain, the barley is put to steep
during 48 hours in water tanks. This operation supplies sufficient oxygen and
water to enable the grain to germinate. The grain's moisture content is thereby raised from 15% to approximately 45%.
During the germination stage which follows the steeping, the grain undergroes
a process of internal desagregation. After germination, the barley has become
malt, known as "green malt" and is then directed to the kiln.

Kilning produces two effects : it stabilizes the malt by stopping diastatic
transformations and imparts a distinctive aroma and colour.
During a first phase, the germinating process is briefly accelerated, diastatic
activity persists and metabolism remains active up to approximately 45°C.
Above 60°C, sugars and nitrogenous compounds give rise to complex substances
known as melanoidins, which contribute to flavour the malt. During a second
phase, diastatic reactions are stopped by curing, using drying air at approximately 80°C, bringing the malt's moisture content down to 5% maximum.

Diastasis must be kept active for brewing ; the drier the malt, the better
diastasis are able to withstand high temperatures. The principle of the malt-
producing process appears very simple, but is in fact based on highly specialized know-how ; interaction between physical factors (temperature, moisture,

timing) are complex, and the biological mechanisms which occur within the living matter are invisible.

A malt kiln is a chamber type dryer. The malt is placed in a fixed layer a meter thick on a perforated deck and crossed through by an ascending hot air flow.

The temperature of incoming hot air follows a carefully predetermined curve, climbing progressively from 55°C to 80°C and predominating at approximately 70°C, while moisture content of the grain is lowered from 45% to approximately 5%.
During this cycle, exhaust air from the layer of malt is at first almost saturated, then, about halfway through the cycle, dryer and dryer as it becomes hotter.

2 - THE KILNS AT MAUCLAIRE MALTING

The twin kiln system (fig.1) at MAUCLAIRE malting makes full use of the air's drying potential by combining two kilns working a semi-cycle apart.(fig.2).

On the "wet kiln", the grain is dried to decrease moisture content from 45% to approximately 15%. The air on the "wet kiln" being the air off the "dry kiln". At the end of the cycle, a flap system places the "wet kiln" (then called "dry kiln") in the hot airflow coming from the burners.
The dry malt is unloaded, the deck is reloaded with green malt and becomes a "wet kiln" placed in the airflow off the "dry kiln", (see fig.1a and 1b).
Exhaust air from the system is thus permanently kept at approximately 30°C, 90% relative humidity.
Size of treated airflow varies during the drying cycle from 98.000 to 110.000 kg/hour of dry air.

Today, MAUCLAIRE malting operates 24 hours a day, year round ; three periods
can be distinguished in the supply of the kiln's thermic requirements :

1 - 90 days at the year's end (beetroot campaign), during which MAUCLAIRE mal-
    ting uses waste hot water at 95°C from a neighbouring sugar factory, which
    covers most of the thermic energy required for kilning.

2 - 105 days at the beginning of the year, (syrup campaign) : hot water is
    still available, but at 60 - 65°C. Heavy topping up is required, and sup-
    plied by butane gas torch burners.

3 - 170 days during which no waste is available from the sugar factory ; the
    burners cover all thermic energy requirements.

3 - <u>THE PROBLEMS : ENERGY AND PRODUCT QUALITY</u>

Malting is a low added-value industry ; a large part of which concerns energy
expenses. This industry is therefore constantly searching for means of reducing energy spending. Moreover, standard torch burners pose the problem of NOx
contained in the combustion fumes ; as more stringent legislation is pending,
maltsters are searching for "indirect" drying solutions, which would eliminate
direct contact between malt and fumes.

Several solutions can be considered in this light :

- the air can be indirectly heated by hot water exchangers. This system solves
the nitroaminate problem, but common boiler efficiency lower than 90% causes
high energy consumption in spite of increased drying power.

- burners with a low production of Nox ("LoNOx") can partially solve the problem of nitroaminates, but are without energy saving.

- recovery by cross-flow exchangers : new air is heated by the exhaust air. Economy would average 20 to 30%, according to the air's enthalpy. This wholly passive system has the advantage of being a comparatively cheap investment, but can make higher recovery rates difficult to achieve ; moreover, because topping up is always necessary, the problem of direct or indirect drying arises once more.

- heat pumps : they enable heat from the exhaust air to be recovered for use at a higher level to heat new air. The system can be dimensioned to cover total drying requirements, thereby, eliminating the need for topping up and related problems. Consumption of primary energy is cut down to a third, and the only change in operation is a slight decrease in the flow of treated air due to the higher drying power (indirect heating replaces a torch burner) ; the heat pump remains outside the drying process.

MAUCLAIRE malting chose the heat pump solution, which solves problems posed by topping up, including the problem of nitroaminates and is the most economical solution.

4 - THEORETICAL REMINDER - CHOICE OF SOLUTIONS

4.1. Theoretical reminder - possibilities of improvement

A heat pump is a machine which carries heat from a cold source to a hot one. A refrigerant fluid evaporates while taking heat from the cold source, and condenses after compression, delivering the heat at a higher level to the hot source (fig.3).

The liquid refrigerant leaves the condenser at condensing temperature. Instead of this hot fluid being sent directly into the evaporator, it can be directed through a sub-cooler before expansion to extract part of its sensible heat.

The percentage of liquid entering the evaporator will then be higher, making better use of the vehicled fluid.

In order to obtain a better Coefficient Of Performance (COP) it is advisable to connect heat pumps in a series : for any given variation in the temperature of useful fluids, (exhaust air - outside air) temperature differences between the hot and cold sources from each machine decrease in proportion to the increased number of machines : the more machines, the smaller the temperature difference. (fig.5).

COP is also increased by driving the compressor with a thermic motor with cooling heat and exhaust gases recovery ; as the final necessary degrees are supplied by recovery, condensation temperature (Tc) can be lowered, which raises the $\dfrac{Tc}{Tc - Tf}$ rate (fig.5).

If the recovered heat is latent heat, great quantities of heat are made available by moderate cooling of the cold source. The evaporator's entire exchange surface can work at a temperature which approximates that of the effluents, giving a good Coefficient Of Performance. When applied to drying, the heat pump is therefore an ideal instrument, thanks to the great quantities of latent heat contained in the waste matter (evaporated water).

4.2. Choice of solutions

The solution chosen for MAUCLAIRE malting heat pump system includes two heat pumps in a series, driven by gas motors.
The circuits are placed in the exhaust and new air flows in such a way as to have the hottest evaporator correspond to the hottest condenser, (reducing the temperature difference between condensation and evaporation). Heavy sub-

cooling and a refrigerant fluid which is adapted to each circuit complete
the above dispositions, with the aim of increasing the plant's Coefficient Of
Performance.

4.2.1. Compressor drive

The high temperature level required for malt drying (up to 80°C) makes it
particularly recommendable to drive the heat pumps by means of thermic motors
using heat recovery on exhaust gases and cooling water.

As compared to electric drive, recovery on thermic motors reduces both the
power required to supply the compressors and the final temperature to be reached by the heat pump proper, by approximately 20%, which considerably improves COP.

Taking average values :
- new air to be heated to 71°C
- exhaust air available at 32°C

the theoretical COP rises from 8.8 to 12.3, i.e. an improvement of 40% ; recovery from the motors heats the new air from 59°C to 71°C.

An additional advantage of the thermic motors solution is the possibility of
choosing high performance refrigerant fluids, thanks to the less stringent
operating conditions (lower condensing temperature).

A heat pump of identical power (and therefore, identical COP) would require
a top up of 20% of total power if an electric motor is used. The power balance would then show predominant use of badly valorised fossil fuel.

Economically speaking, the thermic motor solution is mainly workable with natural gas, which is cheaper than other forms of energy. In the future, as the
cost gap closes between the gas kWh and that of electricity, the thermic motor solution might become less interesting as gains on performance would be
in part absorbed by rising cost due to the fact that gas motors are more ex-

pensive to maintain than electric motors. The thermic motor must therefore be chosen in the industrial range, in order to keep maintenance costs low.

From an investment point of view, volumes are comparable when speed variation is necessary (heavy variations of the power to be supplied) ; otherwise, electric motors have a clear advantage.

The final choise was therefore a gas powered thermic motor, primed by sparking (Beau de Rochas cycle). These motors have a high output at full load (35%), and output is relatively stable up to half-load. Heat is recovered at the rate of 100% from the cooling water circuit and at 80% from exhaust gases, with a global efficiency of approximately 88%.

4.2.2. Dimensioning

To adapt to thermic needs required for drying, major power variations are necessary, during a cycle and depending on atmospheric conditions (1 to 2 ratio between maximal and minimal power to be supplied). A heat pump constantly delivering minimal power would therefore require heavy topping up.

The selected sub-cooling system (patented by Mr.AZNAVORIAN) reduces the gap between extreme values : the lower the outside temperature, the more thermic power the system produces ; under these conditions, condensation temperatures remain more or less stable year round.

A heat pump capable of supplying full requirements only requires minor over-dimensioning ; during operation, COP is scarcely dependant on outside conditions at all.

The system was therefore dimensioned to cover entire thermic requirements year round, with the exception of the period during which waste hot water is available at 95°C, when the heat pump is stopped. This formula gives the best results where energy-saving is concerned.

4.2.3. The compressors

Power to be supplied by the installation is suitable for piston compressors which have the added advantage of facilitating regulation of quantities vehicled by by-passing the cylinders.

These materials, whose technology is similar to that of piston engines, have proved to be highly dependable in many refrigerating installations.
Solutions to prevent liquid from going to the compressors have been thoroughly mastered.

The compressors are coupled directly to the thermic motors, without any power loss as far as driving problems are concerned. This arrangement which links two piston machines required special design to avoid vibration problems, particularly during cylinder by-passing.

Power is regulated by controlling the motor's rotation speed, after approaching the compressor cylinders by by-pass, allowing highly accurate regulation.

4.2.4. Air circuits

Heat insulated leak-proof ducts carry the treated air through the various exchangers. Sub-coolers, recovery condensers and batteries are placed at level + 4 m, with a by-pass system through the hot water batteries to enable use of waste from the nearby sugar factory. The evaporators are placed in a terrace at level + 17 m, over a tank to recover condensates. Motor exhaust gases are recovered up to 120°C, then mixed with the exhaust air upstream from the evaporators.

Exchangers are using fined tubes ; evaporators are made of stainless steel due to the low pH of exhaust air.

5 - BALANCE AND RESULTS

The thermal balance of the heat pump system represented in figure 6 is inten-
ted under average climatic conditions : 10°C outside temperature, with a moisture rate of 6.5 g/kg.

The heat pumps proper supply the hot source with 80% of thermic requirements
(of which 23.5% from the sub-coolers), the remainder being supplied by recovery from the thermic motors.

Thermic power of the sub-coolers plus recovery exchangers is higher than that
of a heat pump condenser.

Actual COP at the shaft of the heat pumps (power furnished to condensers and
sub-coolers over motor power at shaft) is approximately 7.

The installations energy output (power supplied at the hot source over motor
consumption) is 2.8. This figure takes primary fuel consumption into account
and cannot be compared to a classic Coefficient Of Performance ; it makes it
possible to evaluate the perceptible decrease in fuel consumption as paid by
the user ; fuel consumption will be divided by 2.8.

Mechanical output of the motors varies according to load.
However, recovery from the water circuits and exhaust gases leads to high global output which hardly varies with load. The thermic balance of the chosen
motor is summed up as follows :

        - power at shaft   : 35%

        - recovery from cooling
          water   : 36%

        - recovery from exhaust gases
          (cooling at 120°C)   : 18%

        - losses   : 11%

A Sankey's diagram (fig.7) illustrates the general balance of the installation.

The heat pumps will be idle 90 days per year during the period when waste
from the sugar factory is available at 95°C ; heat pumps will therefore operate 6 500 hours per annum. Consumption per annum is detailed on figure 8.
Annual savings will average 7 230 MWh or 622 tep.

The introduction of heat pumps will significantly decrease specific consumption of the twin kilns ; consumption will be brought down from about 800 to
less than 300 Wh per kg of evaporated water.

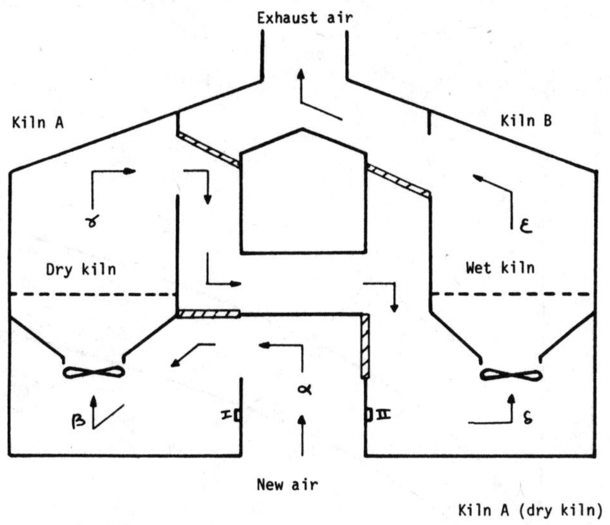

I - Kiln A dry kiln - Burner n°1 heats the new air (α) before it
is directed to kiln A (β)

Exhaust air from kiln A (γ) is directed to
kiln B (ε) at the beginning of the drying
process

Temperature at (δ) is controlled by adding
heat (burner II) or fresh outside air.

II - Kiln A dry kiln

Malt contained in kiln A is dry. Burner I is stopped - the air
heated by its passage through the dry malt is directed straight
at kiln B.

Figure 1a : Twin kilns in operation

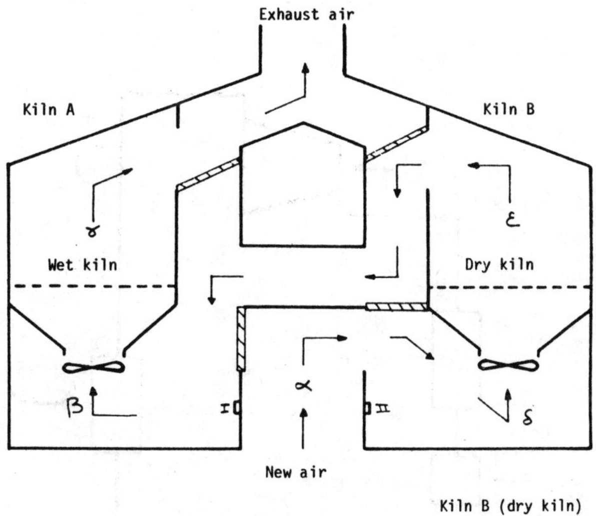

III - Kiln A : unloading operations and placing of a new load

Kiln B : drying cycle is continued with new air heated by
burner II ; exhaust air is released to the chimney

IV - Kiln B (dry kiln) - beginning of drying process, Kiln A

V - Kiln B (dry kiln) - cooling

VI - Kiln A (dry kiln) - unloading / loading of B

Figure 1b : Twin kilns in operation

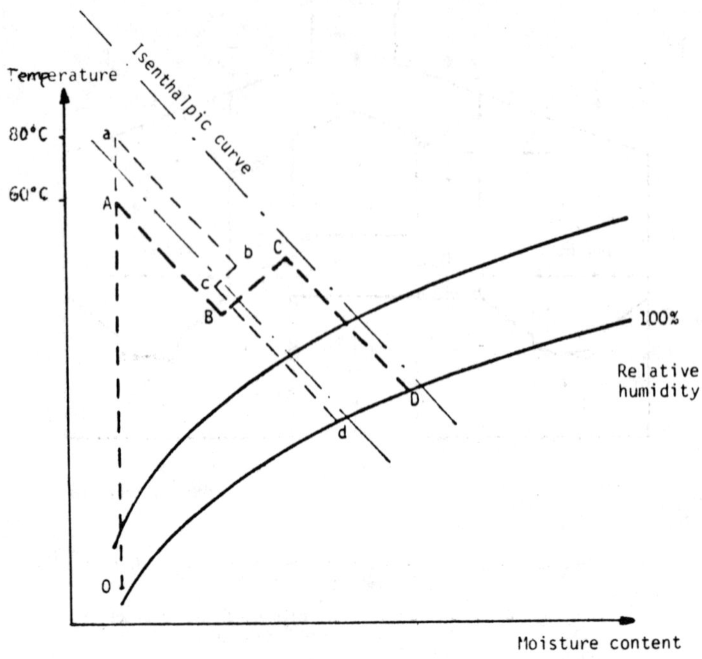

ABCD   : drying with additional heat top-up between decks
abcd   : drying with additional new air between decks

0   : atmospheric air
OA(oa) : heating of new air
A(a)   : kiln entry
B(b)   : exit 1st deck
BC(bc) : additional heat top-up (dilution with fresh air)
C(c)   : entry 2nd deck
D(d)   : exit from kiln

Figure 2 : drying diagram

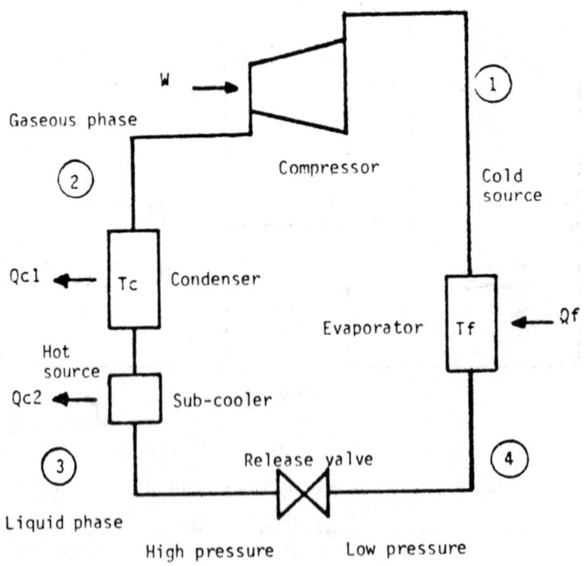

$$Qc = W + Qf = Qc1 + Qc2$$

$$COP = \frac{-Qc}{W} = \frac{Tc}{Tc - Tf}$$

Figure 3 : Principle of a heat pump
with refrigerant fluid

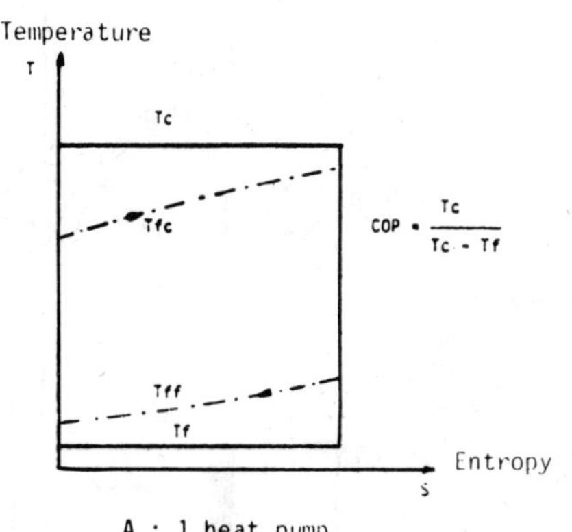

A : 1 heat pump

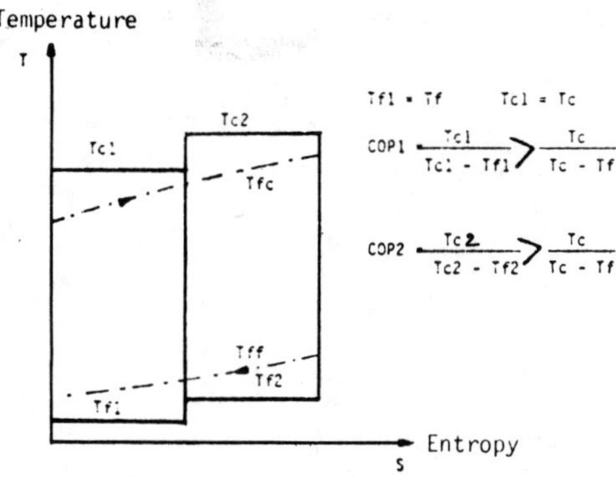

B : 2 cascade - connected heat pumps

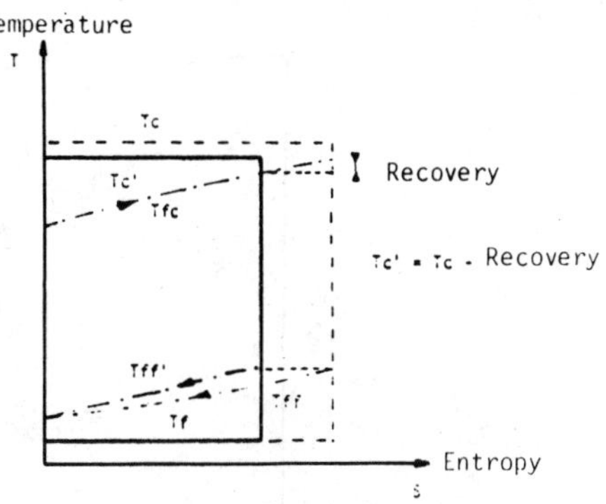

C : Heat pump with recovery on thermic motor

Figure 5 : compared performances

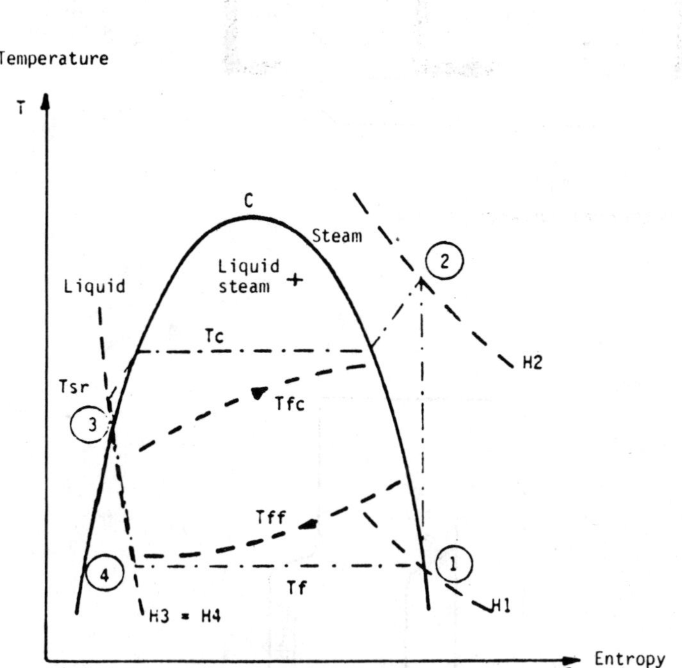

Tfc : evolution of fluid to be heated
Tff : evolution of fluid to be cooled
Tc : condensing temperature
Tsr : sub-cooling temperature
Tf : evaporation temperature

Figure 4 : compression heat pump

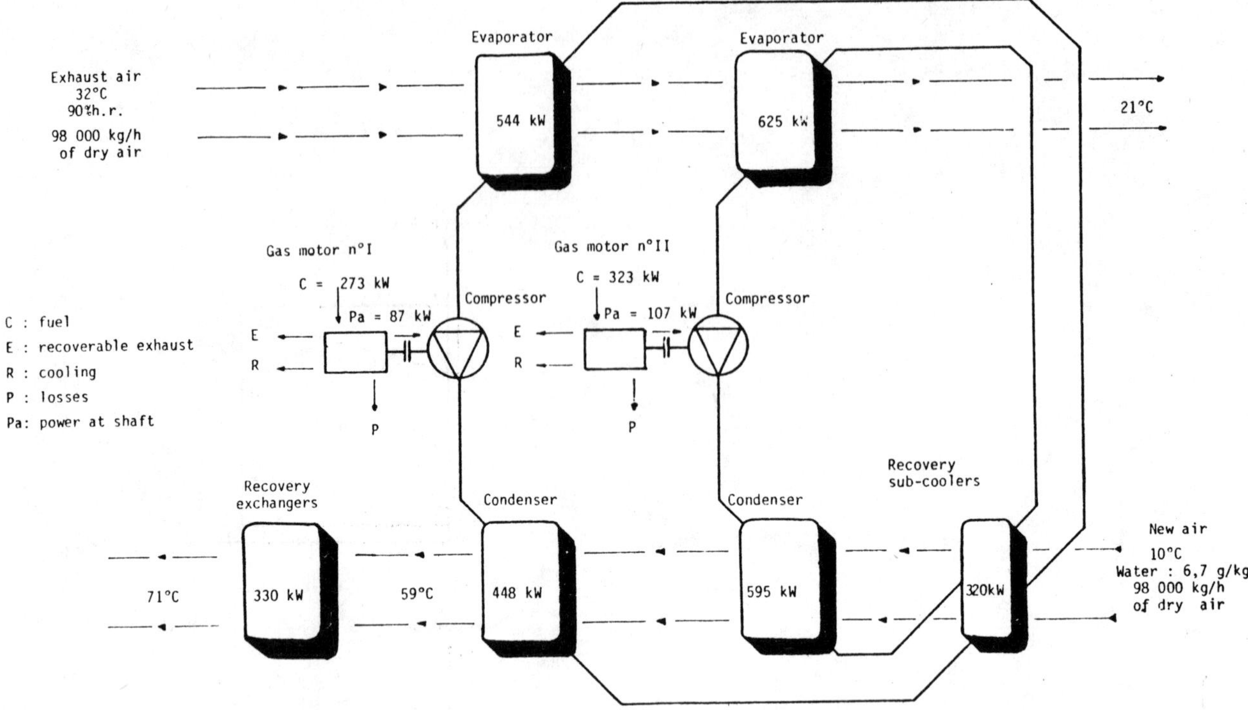

Figure 6 : thermal balance

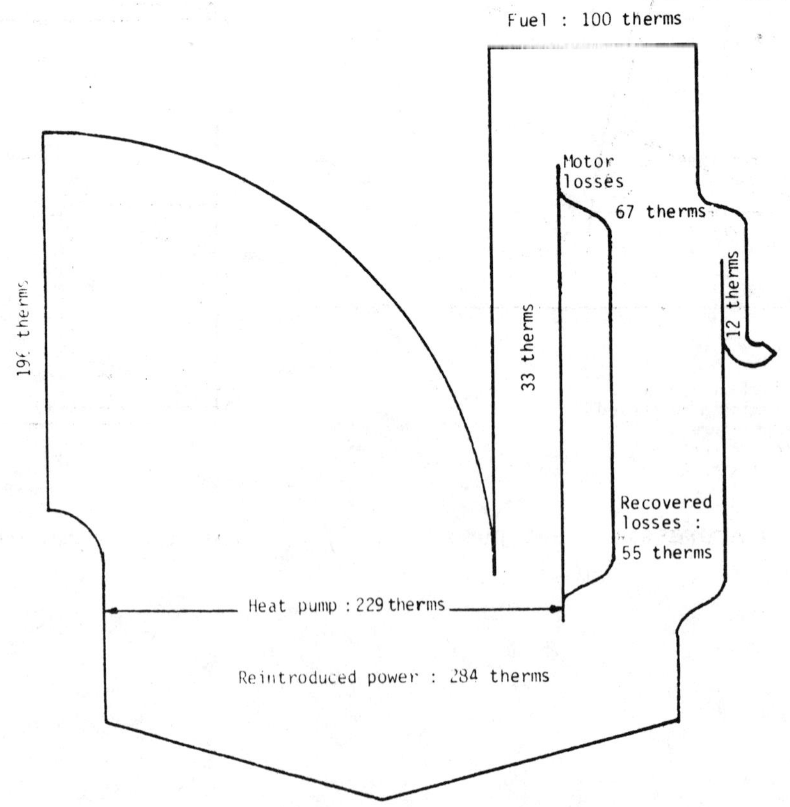

Figure 7 : SANKEY's diagram

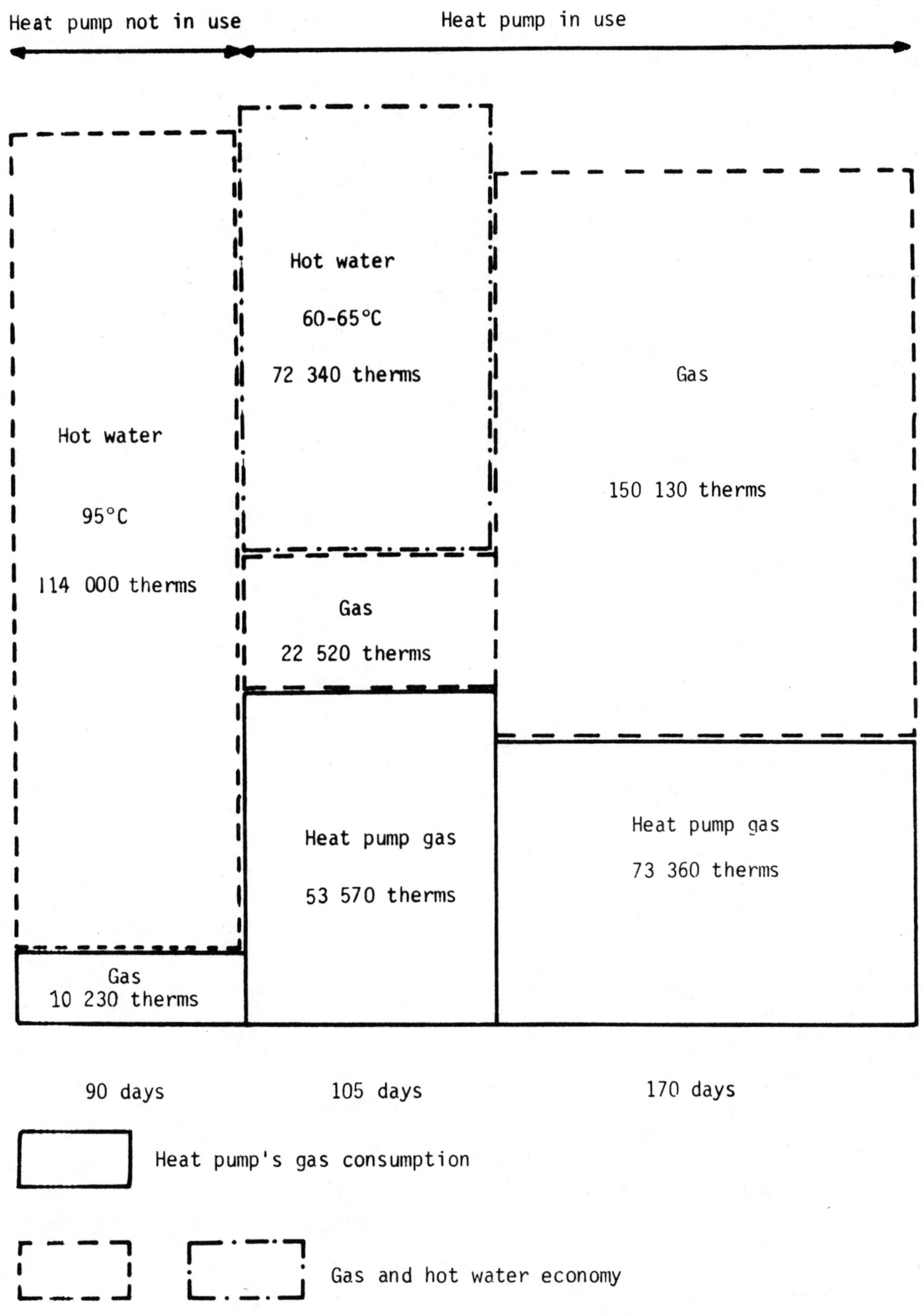

Figure 8 : Annual heat consumption rates

EXPERIENCES IN THE APPLICATION OF INDUSTRIAL HEAT PUMPS – CONCLUSIONS FROM 80 CASE STUDIES

H. Schnitzer and F. Moser

Technical University Graz, Austria

Summary

The application of heat pumps in industry is widely discussed but the actual technological and economic potential of the possibilities for their application in industry is ascertained. The authors have carried out a detailed study of the technical and technological implications of the application of heat pumps in the process industries as well as an estimation of the market potential of heat pumps in Austria and the resulting primary energy reduction.

In the study all types of heat pumps i.e. compression heat pumps, absorption heat pumps (including heat transformers) and open cycle heat pumps (mechanical vapour recompressions) were considered.

Case studies in about 40 companies have been carried out, covering application in 16 branches of industry like chemical industry, metals and wood processing, food industry and others. It was important for the valuation of the heat pump application to differentiate between internal process and external process applications. Internal process applications are evaporation, distillation, drying and internal heat recovery. External process applications are heating purposes in general. The valuation of the heat pumps in the case studies has been carried out through thermodynamical parameters (e.g. COP), energetic parameters (e.g. primary energy ratio) and economical parameters (e.g. DCF). The case studies lead, besides others, to the following conclusions:
- the application of the heat pump has to be studied within the process if possible
- the application of heat pumps to existing processes is difficult and often uneconomical if the operating conditions can not be changed.
- the application of heat pumps demands a change of the operating conditions in general, which will result in a new design concept.

The possible saving in energy by the applications of heat pumps in the industry is 5.5% in 1980 related to the present consumption and will increase to 8.1% at a rise in the price for the energy of 100%. In the same way, as the potential for saving primary energy has been evaluated, the necessary investment has been calculated at 11 Mrd. Austr. Schillings or 14 Mrd. AS respectively.

Held at the University of Warwick, U.K.
Symposium organised and sponsored by
BHRA Fluid Engineering

Nomenclature and abbreviations

a	year
AHP	absorption heat pump
AS	Austrian Shillings (100 AS $\sim$ 3,3 £)
BEP	break even point
CHP	compression heat pump
COP	coefficient of performance
DCF	discounted cash flow
DCFRR	discounted cash flow rate of return
HP	heat pump
ORC	organic rankine cycle
P	power
PE	primary energy
PER	primary energy ratio
$\dot{Q}$	amount of heat
	efficiency
ΔE	energy saving
ΔE_{PE}	primary energy saving
ε	COP

1. INTRODUCTION

The application of heat pumps in industry should be directed to recover large amounts
of energy in industries which are large consumers of energy. Industry in developed
countries consumes from 30 - 50 % of the total energy requirement of efficiencies
which vary from 9 % for nitric acid processes to 78 % for ethylene production, in
which heat pumps are already included (Ref. 1). The average energy efficiency for
chemical processes is around 30 - 40 %. However, there is a difficulty in the
recovery of energy, mostly in the form of waste heat from such processes as 70% of
this waste heat became available below 100°C, while an equal percentage is required
above 200°C. The problem therefore is: How could we bring waste heat from below
100°C, where it becomes available, up to 200°C or higher, where it is required?

There are in general several ways to improve the efficiency of processes and to
reduce energy losses. Basically two different ways can be distinguished:

> 1) By internal use of heat within the process
> 2) By external use of waste heat for other purposes

Both cases should be investigated when considering the energy household, but the first
is of course the more important one.
What then are the options for engineers to reduce the energy requirements in industrial processes. They are:

> 1) Heat exchange
> 2) ORC-processes (organic Rankine Cycle)
> 3) Heat pumps

The present situation with regard to these three possibilities can be summarized as
follows on the basis of detailed studies.

ad 1)   Heat exchange should be used whenever possible to the utmost, because it is the cheapest
way to recover heat
ad 2)   The efficiency of ORC-Processes is low (5 - 8%) and only relatively
small machines are available (max. capacity around 2 MW)
ad 3)   Heat pumps - the knowledge about such arrangements is so far insufficient to
make wide spread application possible. More development work seems necessary.

The situation in more detail, based on conditions in Austria in 1980 is presented
below.

2. AVAILABILITY OF INDUSTRIAL HEAT PUMPS

In order to obtain information about availability of industrial heat pumps, contact
was made with various heat pump producers. At present little practical experience
with heat pumps in industrial processes, except with thermocompressors, is available.
Compression heat pumps up to a temperature of 70°C are offered by a great number of
heat pump manufactors. The maximum temperature that can be achieved with compression
heat pumps today is 220°C where water is used as the heat transfer medium. For
organic refrigerants temperatures above 130°C cannot be achieved at present due to the
lack of stability of the refrigerants.
The maximum temperature for the heat delivered in absorption heat pumps is about 110°C
due to the critical temperature of NH_3. In the generator the temperatures should stay
below 190°C due to the chemical instability of the NH_3/H_2O mixture. Using new working
fluids, especially with water as the solute, temperatures above 200°C are expected to
be possible to achieve within a few years. At present, however, there exists
practically no industrial experience with AHPs. Counter absorption heat pumps
(heat transformers) seem to be extremely interesting for industrial applications,
because they could present a system which makes it possible to generate steam from
waste heat. Maximum obtainable temperatures are from 140 - 210°C with COP from 0.25
to 0.50. The heat sink temperatures for these heat pumps are 90 - 140°C. Again
however, there is no practical experience with these heat pumps in industry until now.
Mechanical vapor recompressors and jet compressors are widely used in industry, ex-
specially in evaporation processes. In distillation there are few applications up to
now. The temperature range in which these types of HP are applied is from low
temperature applications up to the range of 200°C. There is no temperature limitation, but applications are limited depending on the medium, which has to have
properties, suited to the use of process compressors and heat exchangers. Toxic,

corrosive or very viscous media may prohibit the meaningful application of MVR.
Non condensable components in the vapour can bring further problems to the use of
MVR in the chemical industry.

<u>Which heat pump - which drive</u>

Industrial heat pumps today are compression heat pumps in general. In a very limited
number of cases absorption heat pumps are in operation. Absorption heat pumps require
higher investments related to CHP s, therefore they are more economical only if the
heat production is very high or of they can be driven by waste heat. Heat transformers (counter-absorption heat pumps) cannot be compared to CHP's or AHP's directly,
since they are a combination of an heat pump and an heat engine. For the comparison
of HP of equal or different design we define a number of characteristics. They all
set up a ratio of the quantity of useful heat delivered and the amount of heat or
power for driving the HP, the coefficient of performance (COP). As there is a great
difference in the thermodynamic value of power and heat, we can find a set of COP
which is not directly comparable. In order to compare different heat pumps system
using drives from different fuel or energy sources among each other or with other
energy recovery systems as e.g. heat exchangers or organic rankine cycles, the
Primary Energy Ratio - PER - is applied.
The PER takes into account not only the heat pump COP but also the efficiency of
conversion of the primary fuel (e.g. oil, gas,coal or water power)into the work which
drives the pump.
The PER is defined as follows:

$$PER = \frac{\text{Useful heat delivered by heat pump}}{\text{Primary energy consumed}}$$

It is often possible to use an alternative definition, when a heat engine with thermal efficiency η_{th} is used to drive a heat pump compressor.

In this case:

$$PER = \eta_{th} \times COP$$

The primary energy saving ΔE_{PE} indicates, how much primary energy (PE) can be saved
by one process (P1) compared with an other (P2). The primary energy saving can be
calculated by

$$\Delta E_{PE} = 100 \quad (1 - \frac{PER_{p2}}{PER_{p1}}) \qquad /\%/$$

with PER_{p1}.........PER of process 1
$\quad PER_{p2}$.........PER of process 2

On the base of this equation it is possible, to define a minimum COP, required to
archieve savings in PE.

$$COP_{min} = \frac{PER}{\eta_{th}}$$

with η_{th} efficiency of conversion of PE to shaft power. This figures can help
us to select the driving device for our heat pump from an energy   point of view.
Thereby,   we have to consider, that also an economical comparison and a
comparison of operating behaviour, the range of operation and the on-stream factor
must be made.
Although the COP of an AHP is much lower than that of a CHP, its PER may be equal or
even better. This is caused by the different efficiencies for the conversion of fuel
to energy in driving the HP. Power, either electricity or shaft power of a combustion
engine is generated with an efficiency of η_G = 30 -40 %. The efficiency for
generating heat for the absorption heat pump is about η_G is 80 - 90 %. In that way
PER, the product of the COP and this efficiency

$$PER = \eta_G \times COP$$

turns out to be about the same for both cases as e.g.

                for CHP   PER = 0,30 . 4,0 = 1,2
                for AHP   PER = 0,85 . 1,4 = 1,2

(comp. also fig.   1   ). PER can however be improved if the additional use of the
waste heat of gas- or diesel motors is used to drive the motor. Using steam turbines
for driving the heat pump gives about the same figure as the electrical driven unit
does, for the small turbines have about the same efficiency as the production and
distribution of electricity in larger units will do. Other arguments will arise   if
the turbines are driven by organic media, using waste heat for powering.

Minimal COP due to economy

There is also a lower boundary limit for the COP due to the prices for the heat delivered and the drive.

$$COP_{min} = \frac{price\ of\ drive\ per\ kW}{price\ of\ heat\ per\ kW}$$

In European countrys this COP_{min} can vary widely for different drives. Electricity
prices per kWh can range from 1.1 to 4.0 times the price of heavy fuel oil. In countries with plenty of hydro power this ratio can even fall below one.
This prices for gas are in the range of 1 to 2.0 times fuel oil. The price for heavy
fuel oil divided by the efficiency of the boiler gives the price of heat if
depreciation for the COP_{min}. is not taken into account.
The values for the COP_{min} changed rapidly within the last years. . If one
looks   at the increases of the prices of various energies in the last ten years
(for Austria, 1981).

                crude oil: factor   30 - 40
                fuel oil: factor   10
                gas: factor   4
                electricity   factor   2

one can see that prices for the four energies are steadily coming closer together. Gas
now is the cheapest energy for driving the heat pump, also due to its good COP using
the waste heat of the motor. Nowadays gas prices begin to rise with the oil prices, and
bring, in doing so, electricity into a better situation. Electricity, generated in
hydroelectric, coal-, nuclear - or cogenation plants is generally assumed to be most
stable in price. Diesel oil is generally more expensive than natural gas, but diesel
motors are much cheaper. In many countries it is allowed to use extra light fuel oil
in diesel motors in heat pumps.

3.   THE UNIT OPERATION CONCEPT

In every industrial plant, there is a number of process steps, which appear in many
processes in a similiar way. These are called unit operations. There are about 60
unit operations in chemical plants, as e.g. distillation, extraction, drying, filtration and sedimentation. To make it possible to work out 80 case studies within 15
months it was necessary to apply the unit operation concept to the application of
heat pumps.

Out of the various unit operations in chemical engineering not all lend themselves to
the same degree for the application of heat pumps. A survey /5/ showed, that
                        evaporation
                        distillation
                        drying of solids and gases
                        reaction engineering
                        refrigeration and
                        general heat recovery

are the most likeley unit operations where heat pumps will be usefully applied. Of
course there also will be other applications as e.g. in electrolysis but it seems at
present, that the field of application is limited mainly to the unit operations
mentioned. In all cases we have however to fulfil the following criteria:

A)   For the case of process-integrated use of heat pumps there has to be a heat
     source and a heat sink in the process
B)   The temperature difference between heat source and heat sink should be small
C)   The transferred amount of heat should be large
D)   The on-stream time should be large
E)   Continuous operation is preferred
F)   Costs for conventional heating or cooling should be large
G)   Cooling and heating should be required

In general it is useful to distinguish between process integrated heat pump systems
and HP-systems for general heat recovery (comp. Tab. 1). In order achieve satisfactory results in applying heat pumps to the chemical industry it will in many
cases be necessary to change the process technology. An example for this statement
is given in Fig. 2. where heat from one process is to be used in another process. In
the first a process stream has to be cooled down from 70°C to 50°C and in the second
a suspension is to be evaporated at 100°C. The cooling down of the process stream in
the first process is carried out by heating-up cooling water to 27°C.

If we recover the heat from the cooling water, it is possible to heat it to 50°C and
cool it in the evaporator of the heat pump. With the heat delivered by the heat pump,
steam is generated which can be used in the evaporator (case A). A first step to improve the COP of the heat pump is to cool the process stream by the heat transfer
medium of the heat pump directly. The second step would be to transfer the heat from
the heat pump directly to the evaporator, avoiding the steam cycle. Doing so, an
improvement in the theoretical COP from 4.2 to 5.9 would be achieved, which still
might be too small for an economical application of the heat pump.
A further improvement of the energy utilisation can be achieved by lowering the
operating pressure of the evaporation process. In such a case an operating temperature
of 80°C would occur and thereby a theoretical COP of 8.07 be obtained which would be
high enough for an economically useful installation.

It turned out to be one of the most significant conclusions of the investigation that a
worthwhile application of heat pumps to industrial processes can only be done in the
state of the design and seldom by appliying the heat pump to an existing process.
Further details about the application of heat pumps to evaporation, distillation, and
drying processes as well as to chemical reactors are given in /2/ and /3/.

4.   CASE STUDIES

There is so far only a very limited number of practical applications of heat pumps in
industry. Most widely used are at present mechanical vapour recompression heat pumps
and jet compressors. In order to see whether a heat pump installation is economical
dynamic methods like discounted cash flow rate of return (DCFRR) or static or dynamic
break even point (BEP) can be used.
For the case studies, carried out during the work on our market study, we used both
methods DCFRR and static BEP. The correlation between these two is shown in fig. 3.
It can be seen, that a break even point of five years corresponds to a DCFRR of about
15 %. From all the examples there could also be drawn a figure, reflecting the influence of the most imortant parameters to the BEP (fig. 4). For a comparison with a
basic BEP_o five factors can be found in the diagram to correct this value for the
actual case. The basic data for BEP_o

C_1:   price of the heat produced, basis: 0,3 AS/kWh
C_2:   price of the driving power, basis: 1,0 AS/kWh
C_3:   price of the waste heat, basis: 0,0 AS/kWh
C_4:   annual working time, basis: 6000 h
C_5:   factor for investment costs, bais: carbon steel, price of plant = 3,2 times
                                               price of equipment

Heat pumps in cogeneration plants

A special problem arises in using heat pumps in plants with cogeneration units. In
this case two essential facts have to be considered
          -   whether there is a condensation turbine or not
          -   whether there is an electricity surplus or not

This gives four possible cases to be distinguished. In general it is found, that
saving steam at a low temperature level (T 150 C), which is the range where heat
pumps will operate, will reduce the amount of electricity generated in the back pr
pressure turbine. On the other hand, more electricity is needed (or at least power) to
drive the heat pump. This electricity has to be generated in the condensation step of
the turbine, if there is no surplus of power, or it has to be bought. Condensation
steps have a rather low efficiency (15 %) if therefore the COP_{min}, due to energy has
to be rather high (COP_{min} = 0,35/0, 15 = 5,7). Prices for electricity and heat in
cogeneration units are set up rather arbitrarily in general, so that there can be
economic applications, in spite of small or even no energy savings. Case studies have
been carried out in 15 branches of industries, as listed in tab. 1. All the examples
could be traced back to the six unit operations, discussed before. In this way it was
not necessary to learn the technology of all the processes looked at. However the
technological restrictions are of a significant influence to the heat pump application.
It is our conclusion that out of all these industries the chemical and food industry
lend themselves best to the application of heat pumps. Here we have the lowest
operating temperatures. Almost every process in the food industry - cooking,
pasteurisation, sterilisation, drying, washing - is operated at temperatures below
100°C. On the other hand in these cases we do not have in general large companies
operating cogeneration plants.
Chemical industries operate many evaporation and distillation processes, so that we
can use MVR-heat pumps frequently. This gives possible applications in spite of the
presence of cogeneration plants.
Very few applications could be found in steelworks, because of the high temperatures
here. Yet some applications can be found in flotation and waste treatment.

Smaller companies, like workshops and laundries, often offer good processes but have
only small on stream time in general. They work only 2000 h/a and good economy cannot
be found therefore.

5.   DELPHI STUDY

To reinforce our calculations we carried out a delphi study about the present and
future application of heat pumps in industry. About 100 persons out of the Austrian
industry were confronted with 14 questions about heat pumps and their applications.
After analysing the answers, all of them have been shown the results and have been
asked to think over their answers once more. The whole questionnare is cited in /3/.

Many people believe that possible savings in primary energy through heat pumps range
up to 15 %. although they themselves do not believe that they can make use of them.
Most of the persons asked who believe that they could apply heat pumps, think that they
would do it before the year 1985. Of course the best form of support from the
government to industry to induce the installation of heat pumps will be some sort of
investment assistance. This seems much more favourable to the companies than a policy
of modified interest rates. The possibility of premature depreciation of plant equipment could also be a way to encourage industry to invest in heat pumps.

6.   MARKET POTENTIAL

In order to evaluate the potential of energy savings through the use of heat pumps, a
scale up from the 80 case studies in the 40 companies has been made, knowing the
energy consumption and the number of emplcyees in every company and every branch of
industry. Fig. 5 shows the results of these calculations. Greatest savings can be
achieved in the food industry and in the chemical industry. This is caused by the
relatively low operating temperatures of the processes used in these industries.

For all the industries, it would seem that, there is the possibility to save 9.2 % of t
primary energy consumed today with 5,5 % as economic implementations now. This value
can be recalculated by the following simple calculation with the following assumptions

- 30 % of the energy-input of the industry is at temperatures lower than 200°C
  (the maximum temperature for heat pumps)
- 50 % of the possible applications are economical
- the COP of a economical feasible heat pump is at least 4,0

Now the PF-saving of one heat-pump application can be calculated by

$$\Delta E_{PE} = 100 \times (1 - \frac{0,85}{4 \cdot 0,3}) = 29,2 \text{ \%}$$

and the total national saving by

$$\Delta E_{PE,\ddot{o}} = 0,292 \times 0,5 \cdot 0,3 \times 100 = 4,38 \text{ \%}$$

These results are about the same as found in the exact calculations. These figures are as it was assumed, typical for an industrialised developed country. In a similar way, the possible savings in primary energy were found, by calculating the necessary investment costs for these heat pumps (comp. fig. 6).

7. CONCLUSIONS

For the case studies carried out during the work on this market study, 40 companies, were contacted resulting in about 80 possible applications.
Mostly the discussions started speaking about waste heat streams as possible heat sources. This turned out to be the wrong way to look for improvements. In a very few cases waste heat was really the problem for the industry. In all cases it is the heat input which created costs and which has to be diminished.
Therefore in looking for possible heat pump applications, we tried to find low temperature heat consumers at first, and looked for heat sources as the second step. Thereby frequently other possible ways to reduce the energy input could be found. In these cases, the heat pump, although an economical solution itself, frequently will not be used, because other energy saving techniques can been found through the energy analysis. In a great number of cases heat pumps could not be applied to the existing process, because the operating conditions did not allow a meaningful application of the heat pump. Changing pressures, temperatures or humidities in the process frequently can reduce energy consumption even without heat pumps or allow heat pumps to operate economically.
These facts pointed out clearly that a wide use of heat pumps will only be achieved if manufactors of various equipment, as e.g. distillation and evaporation units, crystallers, pasteurizers, dryers, industrial washing machines and others, will include heat pumps by themselvs. For the suppliers of heat pumps this means:
Selling heat pumps in not selling hardware, as it would be by selling heat exchangers. Selling heat pumps in selling services, including the full range of know-how of the specific process. Therefore - as was said in the introduction - an improvement of "Heat Pump know-how" with users and suppliers seems required to improve their application and top the vast potential in energy saving, which so far seems unexplored.

REFERENCES

/1/ Riekert, L.: "The efficiency of energy utilization in chemical processes". Chem. Engng. Sci 29 (1974) 1613

/2/ Moser, F., Schnitzer H.: "The application of heat pumps in the chemical and process industries. Professional development course, 2nd World Congress of Chem. Engng. Montreal, Canada Sept/Oct 1981

/3/ Moser,F., Schnitzer H.: "Primary energy reduction in industry through the application of heat pumps". (Energieeinsparungen durch Wärmepumpen in der Industrie). Energiepolitische Schriftenreihe, Springer Wien/New York, in preparation

/4/ Ortner P., Schnitzer H.: "Economics of energy saving equipment". 14th European Symposium 'Computerized Control and Operation of Chemical Plants'. Sep. 81, Vienna EFCE Pub. Series 17

/5/ Eder, W.,Moser F.: "The heat pump in chemical engineering" (Die Wärmepumpe in der Verfahrenstechnik), Springer Wien/New York 1979

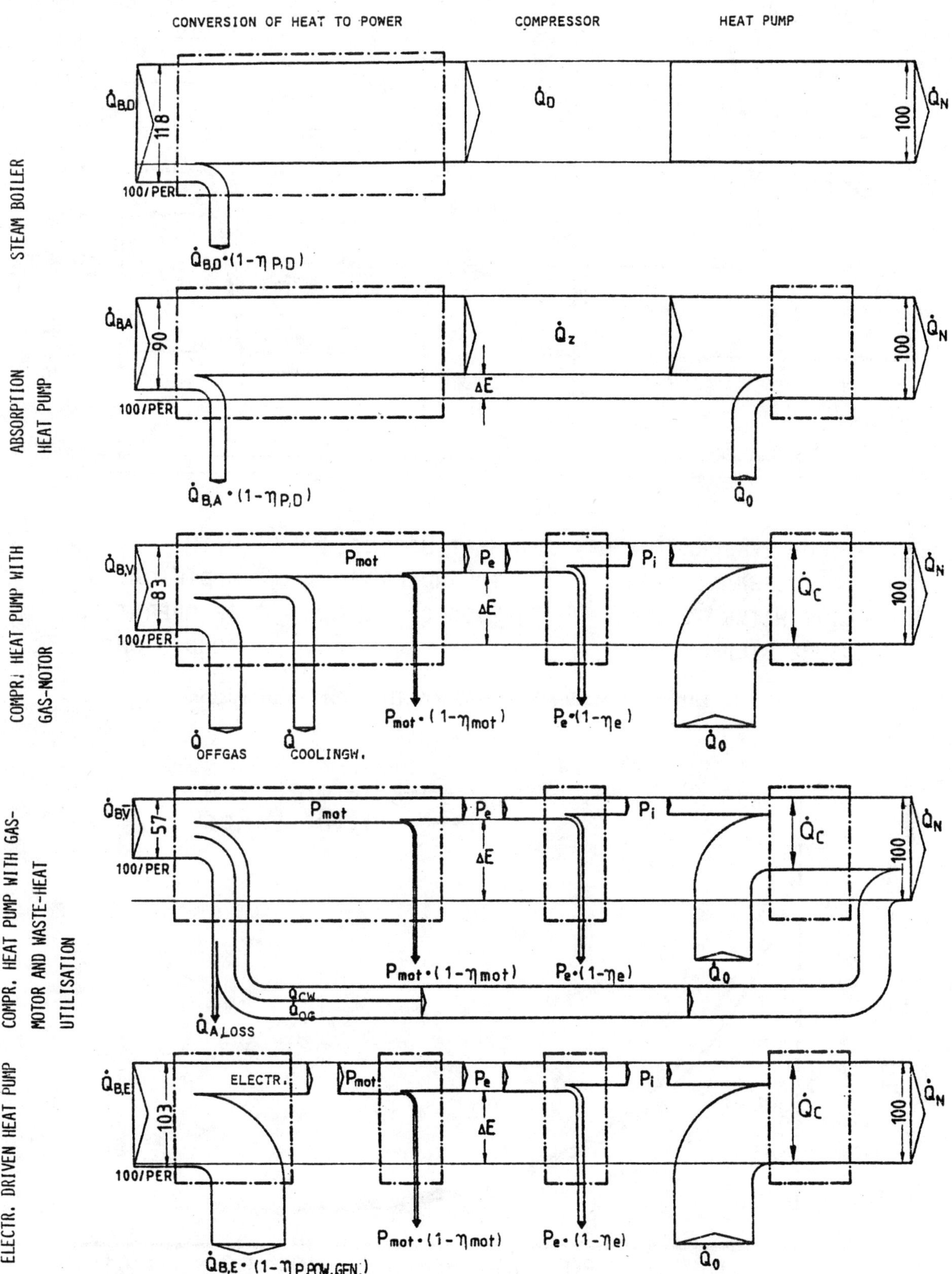

Fig. 1 Energy Flow Diagram for Different Ways to Generate Heat
$(COP_{CHP} = 3.5, COP_{AHP} = 1.3)$

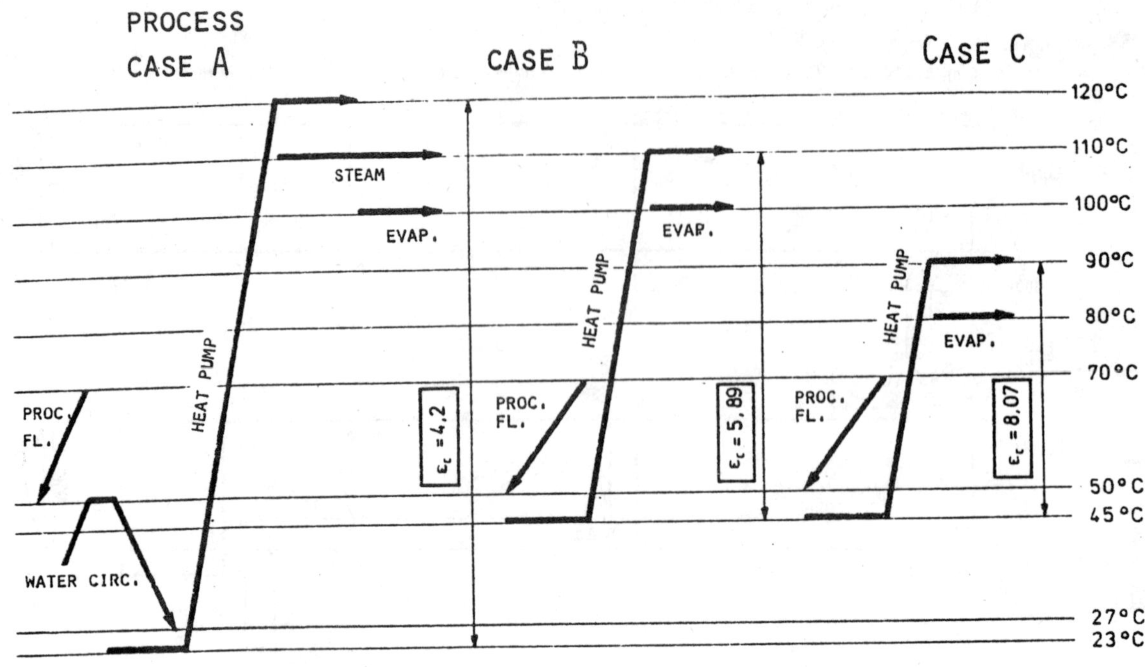

Fig. 2 Different Modes to use Waste Heat for Evaporation

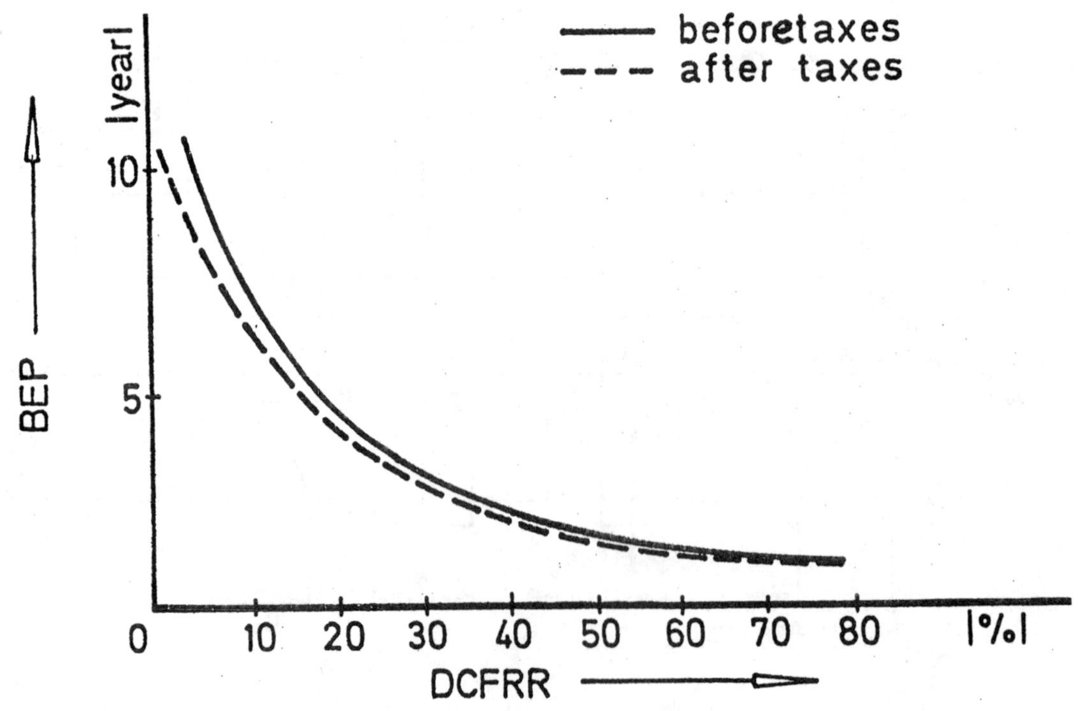

Fig. 3 Coherence of the Break even and the Discounted Cash
Flow Rate of Return for Heat Pump Installation

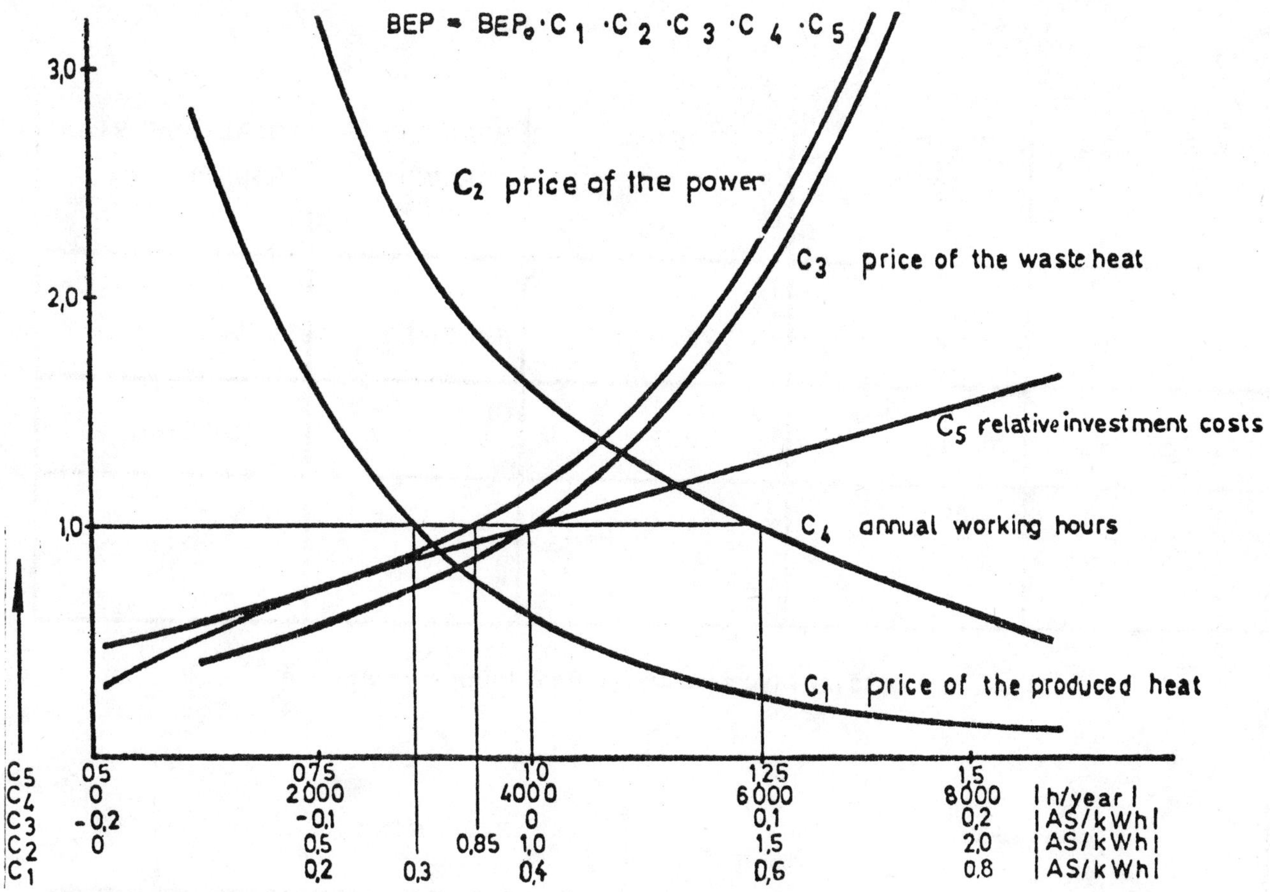

Fig. 4 Correction Factors for Practical Heat Pump Application

SAVINGS : (TJ/a)	COMPANIES ANALYSED	USED FOR SCALE UP	IN BRANCH OF INDUSTRY	TOTAL IN AUSTRIA	SAVINGS IN %
POSSIBLE	4064	4060	25 452	30 300	9.2
ECONOMICAL in 1980	3814	3810	15 224	18124	5.5
ECONOMICAL IF PE-PRICE DOUBLES	76	76	7 277	8 663	2.6

Fig. 5 Possible savings in primary energy through the
use of heat pumps

INVESTMENTS FOR :	COMPANIES ANALYSED	COMPANIES USED FOR SCALE UP	BRANCHES ANALYSED	TOTAL AUSTRIAN INDUSTRY
HPs possible	780.7	773.4	14 234.	16 945.
HPs economical	674.7	668.7	9 324.7	11 101.
HPs economical if PE price doubles	32.1	32.1	2 505.7	2 983.

Fig. 6 Necessary Investments for Heat Pumps in Austria

HEAT SOURCE	HEAT SINK	
ANY WASTE HEAT STREAM EXAMPLES:	ANY EQUIPMENT CONSUMING HEAT	
OVERHEAD VAPOUR OF DISTILLATION VAPOUR OF EVAPORATION UNITS HUMID AIR FROM DRYERS	REBOILER EVAPORATOR DRY AIR TO DRYER	PROCESS INTEGRATED SYSTEMS
COOLING WATER FROM COMPRESSORS, ENGINES,REFRIGERATION UNITS,REACTORS AMBIENT AIR WASTE AIR FROM PROCESS UNITS SOIL, RIVERS,GROUND WATER FUMES FROM PROCESS UNITS SOLAR ENERGY WASTE WATER	HEATING IN GENERAL HEAT PROCESS STREAMS HOT WATER HOT AIR HEAT ROOMS	GENERAL HEAT RECOVERY

Table. 1 Possible Application of Heat Pumps

HEAT RECOVERY FROM WARM WASTE WATER
AT DYEING PROCESS BY ABSORPTION HEAT PUMP

S. Kannoh

Osaka Gas Company Ltd., Japan

Summary

In Japan, technology for energy saving is the primary subject in many factories and buildings, and exhaust gas heat recovery and hot water recovery are practiced popularly.

In textile dyeing processes, bottle washing processes and in public baths, a large quantity of warm water, 293 K to 323 K, is discharged but almost all of such water is discarded because the temperature is low.

In October, 1980, we completed a system to recover the heat of 1.39×10^9 J/hr. out of the waste water from dyeing process of 303 to 323 K discharged at the rate of 60 m^3 per hour using an absorption heat pump, and obtain the hot water at 353 K in the rate of 3.48×10^9 J/hr. The system was installed at a textile dyeing plant.

The details of the system and the effect of energy saving during operation are reported below.

Held at the University of Warwick, U.K.
Symposium organised and sponsored by
BHRA Fluid Engineering

1. CONVENTIONAL SYSTEM

In this dyeing plant for cotton woven fabric, dyed cloth is dipped
into a dyeing bath and is washed with warm water for color fixing.
The washing water temperature differs by each bath in the range
from 333 K to 363 K. Waste water after dyeing is alkaline because
of the dyestuff and is used for desulfurizing the exhaust gas from
oil fired boilers.

After the desulfurizing process, the water is discharged through
a waste fiber eliminating filter. A heat exchanger is installed to
utilize the heat for feed water pre-heating since the temperature of
the discarded water is about 328 K. (Fig. 1)

Even after the heat exchanger, however, the waste water is still at
about 308 K and the rate of discharging is 60 m^3 per hour.

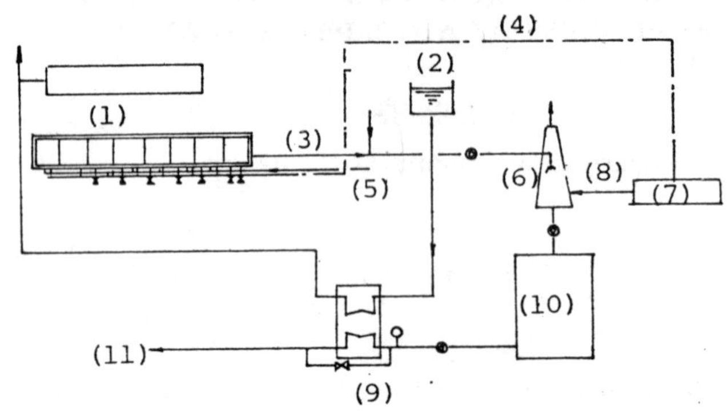

(1) Dyeing bath
(2) Elevated water tank
(3) Discharged water
(4) Steam
(5) Feed water
(6) Desulfurizing
(7) Boiler
(8) Exhaust gas
(9) Heat exchanger
(10) Waste fiber filter
(11) Waste water treating equipment

Fig. 1 Conventional System

2. DEVELOPMENT TO THE COMPLETION OF HEAT RECOVERY SYSTEM

2-1. Selection of Heat Pump Type

High temperature water is taken out of low temperature water by
heat recovery with heat pump cycle, which can be either
compression type or absorption type. In selecting either one of
the two, we took the following conditions into consideration.

From users' point of view, the higher the hot water temperature,
the easier is the water to use. From the point of heat recovery,
on the other hand, it is more advantageous to lower the
temperature of waste water. It is better, therefore, to increase
the temperature difference of refrigerant between condensation and
evaporation. For this requirement, the followings are regarded
important.

i) Efficiency comparison of cycle
ii) Reliability of equipment

By the technology at present in Japan, H_2O - LiBr absorption type heat pump and Freon refrigerant compression type heat pump are practical and the two types are compared.

First for the compression type heat pump, the theoretical efficient (ηth) is given by inverse Carnot's cycle as shown below.

$$\eta\,th = \frac{tc}{tc - te} \quad\dots\dots\dots\dots\dots\dots(1)$$

      tc: Condenser temp.

      te: Evaporator temp.

Accordingly, the efficiency is lowered in hyperbolic manner as the temperature difference increase.

For absorption type heat pump, on the other hand, the theoretical efficiency is given by:

$$\eta\,th = \frac{tg - tc}{tg} \times \frac{tc}{tc - te} \quad\dots\dots\dots\dots(2)$$

      tg: Generator temp.

  (tc=ta tc: Absorber temp.)

With H_2O-LiBr absorption type, however, the efficiency differs greatly from the above and actual becomes as shown in Fig. 2.

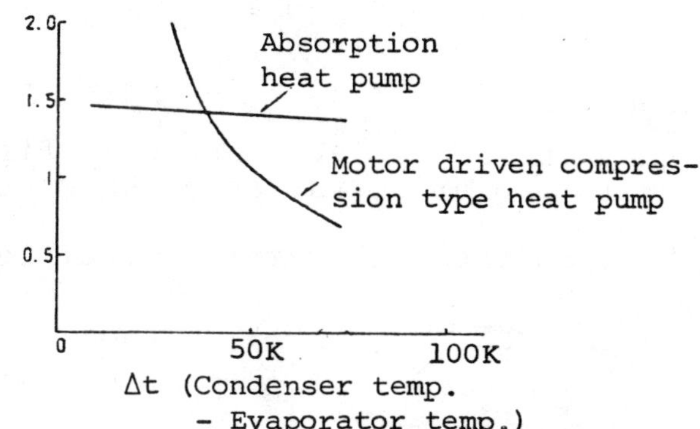

<u>Fig. 2</u>

The efficiency is calculated by primary energy conversion while taking generation and transmission efficiency of electricity into consideration because electric motor driven type compression heat pumps are ordinarily used in Japan.
(The generation and transmission efficiency is about 35% in Japan.)

As the reliability of equipment, the problem is in the durability with Freon refrigerant compression type since the thermal stability is generally lowered when refrigerant gas temperature exceeds 353 K.

We thus came to conclude that absorption type is better for this heat recovery system.

2-2. Consideration of the Corrosive Nature of Waste Water

The waste water is fairly corrosive as a result of the presence of dyestuff and from the exhaust gas desulfurizing process.

For the heat exchanger, SUS-316 was employed as it was found resistive on the corrosion test of various kinds of metals carried out over several months and the price is reasonable.

Table 1 Analysis of Waste Water

Item	Analysis value mg/ℓ	Item	Analysis value mg/ℓ
Cd	0.02	Na	518.8
Cr	0.03	K	25.00
Cu	0.14	Ca	11.19
Fe	1.04	Mg	3.79
Mn	0.09	Al	1.23
Ni	0.13	Si	10.88
Pb	0.26	P	2.13
Zn	0.10		

Note) Specific resistance: 195   -cm, pH: 9.0

Cl^-: 979 ppm, SO_4^{2-}: 713 ppm

3. SYSTEM OUTLINE

3-1. Outline of Absorption Heat Pump

The heat pump is designed in the cycle as shown in Fig. 3 and the concentration of LiBr water solution varies as shown in Fig. 4.

Fig. 5 indicates the heat balance under rated condition.

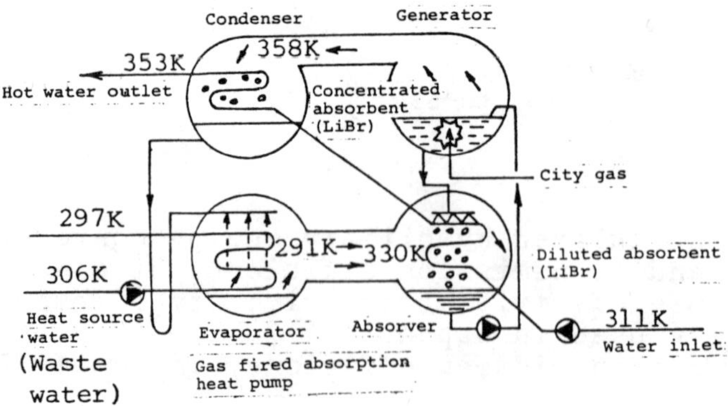

Fig. 3 Cycle Flowchart

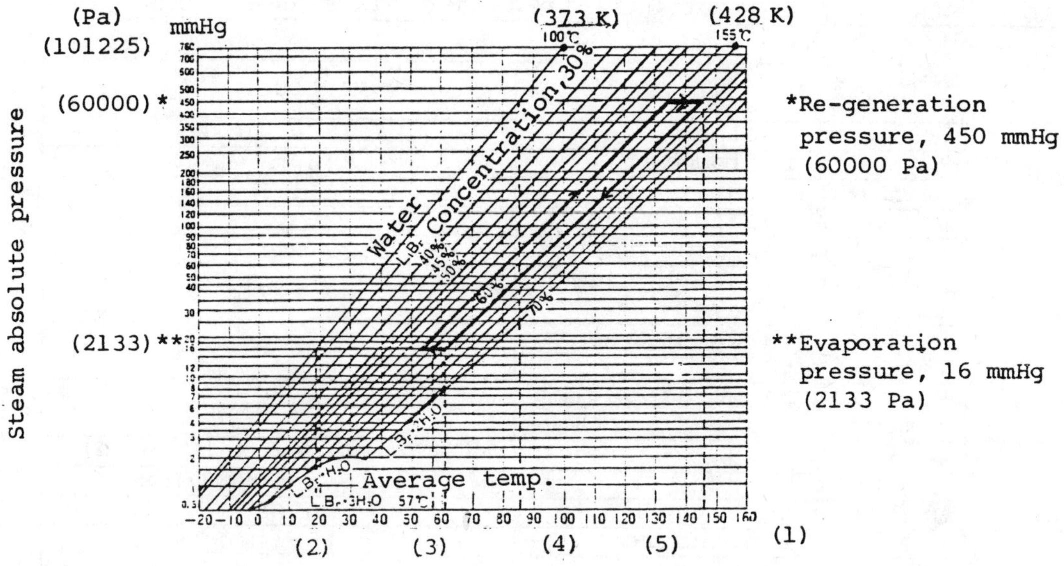

(1) Temperature °C (K)
(2) Evaporator temperature 18°C (291 K)
(3) Concentrated absorption liquid temp. 62°C (335 K)
(4) Condenser temperature 85°C (358 K)
(5) Re-generation temperature 145°C (418 K)

Fig. 4 LiBr Water Solution Pressure-concentration Diagram

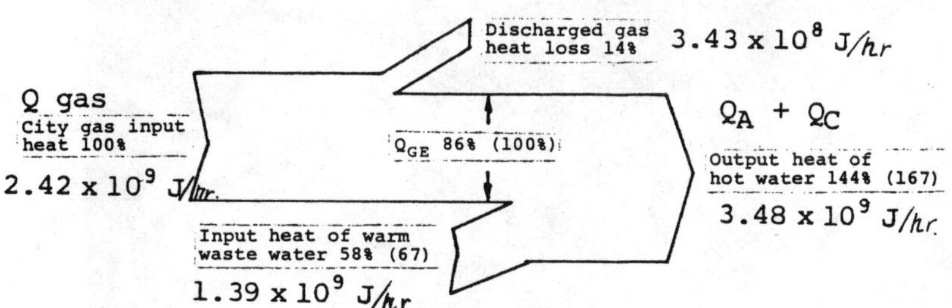

Fig. 5 Heat Balance

The city gas consumption is 58 Nm^3/hr.
(H1 = 4.16 x 10^7 J/Nm^3) and the efficiency (η)
based on this consumption is:

$$\eta = \frac{QA + QC}{Qgas} = 1.44 \quad\dots\dots\dots\dots\dots(3)$$

Table 2 shows the specifications of the heat pump installed.
It was manufactured by TOKYO SANYO ELECTRIC CO., LTD.

Table 2 Specification of Gas Fired Absorption Heat Pump

Model		GT-110HP
Warm water system	Hot water output	3.48×10^9 J/h
	Warm water flow rate	19.8 m³/h
	Warm water inlet temp.	311 K
	Hot water outlet temp.	353 K
Heat source water system	Heat source water input	1.39×10^9 J/h
	Heat source water flow rate	37 m³/h
	Heat source water inlet temperature	306 K
	Heat source water outlet temperature	297 K
Fuel system (City gas)	Consumption	58 m³/h (2.42×10^9 J/h)
	Low heat volume	4.16×10^7 J/Nm³
	Specific gravity	0.65
	Design gas pressure	1961 Pa
	Effective heating surface area (Generator)	12.5 m²
Dimensions	Length	4.305 m
	Width	2.265 m
	Height	2.650 m
	Weight	8000 kg

Photo-1 Outside View of Gas Fired Absorption Heat Pump
Installed for This System

3-2. Overall System Flowchart

As simple description of the whole system, the waste water
discharged from the dyeing bath is discharged at about 328 K to
be used for desulfurization of the exhaust gas from the boilers
as described before. After going through the filter, the warm
waste water is used for feed water preheating by the 1st heat
exchanger and is further subjected to the heat exchange with the
heat source water of the heat pump by the 2nd heat exchanger for
heat recovery by the absorption heat pump and is cooled down to
about 300 K to be discharged. While the absorption heat pump
uses city gas as the drive source, receives the heat from the
waste water, and heats the warm water preheated to 311 K by the
1st heat exchanger up to 353 K for supplying the dyeing bath.
The flowchart is shown in Fig. 6.

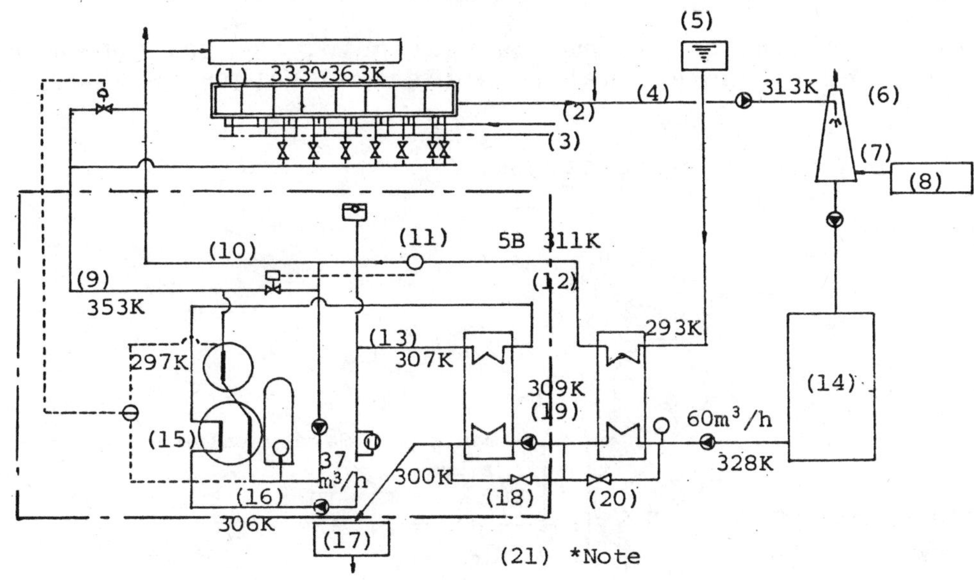

Fig. 6   System Flowchart

 (1) Dyeing bath
 (2) Feed water
 (3) Steam
 (4) Waste water
 (5) Elevated water tank
 (6) Desulfurizing
 (7) Exhaust gas
 (8) Oil fired boiler
 (9) Hot water
(10) Automatic valve for circulation
(11) Flow switch
(12) Feed water
(13) Heat source water
(14) Filter
(15) Gas fired absorption heat pump (GT-110HP)
(16) City gas
(17) Waste water treating equipment
(18) 2nd heat exchanger

(19) Warm discharge water
(20) 1st heat exchanger (existing)
(21) *Note: The temperature at respective parts changes
             according to the season; the flowchart shows the
             average temperature.

By this system, hot water from the heat pump is supplied to one
dyeing bath. However, the hot water in the bath must be replaced
several times a day because the kind of dyestuff is changed.

At the replacement, therefore, the differential pressure valve is
opened to feed the hot water to other systems since the hot water
load is eliminated.

When the hot water load of the whole system is lowered, the
automatic valve is opened to circulate the hot water.

As for the control method of the heat pump, the gas consumption
is controlled by the hot water outlet temperature as shown in
Fig. 6.

Fig. 7 shows the characteristics.

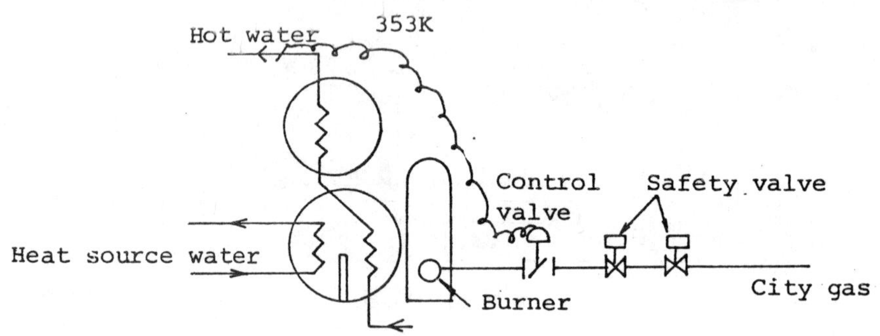

Fig. 7   Gas Fired Absorption Heat Pump

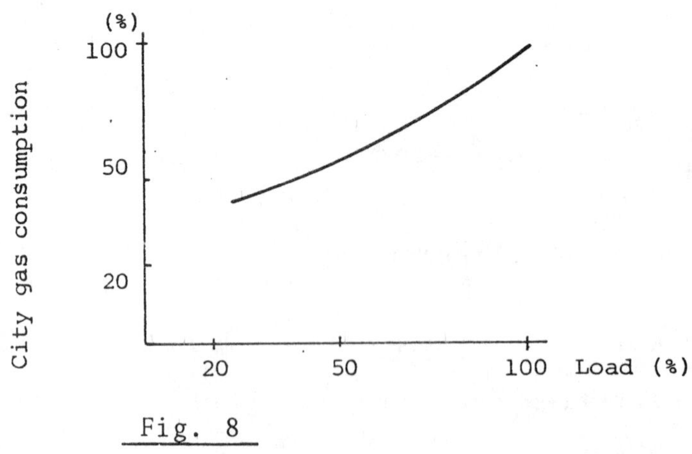

Fig. 8

4. EFFECT

4-1. Operational Condition

Fig. 9 and Fig. 10 indicate the operating data of average days
in August and December.
The heat source water temperature (waste warm water temperature)
differs by about 10 K between summer and winter, but the warm
water inlet temperature is less susceptible to seasonal change
because of the preheating by the 1st heat exchanger. The hot
water outlet temperature, therefore, is stable throughout the
year.
The operating time is longer in summer because of higher
production.

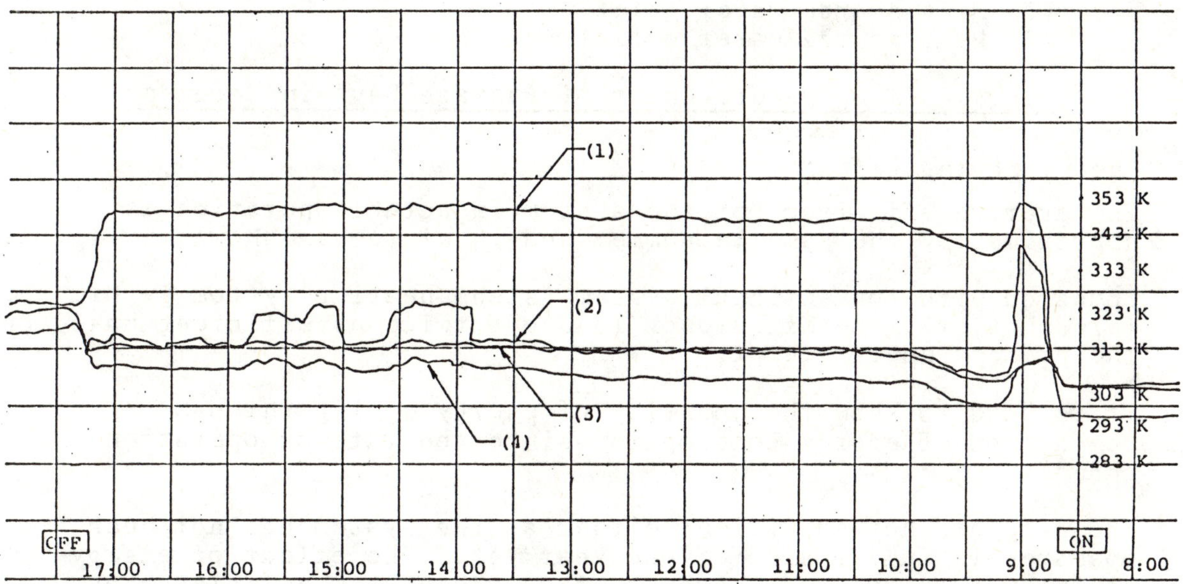

(1)   Hot water outlet   (3) Heat Source water inlet

(2)   Warm water inlet   (4) Heat Source water outlet

<u>Fig. 9   Temperature Chart of Average Days in August</u>

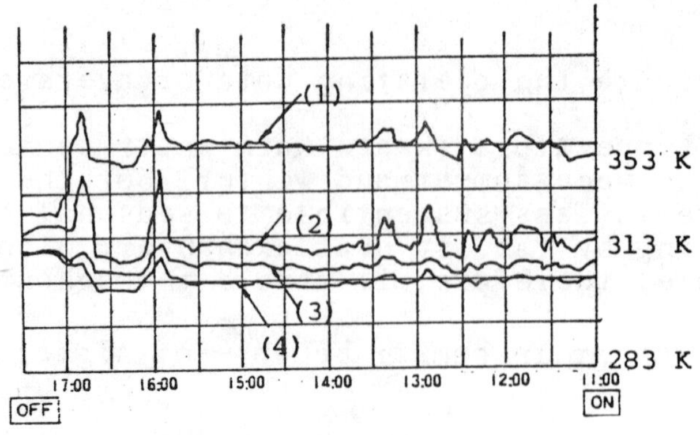

(1) Hot water outlet
(2) Warm water inlet
(3) Heat source water inlet
(4) Heat source water outlet

Fig. 10 Temperature Chart of Average Days in December

4-2. Energy Saving Effect

The average efficiency of the system in actual operation is approximately the same in summer and in winter as shown in Fig. 11.

Compared with conventional system (steam heating system by oil fired boiler), the efficiency is about twice as efficient based on the primary energy input.

It is hard to know the quantity of energy saving through comparison of energy consumption since the rate of operation changes.

The data are not sufficient either as the system has not been operated in full scale for one year yet. The effect of energy saving was calculated, however, from the system efficiency and rate of operation based on the operational data up to this date. The typical results in Summer and Winter seasons are as shown in Table 1 and Table 2.

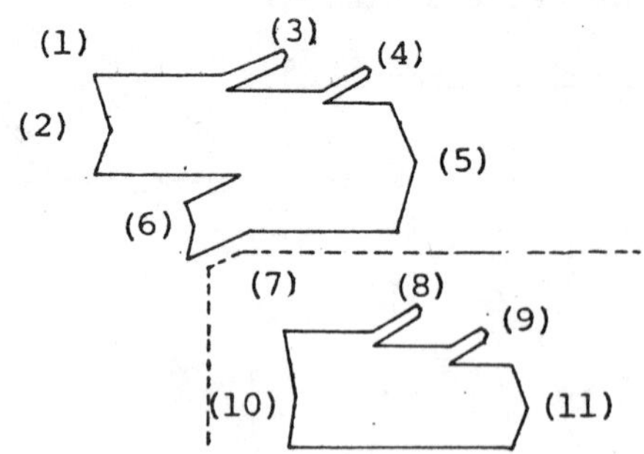

(1) Gas fired absorption heat pump
(2) City gas, 100%
(3) Exhaust gas loss, etc., 14%
(4) Heat loss from piping, 5%

    (5) Hot water output, 141%
    (6) Warm waste water, 60%
    (7) Conventional system (steam heating by boiler)
    (8) Exhaust gas, etc., 18%
    (9) Heat loss from piping, 10%
   (10) Oil, 100%
   (11) Steam output, 72%

Fig. 11   System Efficiency

From the above results, the annual energy saving is presumed to be 3.86×10^{12} J (100 kl of fuel oil).

Table 3   Effect of Energy Saving in August

	Absorption heat pump system	Conventional system (Boiler system)
Hot water output heat	5.44×10^{11} J	
Heat recovery	2.32×10^{11} J	———————
Energy consumption	3.86×10^{11} J (Gas consumption, 9280 m³ 4.16×10^7 J/m³)	7.56×10^{11} J
Efficiency	1.41	0.72
Energy saving	3.7×10^{11} J (Converted into fuel oil 10,000ℓ)	———————

Table 4   Effect of Energy Saving in December

	Absorption heat pump system	Conventional system (Boiler system)
Hot water output heat	4.01×10^{11} J	
Heat recovery	1.70×10^{11} J	———————
Energy consumption	2.84×10^{11} J (Gas consumption, 6840 m³)	5.57×10^{11} J
Efficiency	1.41	0.72
Energy saving	2.73×10^{11} J (Converted into fuel oil 7,700ℓ)	———————

5. CONCLUSION

The unit cost of fuel oil at present in Japan is about £0.2/1 (3.56×10^7 J/1) and that of city gas is about £0.28/m^3 (4.16×10^7 J/m^3). This system which mainly consisted of the gas fired absorption heat pump unit, a heat exchanger, three pumps and the piping cost about £57,000. Annual means the about £18,000 is therefore expected which means the system can be paid back in about three years. This is significant not only for energy saving but also for cost saving.

This system is suitable for the users with ample warm water load. Being already operated as the system to obtain the hot water output of 6.33×10^9 J by recovering the heat of 2.53×10^9 J from waste warm water of public bath and as the system to use the heat of 9×10^8 J for room heating by recovering the heat of 30RT (3.8×10^8 J) from computer room cooling system, this heat recovery system is realized a substantial effect of energy saving.

THE POSSIBILITIES OF WASTE HEAT RECOVERY BY HEAT PUMPS
SHOWN ON THE BASIS OF SEVERAL STUDIES AND PLANTS,
REALISED IN SWITZERLAND

R. Gfeller

Sulzer Brothers Ltd, Switzerland

Summary

This paper describes some possibilities of using waste heat of industrial and other origin for heating purposes by the means of heat pumps.

It also shows, that when it is not possible to use the recovered heat in full quantity and at the temperature level offered by the heat pump in the own works, it may be interesting to look outdoors for possible uses.

Held at the University of Warwick, U.K.
Symposium organised and sponsored by
BHRA Fluid Engineering

<u>Nomenclature</u>

t	metric ton
m3	cubic meter
/a	per annum
kW	kilowatts (1 kW = 1 kJ/s)
kWh	kilowatthour (1 kWh = 3,6 MJ)
Fr.	Swiss franks (1 Fr. = ca. 0,3 £)
COP	coefficient of performance

The possibilities of waste heat recovery by heat pumps shown on the basis of several
studies and plants realised in Switzerland

1. Introduction

Because of the enormous growth of fuel-consumption during the last three decades
and not the less because of the dangerous dependence of Switzerland from the im-
portation of oil, it becomes more and more important for our country to look for
possibilities of recovering some of the wasted heat in industrial processes and
other ranges of application.

During the years of plentiful, low cost energy, there was very little interest
in the investment of money for heat recovery purposes.

Since the price-shocks of 1973 and 1980, offered by the OPEC-organisation to the
rest of the world, men began to get aware of the dangerous dependence of their
economic system from an energy source quickly becoming rare.

In Switzerland, where the technology of heat pump application was developed to a
quite high standard during the last world-war in order to economise the coal,
short for heating-purposes, and to replace it by environmental waste heat and
hydro-electric power, one began now to re-think these possibilities, nearly for-
gotten during the time of cheap fuel.

Today, everybody speaks of heat pumps and a lot of firms offer the installation
of such plants for small family and appartment houses. Some few firms are able
to offer and to install also plants for industrial purposes and with much larger
thermal capacities.

The aim of my paper is to show on the basis of several projects and plants realised in Switzerland some possibilities of waste heat recovery by means of heat
pumps.

Not all of these projects are for industrial purposes only, but we assume, that
the recovery of waste heat from an industrial application or plant and the use
of it at a higher temperature level for room heating purposes within the plant
or even outside of it may also be of some interest.

In the following, I shall present data and schematic lay-out of 5 different
plants, each representing some special particularities:

- Large central heating plant, using waste heat from the generator cooling of a
  hydro-electric power plant for about 1400 flats and 2 schools at Birsfelden,
  near Basle.

- Heat pumps installed at Airolo, end of the largest motor-way tunnel of the
  world, the Gotthard-tunnel.

- Heat pumps in combination with the cooling system of artificial ice rinks.

- Heat pump and combined refrigerating plant in the factory producing the famous "Swiss Army Knife" in Ibach in Central-Switzerland.

- Central heating with a gas-engine-driven heat pump using waste heat from a
  factory for school buildings in the neighborhood.

2. <u>Description of some heat pump plants</u>

2.1. Heat pump plant for alternative heat supply for the district of "Sternenfeld" at Birsfelden, near Basle

The district of "Sternenfeld" consists of about 36 large buildings with flats, belonging to 10 different owners as for example some super-annuation-funds of industrial companies, co-operative societies of personnel of the Swiss railways and governmental personnel and some private owners (fig. 1). In addition, there are two school complexes and some other buildings for general purposes as a general shop and a restaurant. This district was erected in the years 1960-1970 and was heated by about 12 differentcentral heating plants with oil-fired hot water boilers, producing also the domestic hot water. Different heating systems were realised in the various buildings, as for example low-temperature panel heating, heating with radiators and convectors, air heating for special purposes, etc. The annual fuel oil consumption went up form 2300 to 2500 t in total.

Our first idea was, to use waste heat from the underground water stream heated up from waste cooling water of several industrial plants in the neighborhood. Then we studied using heat from the cooling water of the generators and transformers of the near-by hydro-electric plant of Birsfelden. These alternative solutions with heat pumps found quickly some interest by the building owners because of the present high heating cost.

We were then charged with a general study of the possibilities to realise such a system, of the investment and running costs and of the economy compared with the existing heating.

Different possibilities were examined:

a) One large central heat pump station in the neighborhood of the hydro-electric plant serving as principal waste heat source and a heating water distribution system at 60°C to each of the various existing boiler houses from where the distribution to the buildings would be by the existing piping.

a1 The heat pumps driven by electric motors.

a2 The heat pumps driven by gas (diesel) engines with recovery of the waste heat from the exhaust gases and the cooling water.

a3 Different types of heat pump compressors as for example a large number of piston compressors or only 3 to 4 screw or turbo compressors.

b) One large central station with a distribution network to six of the existing boiler houses serving as stand-by and peak-load capacity. From these boiler houses the heating water would flow through a secondary distribution system to the substations in the various buildings.

c) Two central heat pump stations,placed near the two groups of buildings, with the waste heat to be transported over a greater distance. Waste heat for one station from the hydro-electric plant, for the other one from ground water heated up by industrial processes.

The comparison of these possibilities showed clearly, that the solution "b" with one large heat pump station near the source of waste heat, equipped with electrically-driven turbo compressors is the most advantageous.

In closer discussion was also the solution of turbo compressors, driven by
gas engines, but it could not be considered because of the much higher investment and maintenance cost and as the gas price does not seem to be as
stable as the price of electricity.

The following diagram represents the finally chosen solution: (fig. 2)

The cooling water going out of the hydro-electric plant with 14 to 20°C is
collected just before its entry into the river Rhine and pumped to the central building, where it serves as heat source for the heat pumps.

A second heat source is given by an underground water well, serving as stand-
by and as additional source, when the cooling water quantity is reduced because of the low water level of the Rhine.

Three groups of heat pumps, each with a thermal capacity of 1500 kW will heat
up the heating water of the buildings from 51 to 60°C. The average COP will
be about 3.2. (fig. 3).

Through an underground piping system of well-insulated tubes, placed directly
into the ground or passing as much as possible in the basements and car parks
of the district, the water enters six different boiler houses serving as sub-
stations, where it is injected into a secondary distribution system going to
each building served by the new plant (fig. 2 and 4).

As shown in the following diagram (fig. 5), the 3 heat pumps can supply about
40% of the peak heat load of 11 MW at the lowest outside air temperature of
-11°C. As these low temperatures are rare, it will be possible to cover about
90% of the annual heat demand.

The total investment for this alternative heat supply system will be about
SFr. 9 600 000.-, all changes in the various sub-stations, as controls,
additional storage vessels for domestic hot water, piping, etc. included.

An exact cost calculation has shown that the capital and running cost are
balanced under the following consumptions:

Fuel oil price   SFr. 570.-/ton

Capital interest   5 %

Amortization period
    machinery   25 years
    piping system   40 "
    buildings   50 "

With a rising fuel oil price the above periods will be shorter.

The project, prepared in 1979 to 1981, was accepted in principle by the
owners, so that now the detailed planning is going on. It is foreseen that
the plant shall be running for the heating period of 1983/84.

2.2. Heat pump and refrigeration plant for the maintenance and controll building
     of the new tunnel for motor-cars of the St-Gotthard at Airolo
     ----
During the construction time of the longest motor-traffic tunnel of the world,
the water penetrating the tunnel in very large quantities and with temperatures of 4 to 19°C was a great nuisance. The defence against this infiltration consists in the capture and drainage of the water - still running now -
out of the tunnel.

At both ends of the tunnel are large control and maintenance buildings with offices for control personnel, police and maintenance crew. The very complexe control system with a lot of electronic and electric equipment demands a continous survey of the ambient conditions, which means the continous air conditioning and heating of the buildings. The air conditioning system has to dry the outside air by refrigeration and/or to heat it if necessary.

In order to take advantage as much as possible of the environmental sources, it was decided to use the water from the tunnel as a heat source for the principal demands for room heating and air conditioning. The following diagram shows the functional details (fig. 6).

Three groups of heat pumps are installed:

One each of 130 and 245 kW of cooling capacity with dual service:
Primarily one evaporator in the closed refrigerating circuit, securing the cooling demand of the air condition system and two additional evaporators placed in the tunnel water flow, serving as heat source for the heat pump service. For these two groupes, the refrigeration service has priority, because it supplies the computer cooling, which must never be interrupted.

The second service of these two machines is the supply of heat at different levels for re-heaters in the air conditioning plants, for heating domestic hot water and for 55°C radiator circuits.

Working with the refrigerant R 12, these machines can reach, if necessary, a water temperature coming from the condenser of 70°C, serving a radiator system of higher temperature.

A hot water storage tank of 20 m3 helps to balance heat production and demand and reduces the number of starts per hour of the heat pumps.

A third group is installed for heat pump service only, supplying water at variable temperatures for low temperature heating as for example floor heating and air pre-heaters of the air conditioning plants.

Here too is installed a vertical heat storage tank of 40 m3 for balancing purposes. The thermal capacity of this machine is about 1280 kW with an evaporating temperature ranging from 2 to 15°C and a condensing temperature varing from 45 to 55 $^{\circ}$C. The refrigerant is R 22.

For the peak load and as stand-by in the case of failure of one or the other of the heat pumps, several electric water heaters are installed.

The normal contribution of these electric heaters is less than 10% of the yearly heat demand.

This combination of the otherwise useless water from the tunnel with the heating and air conditioning system of the control building permits to economise a great quantity of fuel and contributes to the protection of environment in this beautiful valley of the higher canton of Tessin.

A special feature of this plant are the plate-type evaporators placed in a basin, into which flows the water coming from the tunnel. This type of evaporator has been chosen because of the sand contained in the water and because a water temperature down to 4°C was assumed. In reality, experience shows that this temperature doesn't go below 8 + 9°C.

2.3. The heat pump and refrigeration plant for the sports and recreation centre
     at Wettingen near Zurich

At the beginning of 1970 the town of Wettingen with about 20 000 inhabitants
decided to build a large recreation centre with ice-rinks, indoor and outdoor swimming pools, sporting hall and restaurant (fig. 7).

Preliminary studies clarified the question, in which way the important heat
supply of the whole centre could be assured, taking care of the environmental protection and minimising the running-cost.

The question of environmental protection led to the solution, not to install
an ordinary oil-fired boiler house, but to use as much as possible waste
heat from the ice-making installations of the open ice-rinks and in addition
heat from the underground water system serving the town water supply by means
of a heat pump.

As stand-by plant a liquid-gas-fired boiler was chosen because there was no
district gas supply in the neighborhood.

Detailed calculations showed, that - compared with the cost of an oil-fired
installation - the much greater investment could be paid back in about 12
years with the economies in running cost.

This calculation was based on the very low price of the oil before the important increase of the last years. Today one can say, that the return on
investment is quite higher and the pay-back period considerably shorter.

This result is possible, because the very important heat production of the
ice-rink machines can be reclaimed practically without supplementary cost,
as the heat from the condenser can be used directly at the very low level
of 43°C for the pools and the floor heating.

The following diagram shows the whole system: (fig. 8)
The 3 oil-free compressors for the ice-rinks supply the heat in a first stage
of the heating system at the lowest temperature of 43°C, for the poolwater
heater, the floor heating of the pool and the air conditioning system via
an intermediate heat storage tank of 20 m3 (fig. 9).

When the ice-rinks have no cooling demand, there is a possibility to get the
needed heat from the underground water by means of a separate evaporator.

A second stage of the heating system with a slightly higher temperature of
48°C is supplied by a turbo heat pump, taking the environmental heat only
from the ground water of the town water network.

This heat pump has two different condensers. The first one supplies supplementary heat to the heating circuits of the radiators and to the air conditioning plant.

The second condenser supplies the heat for the domestic hot water, stored in
three vertical tanks of 31 m3 each, at a temperature of about 45°C.

Because this temperature is too low for the kitchen, an electric re-heater
in one of them is installed, permitting to heat up about 9 m3 in the top of
the vessel to 60°C.

For peak load and as a stand-by a liquid-gas-fired heating boiler is installed.

The total electric power for all these machines represents nearly 10% of the
total power,the town of Wettingen is consuming on normal days. It was therefore of great interest to reduce the total electric peak load during the
cooking-time by stopping 3 of the 4 machines one by one.

Therefore, an automatic controller was installed, switching off the machines
as the demand of the town of Wettingen is going up. This apparatus economi-
ses a lot of money for the town, which is at the same time owner of the sport
center and of the local electric utility, buying the power from the national
grid on the basis of a peak load and consumption tariff.

The thermal capacities of these installations are as follows:

total max. heat load	3 000 kW
heat load of air conditioning	2 000 kW
refrigeration load of air conditioning in summer	530 kW
" " " ice-rinks	975 kW
heating capacity ice-rinks	1 250 kW
" " ground water (first stage)	1 560 kW
" " " " (second ")	1 160 kW
" " gas-fired boiler	930 kW
electric peak load in total	1 300 kW

2 indoor pools of 486 and 153 m2

2 ice-rinks   40 x 60 and 30 x 60 m

wardrobes for 860 persons

restaurants with 160 seats.

The whole complex is in service since 1972 and has given full satisfaction.

As the running cost of the whole plant represents a very heavy charge for the
town and while the price of liquid gas and electricity are going up and up,
we were charged to look for possibilities to reduce the energy consumption.

We found some possibilities of recycling energy as for example a heat exchanger between the drained water from the pools (about 10% of the basin
content per day) and the fresh water from the town network.

Another heat recovery system became economic now:
A heat exchanger between the exhaust air and the outside air in the air conditioning systems of the pools. These possibilities have been realised since.

2.4. The combined heat pump and refrigeration installation in the new factory of
"Victorinox", the famous "Swiss Army Knife", at Ibach near Schwyz
----

The owner of this factory became concerned with his mounting heating costs.
He decided, that in his new and much larger buildings, planned 1976 and
constructed 1977-78, he would assure as much heating as possible without the
necessity to buy oil.

On the other hand, he needed for the fabrication a lot of cooling water at a
more or less constant temperature. As this constant cooling temperature could
not be assured by using the different sources of water available from different wells, from a little rivulet or from an underground pit, he was led to
the idea, to combine the refrigeration with the heat pump by recycling the
heat produced in the fabrication process (grinding machines for instance) to
the heating system.

The system is now realised as follows: (fig. 10)

Two multi-compressor heat pumps with a heating capacity of 465 kW each, working with the refrigerant R 12 cool down the refrigeration circuit from between 23-26 to between 16-19 $^{\circ}$C and heat the heating circuit up to 65 $^{\circ}$C.

The eight compressors are of the semi-hermetic type and work with a COP of 3.65 based on the electric power entering the motor (fig. 11)

The two heat pumps deliver the heat directly to the various heating circuits of the radiator heating in the new and old buildings as well as to the ventilating system of the new factory.

In order to balance the heat production - dictated by the refrigeration needs of the production - and the heat demand, there are two vertical storage tanks of 16 and 21 m3 installed and heated up to 65 $^{\circ}$C.

A third vessel of 8,5 m3 is installed for the domestic hot water service.

The two storage tanks provide the heat requirements during nighttime, when production is stopped.

Peak load during cold weather is covered by the existing boiler house with three boilers.

First is used a boiler burning waste wood from the fabrication and in a second stage the two oil-fired boilers serve as stand-by.

In order to reduce even more the oil consumption, a third heat pump will be installed this year with a heating capacity of 414 kW when working on the cold water system of different sources of water coming from outside. This machine can also work on the production cooling system and can then produce about 600 kW at 65 $^{\circ}$C. The two different regimes are assured by two distinct water-tube evaporators, both easily to clean in all water-ways from deposits of dirty water.

This third heat pump will go into service during the weekends and other stops of the production and will contribute to a substantial saving of the fuel-oil still needed today. Once this machine is functioning, the total demand of fuel-oil will be nearly zero.

2.5. A central heating plant with a gas-engine-driven heat pump using waste heat from a food factory for heating several school buildings in the neighborhood
----

As the last of these examples of heat pump installations a project shall be described, which will be realised as soon as the public authorities of two cantons get the needed credits. It shall be shown, that with the good-will of private enterprise-owners on one side and active authorities on the other very interesting possibilities of fuel-saving may be found:

A food-fat and oil factory near Basle has very important cooling needs for its production. The owner installed an important ground-water well, delivering about 144 m3/h of water at 12 $^{\circ}$C, which after process cooling, is heated up to 21 $^{\circ}$C. Until now, this water flows to the sewage treatment plant of the town and causes high cost to the factory owner.

We were asked to investigate, whether it was possible to change this situation and to reduce these expenses. Very soon we realised, that it would not be possible to install something realistic within the factory, because the heating needs for technical processes were at too high a temperature level and the needs for room heating were too small for the capacity offered by the cooling circuit.

The factory owner declared then, that he was prepared to supply the heat of
this cooling water to somebody outside the factory under the condition to get
the cooled water returned for his processes.

Very soon, it was possible to find several buildings of the town, as for instance an engineering school, several other school buildings and an industrial training centre which could use the whole capacity of heat offered during
most of the year.

The discussion with the owners of these buildings (canton, town and private
industry) showed, that all were very interested to use this offered waste
heat in order to reduce the fuel-oil consumption and the environmental charge.
One condition was given by the authorities: the heat pump needed for the heat
recovery had to be driven by a gas engine!

The following diagram shows, in which way the solution was found serving
both, the factory owner with his cooling and ground-water problem and the
owners of the school buildings with their growing heating cost: (fig. 12)

The cooling water is collected in the factory during the working days with
21°C and enters a heat exchanger, where it transfers the heat to an external
transport circuit going to the boiler house of the engineering school. This
heat exchanger has a construction, which can be cleaned very easily on the
primary side from the small oil and fat deposits of the water coming from
the cooling processes. (fig. 13)

On week-ends, when the factory produces no heat, ground water serves as a
heat source.

At full load, 144 m3/h of water at about 14°C flows in underground piping to
the two heat pumps, installed next to two existing boilers in the boiler
house of the school. It enters the evaporators and is then cooled down to
6°C and returned to the factory.

The two gas-engine-driven turbo heat pumps have a heating capacity of 1250 kW
each at 65°C of the water coming out of the condensers and the re-heaters of
the engine cooling and combustion gas heat exchangers (fig. 14).

The heat pumps supply heat to a vertical storage tank which has to balance
the heat production and heat consumption in order to reduce the number of
starts of the machines, when the demand is smaller than the smallest production of one heat pump.

By a new, separate piping system, this low temperature water flows to the
various heating substations of the buildings, where it is directly introduced into the existing heating systems (with one exception, the main building of the engineering school, where a heat exchanger had to be foreseen).

The existing hot water heating system going from the two existing boilers to
the various buildings serves as stand-by and peak load capacity and to heat
up the heating-water of the buildings, when the demand of temperature is
higher than 65°C and the return higher than 55°C. (fig. 15)

The connection of the district heating systems is assured by heat exchangers.

The estimated cost will be as follows:

a) Transformations in the oil factory and water circuit
   to the heat pumps   Fr.   625 000.-

b) Central heat pump station   " 1 830 000.-

                                                             report Fr. 2 455 000.-

                                                    report Fr. 2 455 000.-

    c) Low temperature district heating and
       sub-stations in the connected buildings   "   615 000.-

                                                     total Fr. 3 070 000.-
                                                     ========================

    The heat pump installation will cover about 83% of the annual heat demand of
    the connected buildings. (fig. 16)

    Today, the oil consumption is about 1250 tons, which represents a net pro-
    duction of heat of about 10 500 MWh/a.

    After the installation of the heat pumps 4 000 MWh of recovered heat comes
    from the factory.

    536 000 m3 of natural gas will contribute 4 700 MWh of the demand and for
    peak load it will be necessary to burn about 200 t/a of fuel-oil with a net
    heat capacity of 1800 MWh.

    As the following diagram shows, the final consumption of fossil fuels will be
    reduced from 14 200 MWh/a to 7 900 MWh/a! (fig. 17)

    One can assume, that in addition, there will be a remarkable reduction of
    environmental charge by CO_2, SO_2, NOx and so on.

    The annual running cost, based on the actual prices of fuel-oil and gas, com-
    pleted by maintenance cost will be about Fr. 495 000.- compared with
    Fr. 770 000.- with the actual oil heating. The difference will just cover the
    additional capital cost of the investment, when one admits a return of in-
    vestment in 20 years.

    But when you admit that the relative dearth of oil price compared with the
    gas price will be 2%/a, an additional saving of Fr. 1 717 600.- in 20 years
    will be attained or in other words the investment will be paid back in less
    than 10 years!

3. <u>Summary and conclusions</u>

    The few examples of realised heat pump plants, chosen from several hundred instal-
    lations realised in the years 1940 to 1980 in Switzerland may be representative
    to show the possibilities given by this technique.

    The heat pump technology is not quite new, but it is certainly one of the best
    ways to economise the fossil fuels getting more and more scarce.

    It will not only help to reduce the consumption, but by grading-up waste heat will
    help to reduce the general load of the environment. (fig. 18)

    The three realised plants described have shown some possibilities of application
    by using the recovered waste energy within the own plant.

    Very often it will not be possible to do this, because the amount of waste heat is
    too large to be used at the temperature level, which can be assured economically
    and technically by a heat pump.

    The two projected plants, described as number 2.1. and 2.5. may show, that often
    it can be possible to find needs of low temperature heat in the neighborhood of
    the plant producing the waste heat. This means, that in the future it can be use-
    ful to combine the source of waste heat with an outside user of the graded-up
    heat.

As the installation of a heat pump is in general a quite costly matter, it is necessary to analyse the situation very closely and to seek the most economical solution. Even when the pay-back period of such an investment seems very long under the today's economical situation of fuel and installation cost, it is sure, that with the growing fuel price, the real time of pay-back will be reasonable.

With the experiences of a large number of installations realised all over the world, we propose to use the heat pump technology as much as possible as a means to recover waste, but quite useful energy.

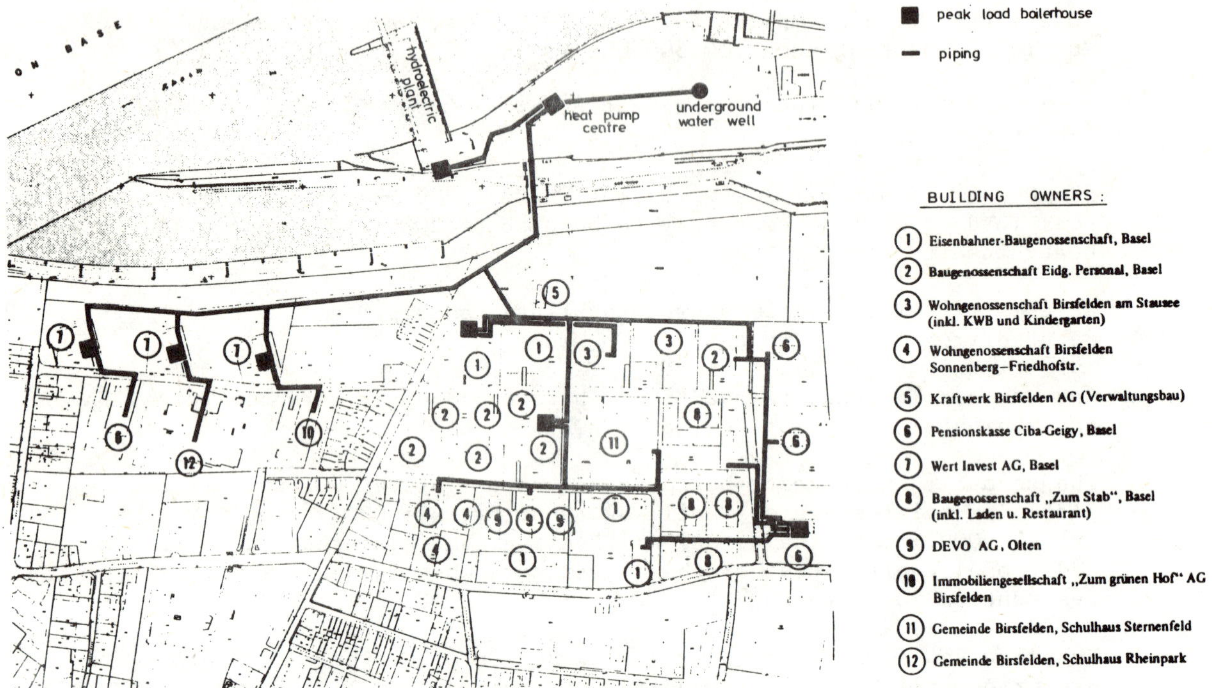

Fig. 1 Heat Pump district heating project "Birsfelden":
- general layout

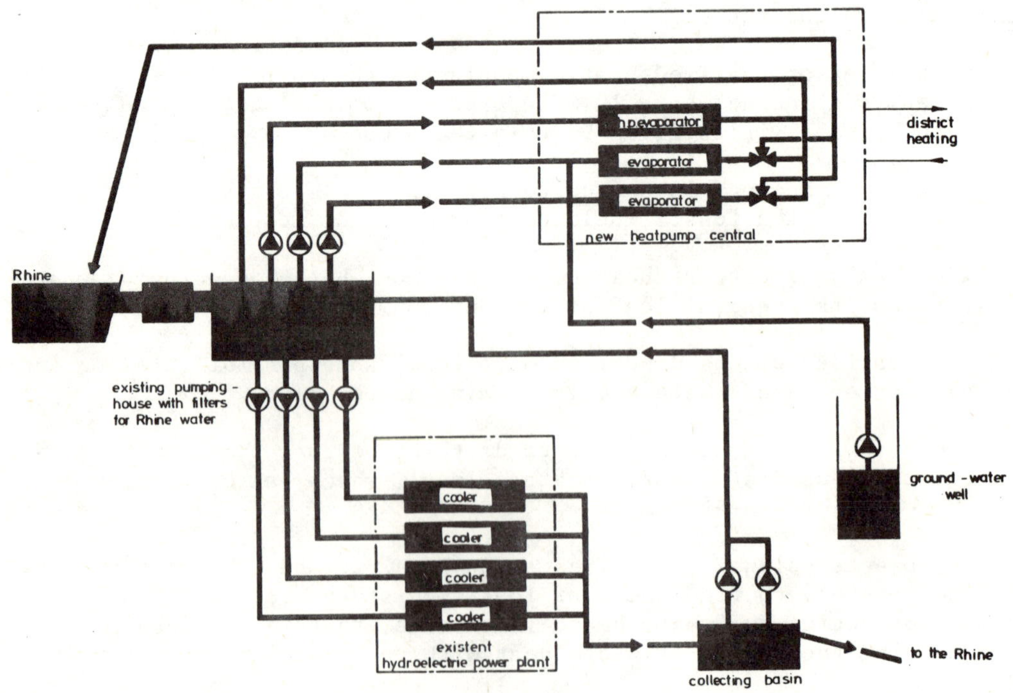

Fig. 2 Heat Pump district heating project "Birsfelden":
- capture of waste heat

on the bottom: evaporater vessel

on the middle: condenser vessel

on top: turbo-compressor with
electric motor

heating capacity 1500 kW at +6/+60°C

Fig. 3 District heating plant at Birsfelden
Proposed type of electric driven turbo-heat pumps

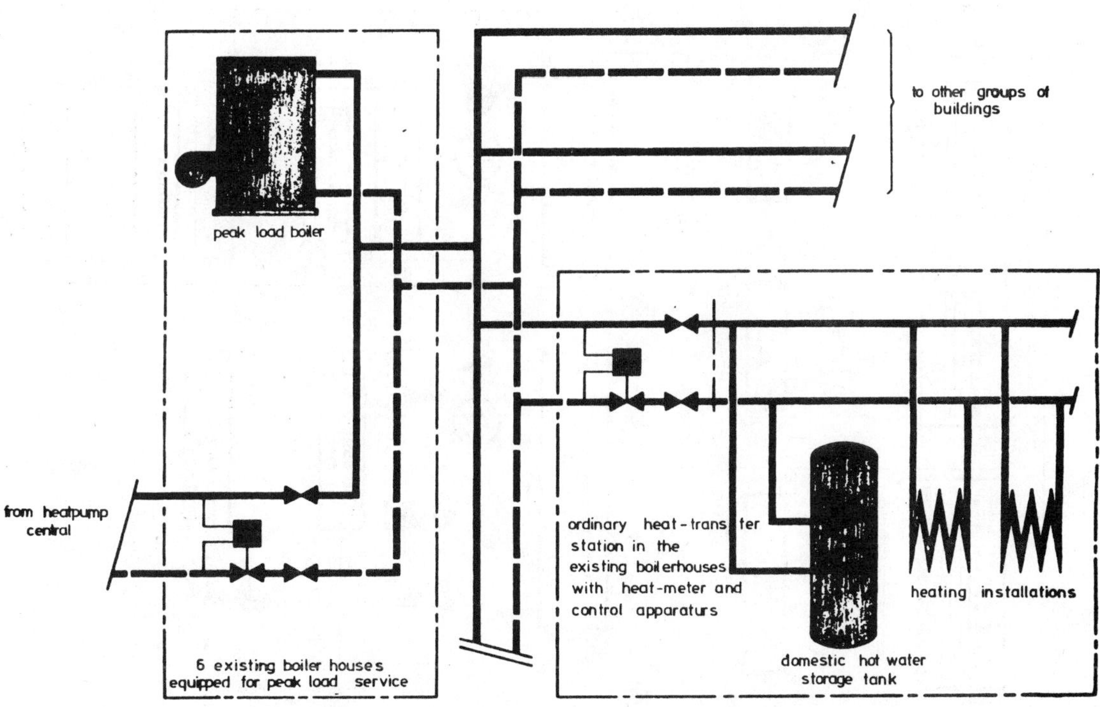

Fig. 4 Heat Pump district heating project "Birsfelden":
- transfer and peak-load station

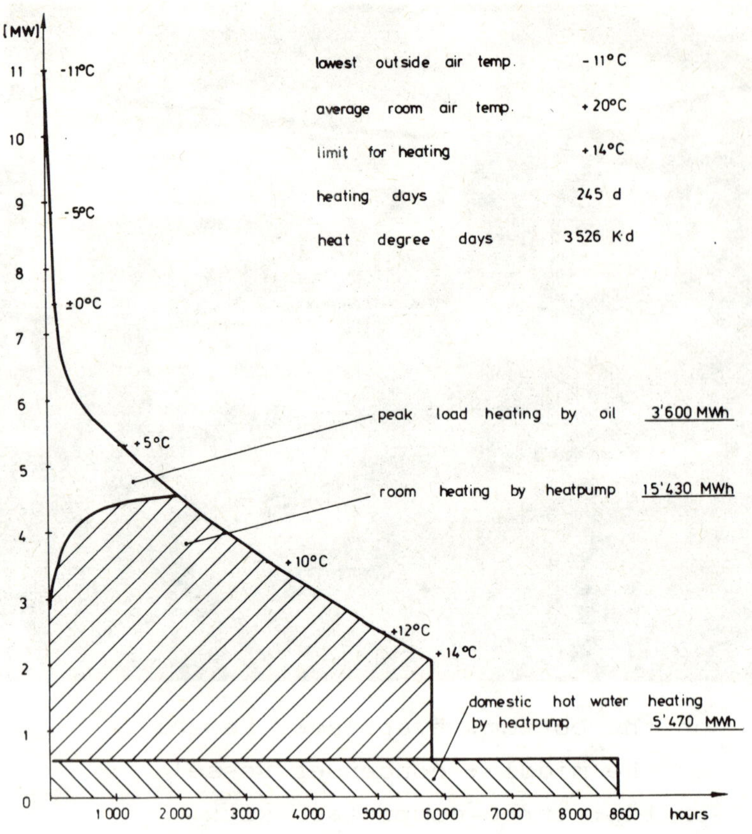

Fig. 5 Frequency curve of yearly heat demand

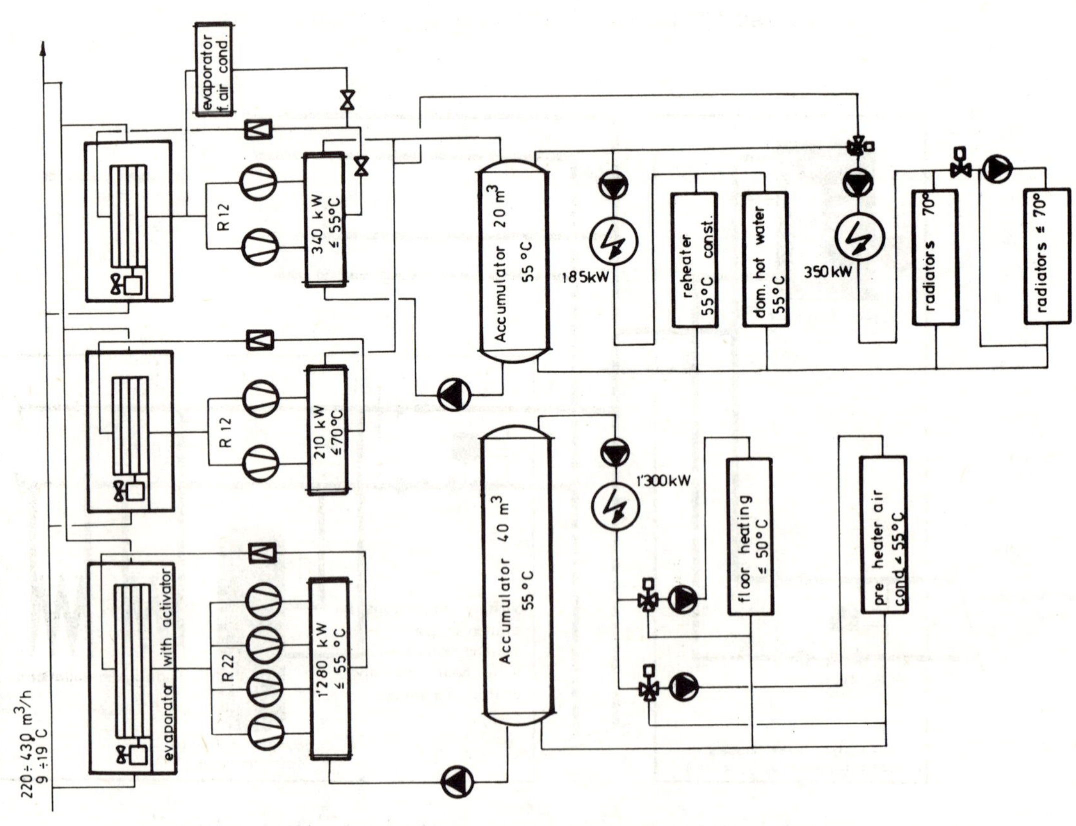

Fig. 6 Motor-car tunnel "St Gotthard":
Heat Pump and cooling plant simplified principle

foreground left: 2 openair swimmingpools

center: indoor swimmingpools, restaurant
hall for general purposes

right side: 2 openair ice-rinks

Fig. 7 Sports Centre at Wettingen - general view

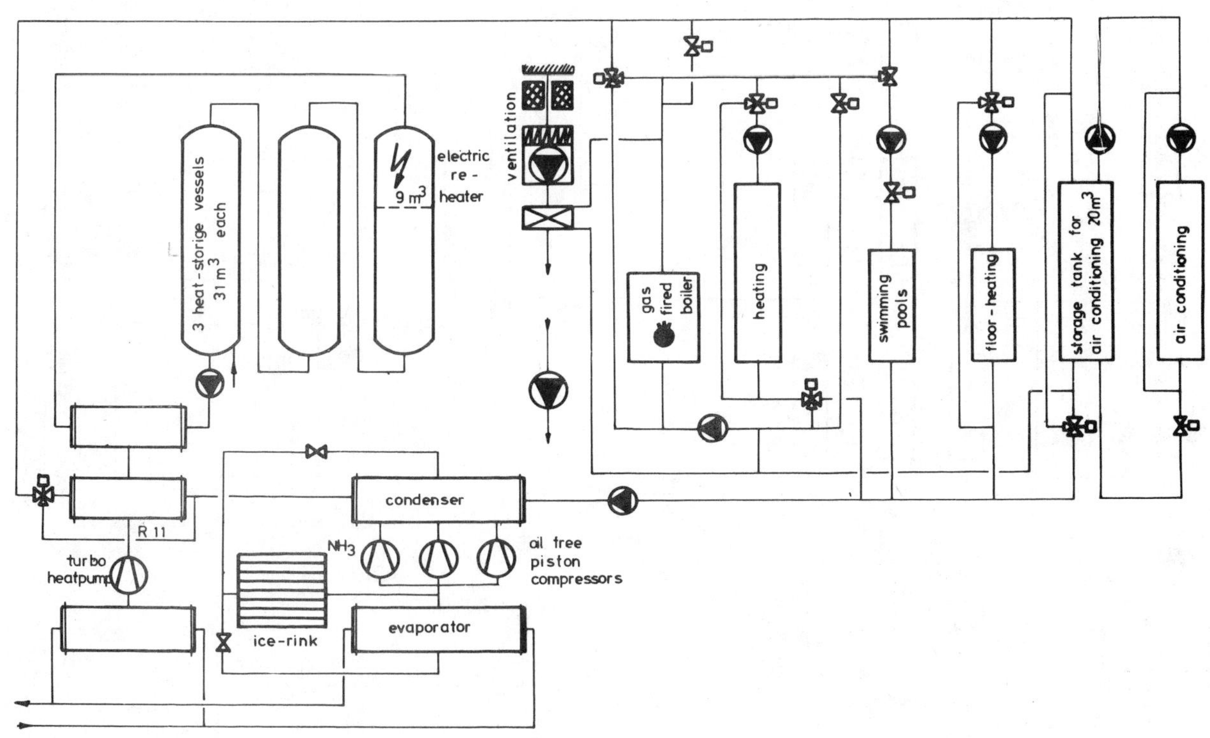

Fig. 8 Sports Centre at Wettingen - heating and cooling plant

155

total refrigeration capacity   975 kW at −7°C
total heating capacity   1250 kW at +43°C
heating capacity with groundwater 1560 kW at +2/+43°C

Fig. 9 Sports Centre at Wettingen -
        3 oilfree piston compressors for ice-rinks

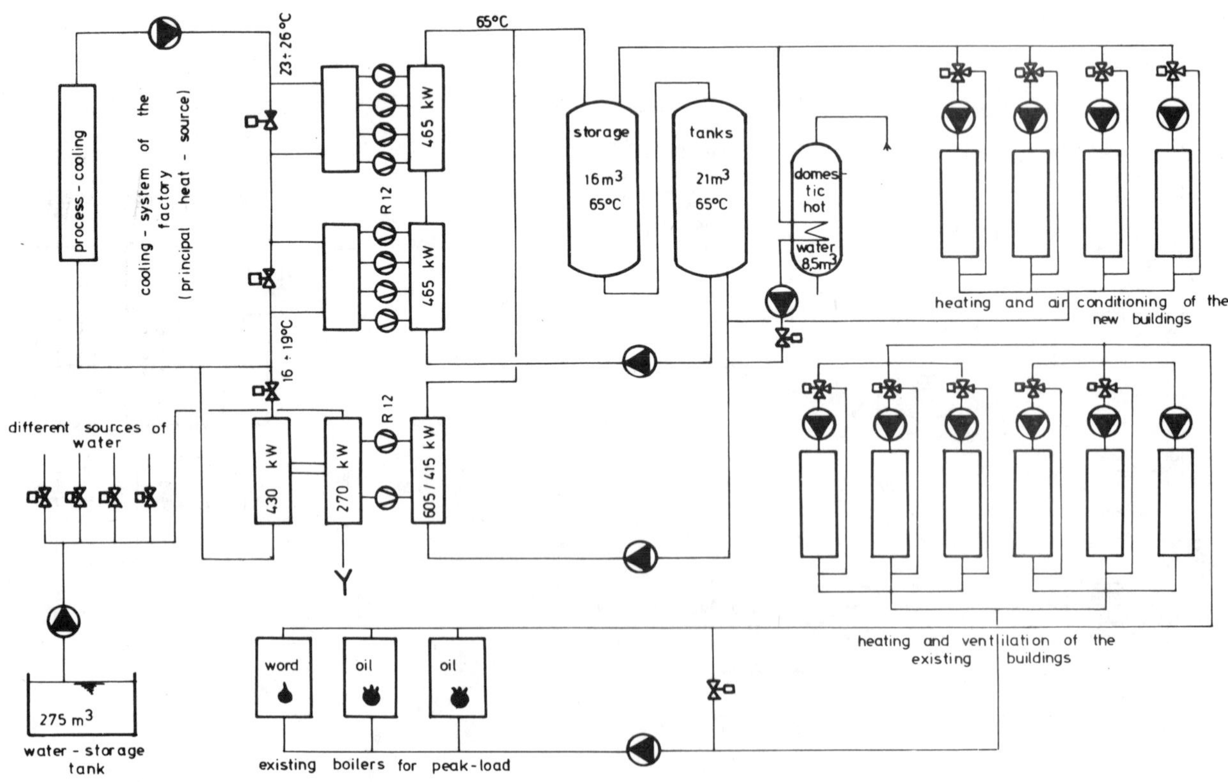

Fig. 10 Victorinox Ibach
         Principle of the heating system

heating capacity 465 kW
evaporating temp. 12 to 18°C
condensing temp. 68°C
refrigerant R12

Fig. 11 Victorinox works at Ibach
Multi-compressor heat pump group Escher Wyss

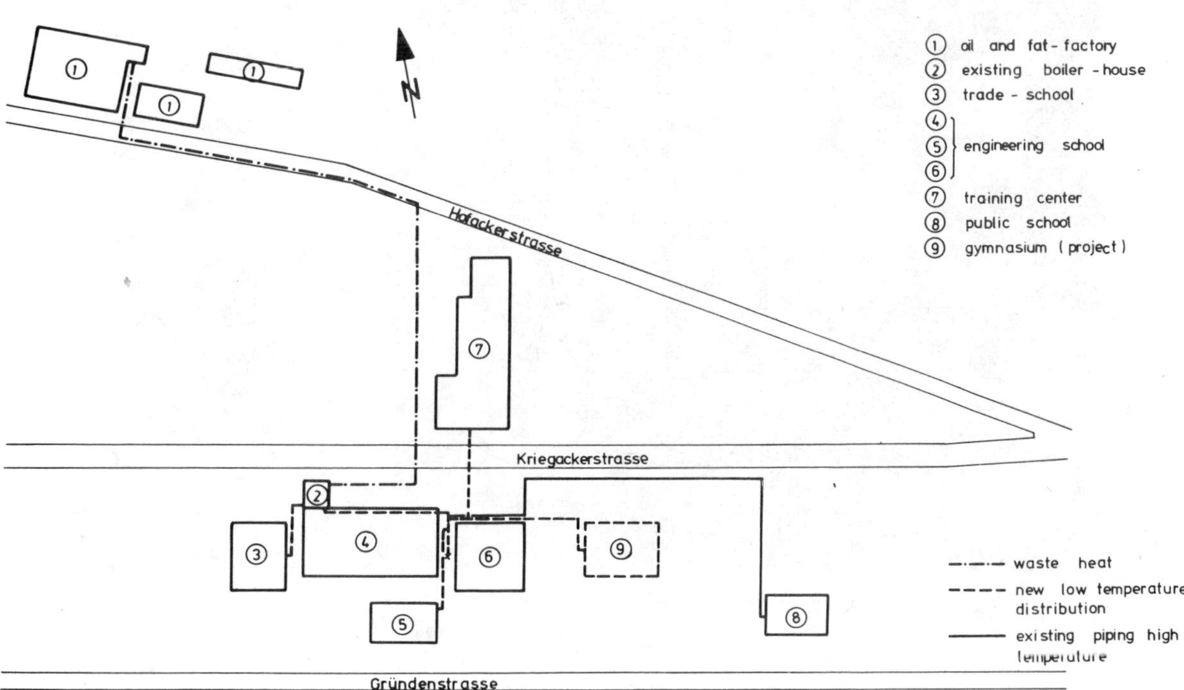

Fig. 12 Centre of Schools at Muttenz -
General situation : scale 1:2 000

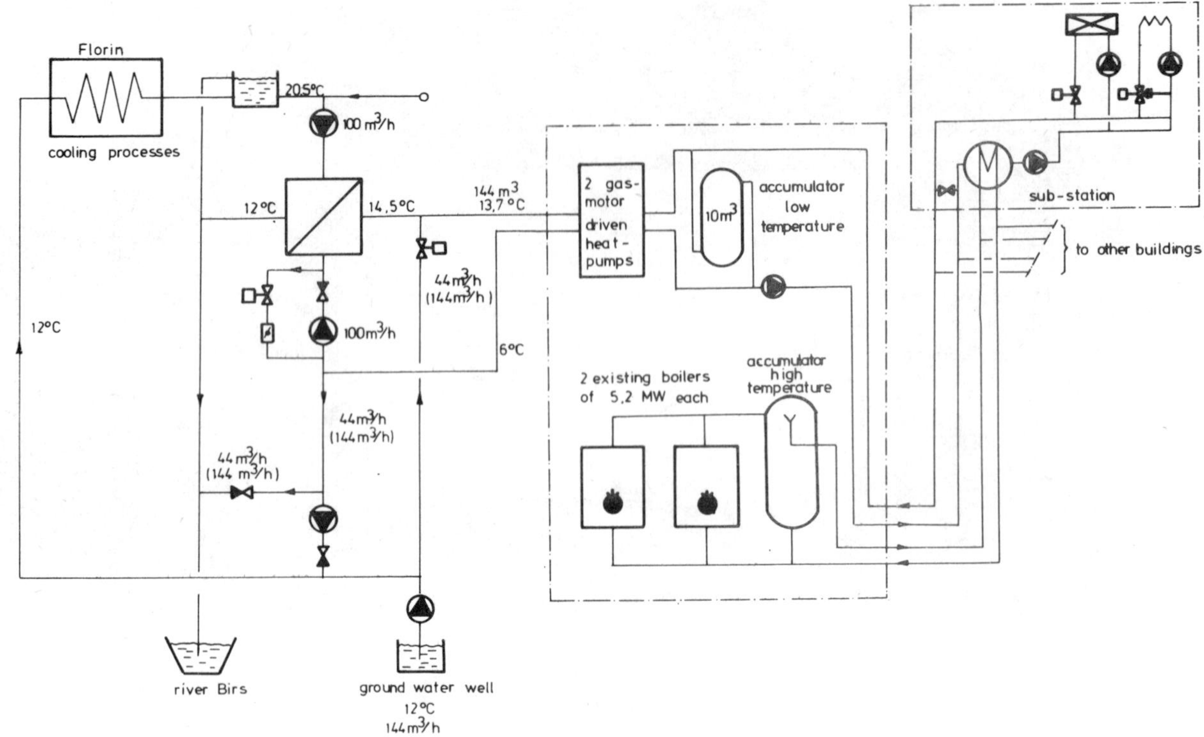

Fig. 13 Centre of Schools at Muttenz -
Principal diagram of heating system

heating capacity 1,26 MW at max 65°C
COP 3.35

Fig. 14 Centre of Schools at Muttenz -
proposed turbocompressor with gas engine
for heat pump purposes

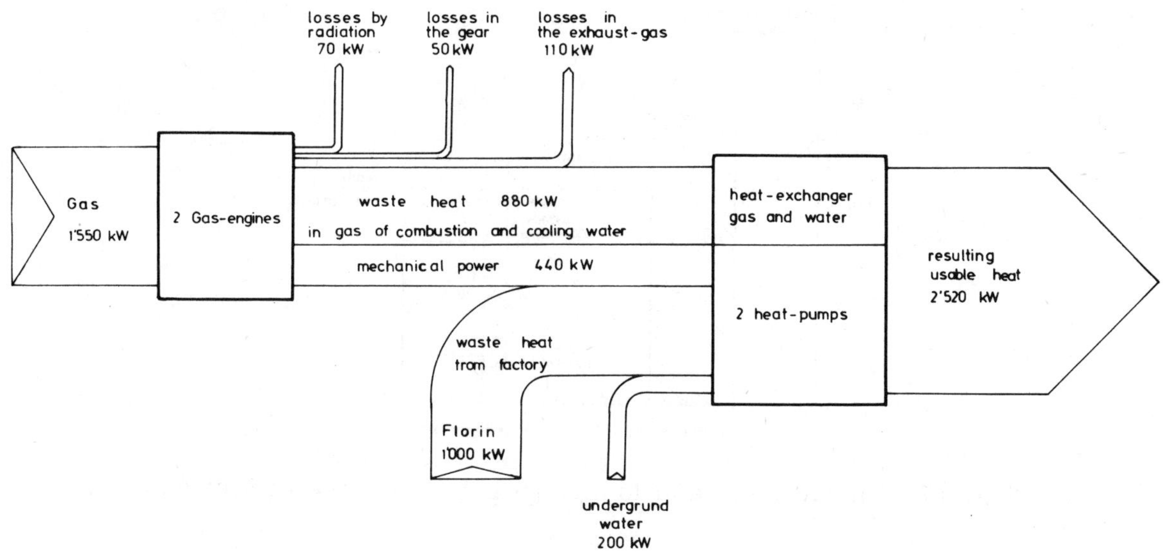

Fig. 15 Energy flow of the gas-motor driven heat pump

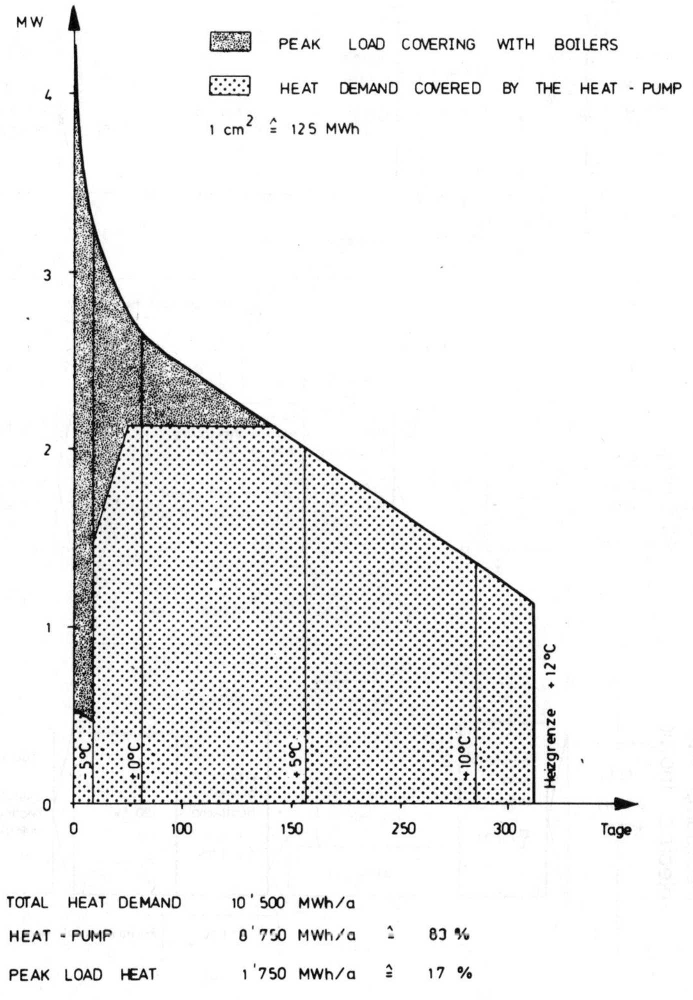

Fig. 16 Frequency curve of yearly heat demand

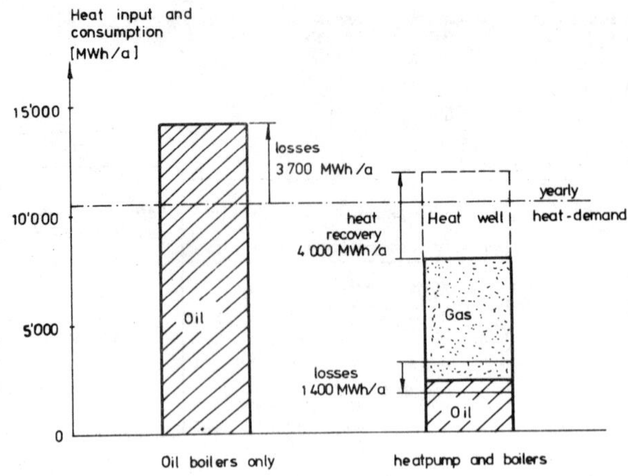

Fig. 17 Consumption and losses in school centre at Muttenz

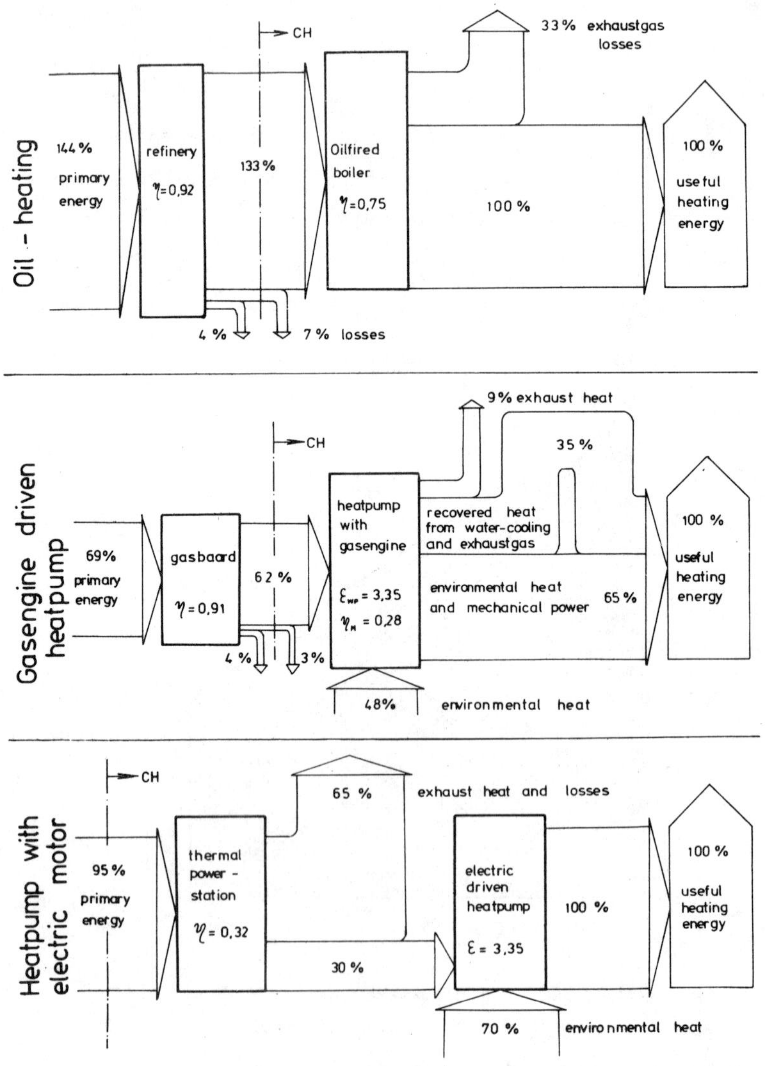

Fig. 18 Comparison of different heat
         production processes

HOT WATER PREPARATION AND HEATING
OF THE PREMISES AT THE METZ ABATTOIR.
REPLACEMENT OF THE BOILERS BY A HEAT PUMP

A. Gauthier

Electricité de France, France

R. Wisshaupt

Ets Quiri, France

Summary

The study describes an efficient electrical system for producing hot water and for heating the slaughtering premises (slaughtering hall and offal section).

This system uses the condenser of the refrigerating units, a heat pump recovering heat from the condenser, supplemented by water storage heating.

The least favourable operating conditions of the use of a heat pump in the abattoir, i.e. in winter, were taken into account in designing this installation.

Held at the University of Warwick, U.K.
Symposium organised and sponsored by
BHRA Fluid Engineering
©BHRA Fluid Engineering, Cranfield, Bedford MK43 0AJ, England.

1 — INTRODUCTION

The size of the heat pump needed for the preparation of hot water is determined by the requirements of the abattoir for space heating of premises and the supply of hot water, taking account of calorific capacity at the condenser of the refrigerating units.

The study was based on the complementarity of refrigeration and heat needs in a working day in the abattoir.

Heat for the premises does not involve any special technique ; it is supplied by the hot gases delivered by the compressors of the refrigerating units.

The additional heating capacity delivered by ammonia to the condenser of the refrigerating units is used to preheat the water needed for the abattoir and to evaporate the R12 in the heat pump.

The condenser of the heat pump heats the water to 65 °C.

The heat pump has two compressors. A possible failure of one of the two compressors preparing hot water is offset by immersion heaters fitted in the storage tanks ; these operate at night.

The nominal slaughtering capacity of the Metz abattoir is 12,000 tons of cattle and 2,000 tons of pigs per year of 260 working days.

2 — SPECIFICATION OF THE HEAT REQUIREMENTS OF THE ABATTOIR

The heat requirements of the abattoir are essentially met by hot water and the hot gases delivered by the compressors of the refrigerating units.

The hot water temperatures at production and during use are distinguished, allowing for losses in the pipes estimated between 3 and 5 °C.

Most of the hot water used in the abattoir is between 50 °C and 62 °C, with a small quantity heated to 90 °C for the offal section.

At the level of the air heaters, the hot gases delivered by the ammonia compressors reheat the air circulating in the slaughtering premises.

In addition to the heating of premises and production of hot water, make-up heat and water are necessary to maintain a constant water level and a temperature of 61 - 62 °C in the pig scalding-tub and to keep the temperature of some offal containers at 90 °C.

The main work stations at which heating capacity is necessary are as follows :

— Pig scalding

— maximum rate of slaughter	80 pigs per hour
— hot water to fill up the tub	5 m³/day
— temperature at production	65 °C
— temperature of use	61 °C
— tub make-up hot water	15 litres per pig
— temperature at production	65 °C
— temperature of use	61 °C
— make-up heat to maintain temperature	2,700 KJ/pig
— daily operating time	4 hours
— average weekly frequency over 52 weeks	3 days
— working hours	7 a.m. to 11 a.m.

— Offal section

— scalding water (3 m³/hour)	24 m³/day
— temperature at production	65 °C
— temperature of use	62 °C
— hardening water for rennets (0.5 m³/hour)	4 m³/day
— temperature at production	90 °C
— temperature of use	90 °C
— make-up heat to maintain temperature	35 kW
— daily operating time	8 hours
— average weekly frequency over 52 weeks	5 days
— working hours	7 a.m. to 5 p.m.

— **Cleaning and other activities**

— hot water (maximum)	92 m^3/day
— temperature at production	65 °C
— temperature of use	50 - 60 °C
— daily period of use	10 hours
— average weekly frequency over 52 weeks	5 days
— working hours	7 a.m. to 5 p.m.

— **Heating of the slaughtering hall and offal section**

  — maximum power requirement for heating :

— offal section	130 kW
— pig slaughtering	100 kW
— cattle slaughtering	100 kW

  — total requirement :

— winter :	between 7 a.m. and 11 a.m.	330 kW
	between 11 a.m. and 5 p.m.	230 kW
— summer :	between 7 a.m. and 11 a.m.	165 kW
	between 11 a.m. and 5 p.m.	115 kW

— maximum daily operating time	8 hours
— winter months	8
— summer months	4
— weekly frequency	5 days
— operating hours	7 a.m. to 5 p.m.

Remark : The changes in requirements for space heating at different times of day arise from the fact that pigs are slaughtered during a four-hour period only.

Temperature requirements for offices are met independently by electrical convectors.

3 — TECHNICAL INSTALLATIONS USED

Air heaters circulating the hot gases delivered by the compressors of the refrigerating units heat the slaughtering hall and the offal section.

Hot water at 65 °C is prepared in two stages ; water is initially preheated from 10 °C to 31 °C in the condenser of the refrigerating units and then heated from 31 °C to 65 °C in the condenser of the heat pump using R12 between 26 °C and 70 °C, recovering the heat from the condenser of the refrigerating units.

Additional heat needed to raise the temperature of 4 m^3 of water in the offal section from 60 °C to 90 °C is provided by night-time water storage heating.

Hot water produced by the heat pump outside working hours is stored in a tank whose size was determined as indicated later.

The heating capacity of the immersion heaters in the storage tanks is determined such as to be able to face up to any failure of a heat pump compressor to operate or in the delivery of the refrigerating power required.

Figure 1 shows the diagram of the water and space heating system of the slaughtering premises. The installed power is indicated for each apparatus, as well as the maximum flow of water heated to 65 °C.

4 — DESCRIPTION OF THE REFRIGERATING SYSTEM

This is a centralised refrigerating system comprising four NH_3 compressor sets, operating between — 6 °C and 35 °C to meet the need for positive refrigeration. The system has the following characteristics :

— refrigerating power installed at the evaporator	650 kW
— heating power installed at condenser level	820 kW
— installed power of the compressors	4 x 55 kW

The refrigerating power demand is only a portion of total installed power :

  — 80 % during 8 hours,
  — 40 % during 14 hours.

In these conditions, the heating capacity dissipated at the condenser of the refrigerating units is as follows :

- 660 kW during 8 hours,
- 330 kW during 14 hours.

The heating capacity which can be recovered at the level of the condenser of the refrigerating units for preparing hot water is governed by the total heating capacity dissipated at the condenser of the refrigerating units and by needs for space heating. The heating capacity which can be recovered in each time bracket is indicated in Table 1. It is such that :

Recoverable capacity = total capacity — capacity needed for heating the premises.

Table 1 — Heating capacity which can be recovered at the condenser
of the refrigerating units in each time bracket

Time	Total heating capacity at the condenser of refrigerating units (kW)	Heating capacity for space heating (kW)	Recoverable heating capacity (kW)
7 a.m. to 9 a.m.	350	330	20
9 a.m. to 11 a.m.	660	330	330
11 a.m. to 3 p.m.	660	230	430
3 p.m. to 5 p.m.	660	0	465
5 p.m. to 7 a.m. (off-peak)	330	0	330

It can thus be taken that recoverable heating capacity in a working day is as follows :

- 465 kW during 2 hours,
- 430 kW during 4 hours,
- 330 kW during 14 hours.

Remark : The recoverable heating capacity was calculated for the worst **winter** weather conditions, when the space heating demand for the premises is maximum. In addition, it was taken that the refrigerating power demand during two hours when the abattoir begins its activity, represented 40 % of installed capacity. Maximum demand for power (i.e. 80 % of installed power) occurs during eight hours between 9 a.m. and 5 p.m. when rapid freezing of carcasses is needed. Refrigerating power demand outside the slaughtering period is 40 % of installed power during ten hours.

5 — DESCRIPTION OF THE WATER HEATING SYSTEM

The water heating system comprises :

- a section of the condenser of the refrigerating units, in which y m^3/h of water are preheated from 10 °C to 31 °C ;
- a heat pump whose evaporator is so placed as to recover the heat from the condensing circuit of the refrigerating units, providing the condenser with additional heating of y m^3/h of water from 31 °C to 65 °C ;
- two water storage tanks fitted with immersion heaters ; the 5 m^3 first tank supplies the additional heat for the offal section water to be heated to 90 °C. The 80 m^3 second tank stores the hot water delivered by the heat pump outside working hours ; it is fitted with immersion heaters which can supplement or even replace the heat pump.

5.1 — Dimensions of the heat pump

Heating capacity supplied to water by the condenser of the refrigerating units :

$$Q_{CRU} = 24.4\,y \quad (kW)$$

Heating capacity supplied to water by the condenser of the heat pump :

$$Q_{CHP} = 39.5\,y \quad (kW)$$

Heating capacity recovered at the evaporator of the heat pump (as per manufacturer's estimate) :

$$Q_{EHP} = \frac{133.3}{174.5}\,39.5\,y \;=\; 30.2\,y \quad (kW)$$

Knowing Q_R, the recoverable heating capacity at the condensor of the refrigerating units, we have :

$$Q_{CRU} + Q_{EHP} = Q_R$$

$$24.4\,y + 30.2\,y = Q_R$$

$$\boxed{\; y = \frac{Q_R}{54.6} \;} \quad (m^3/h)$$

The hot water flow y produced as a function of recoverable heating capacity is indicated in Table 2.

Table 2 — Planned preparation of hot water (65 °C) by the heat pump

Recoverable heating capacity Q_R (kW)	Hot water flow at 65 °C, y (m³/h)	Number of hours	Production per period (m³)
465	8.5	2	17
430	7.87	4	31.5
330	6.06	13.5	81.5
TOTAL		19.5	130

The operating characteristics of the heat pump for each of the two periods of hot water preparation are given in Table 3.

Table 3 — Heat pump operating characteristics

Hot water flow, y (m³/h)		8.5	7.87	6.06
Refrigerating power absorbed at the evaporator	(kW)	257	238	184
Heating capacity delivered to the condenser	(kW)	336	312	240
Power absorbed by compressors	(kW)	79	74	56
Power absorbed by motors	(kW)	86	80	61

Table 4 sets out the installed capacity of the heat pump.

Table 4 — Installed power of the heat pump

	Installed capacity
Evaporator	270 kW
Condenser	350 kW
Motors	2 x 55 kW

The maximum daily rate of operation of the heat pump is 20 hours at the time of the year at which heating requirements are greatest. During intermediate seasons and in particular in summer, recoverable heating capacity is greater and the operating time of the heat pump can be reduced accordingly to below 18 hours.

The heat pump operating in winter will be invoiced for the following electricity tariff periods :

- high-load hours (slaughtering hours)   8 hours
- high-load hours (outside the slaughtering hours)   3 1/2 hours
- low-load hours (outside the slaughtering hours)   8 hours

Consumption of electricity of the heat pump will be :

- high-load hours $2981 . 10^3$ KJ
- low-load hours $1757 . 10^3$ KJ.

5.2 — Dimensions of hot water storage facilities

During working hours, between 7 a.m. and 5 p.m., the heat pump supplies 8.5 m^3/h of hot water for two hours, 7.87 m^3/h for four hours and 6.06 m^3/h for two hours, i.e. a total of 60.6 m^3 of hot water. The abattoir will also have to store all the hot water produced outside these hours, i.e. 69.4 m^3.

A storage capacity of 85 m^3 has been projected, comprising :

- a   5 m^3 tank for offal section water heated to 90 °C,
- an 80 m^3 tank for cleaning water heated to 65 °C.

Each of these tanks will be fitted with immersion heaters whose absorbed and installed power are as follows :
- for the 5 m^3 tank :

  heating 4 m^3 of water from 60 °C to 90 °C

  - electrical power absorbed during 8 hours   20 kW
  - installed power   50 kW
- for the 80 m^3 tank :
  - electrical power absorbed   nil
  - electrical power installed for emergencies   560 kW

Remark :   Installed power was calculated to guarantee heating of the water stored in the 80 m^3 and 5 m^3 tanks, respectively from 20 °C to 65 °C and from 25 °C to 90 °C.

Daily electricity consumption of make-up heat supplied by storage water for the offal section at low-load hours will be :

$$576 . 10^3 \text{ KJ.}$$

6 — CONCLUSION

The lowest energy cost of hot water preparation is obtained by heating water to 65 $^{\circ}$C with the heat pump and then raising the temperature of offal water through night-time storage heating. The cost is 2.74 F/m^3 as against 6.9 F/m^3 using a gas-fired boiler.

The association of a heat pump with make-up water storage heating, as followed at the METZ abattoir, is the only one to yield a substantial saving in energy costs. The installation of the electrical system necessitates a larger investment than would a traditional boiler. However, the investment payback period is 18 months or less.

It should be observed that in the event of a reduction in refrigerating power demand in the abattoir by comparison with the value taken into account in this study, the greater recourse to storage heating would raise the cost of hot water preparation. However, this would never exceed the cost of the other possible methods, even in the most unfavourable case.

Figure 1 — Diagram of the water and space heating circuit using a heat pump and make-up water storage heating

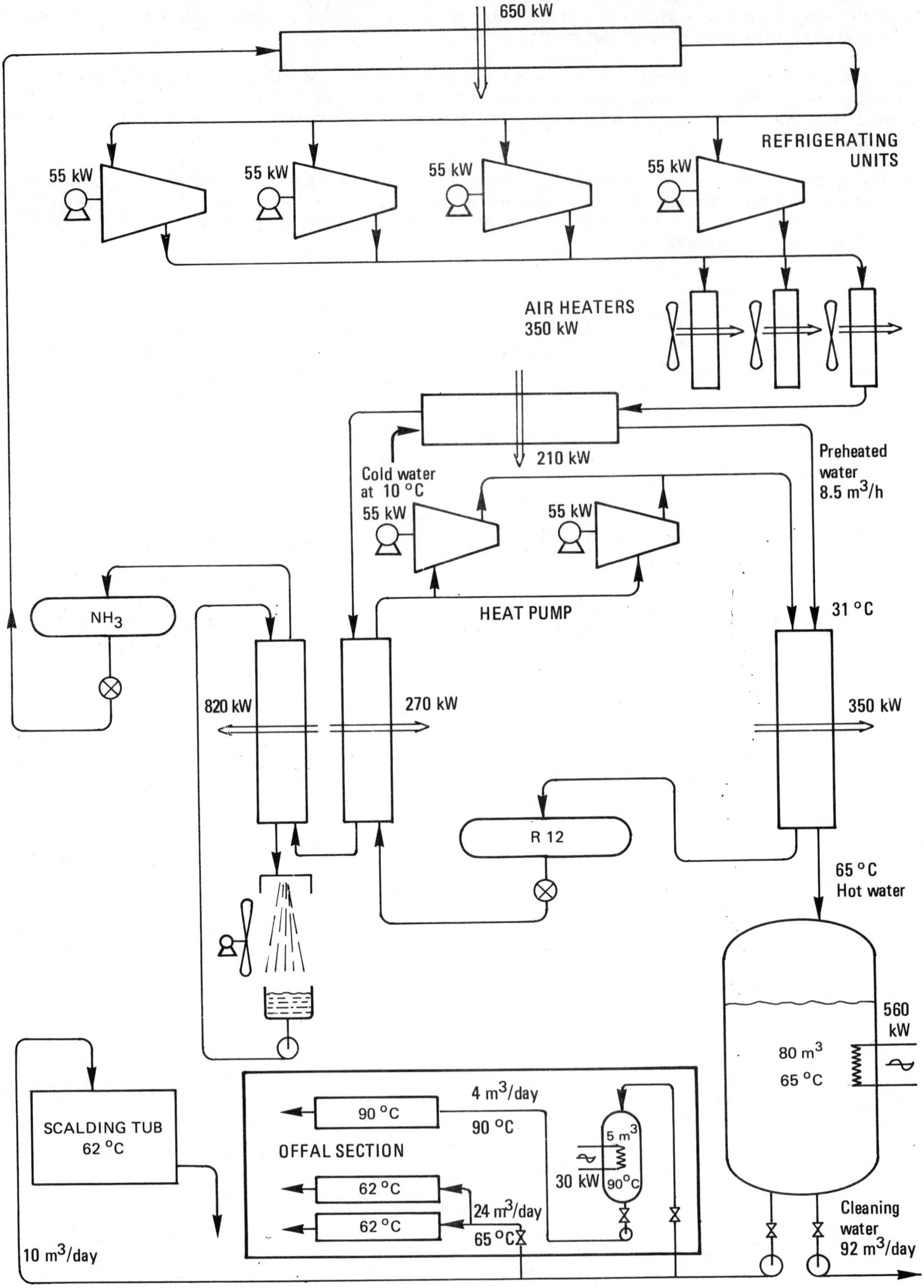

AN OVERVIEW OF THE INDUSTRIAL HEAT PUMP APPLICATIONS ASSESSMENT AND TECHNOLOGY DEVELOPMENT PROGRAMS OF THE GAS RESEARCH INSTITUTE, U.S.A.

E.S. Tabb

Gas Research Institute, U.S.A.

D.W. Kearney

Insights West Inc., U.S.A.

Summary

This paper describes the Gas Research Institute's program to develop advanced vapor compression heat pumps for industrial applications. Three of four projects discussed involve high-performance vapor-compressor concepts for open-cycle systems and the fourth is a broad assessment of the utilization of both open- and closed-cycle heat pump systems in United States industry. The compressor development projects are directed towards development and proof-of-concept testing at industrial sites, and include (1) an advanced centrifugal vapor compressor driven by a gas turbine, (2) a gas engine driven rotary screw steam compressor, and (3) a unique positive displacement epitrochoidal rotary vapor compressor. The assessment of the potential impact of vapor compression heat pumps on the U.S. industrial utilization of natural gas requires specific process knowledge of the potential applications throughout all industrial sectors. To accomplish this, data on process heat needs and waste heat availability were gathered from plant interviews and other sources for over 50 U.S. industries. This information is being evaluated to identify the most promising areas for both closed- and open-cycle heat pump systems in the heat recovery, low-pressure steam production, drying and process mechanical vapor recompression areas.

Held at the University of Warwick, U.K.
Symposium organised and sponsored by
BHRA Fluid Engineering
©BHRA Fluid Engineering, Cranfield, Bedford MK43 0AJ, England.

<u>INTRODUCTION</u>

The Gas Research Institute (GRI) is an independent, nonprofit, scientific organization created in 1976 to plan and implement a comprehensive research and development (R&D) program for the benefit of the gas consumer. GRI itself conducts no R&D. Instead, GRI establishes broad, far-sighted goals and contracts with research organizations, professional services firms, universities, energy companies, engineering firms, and manufacturers to conduct its gas-related R&D projects. Many of its projects are cooperatively funded by the contractor or various government agencies. GRI seeks to expand the technological base necessry to accomplish three objectives: increase gas supplies, improve the efficiency of using gas, and further enhance the quality of service provided to the gas customer, all at lower cost and in an environmentally acceptable manner.

GRI's industrial R&D program is aimed at developing higher efficiency gas-using equipment that will improve productivity while meeting or exceeding all proposed environmental standards. Successful developments will help lower projected consumer costs while saving natural gas. Specific objectives of the industrial program are to: 1) develop large-capacity, gas-fired heat pump systems for recovering the large quantities of low grade heat that would otherwise be lost from industrial processes; 2) develop highly efficient, cost-effective industrial heat recovery systems for multiple applications; 3) develop more effective and less polluting multi-fuel industrial burner systems for direct and indirect combustion applications; and, 4) develop new equipment for improving the utilization efficiency of existing energy and capital intensive processes.

Large quantities of low-grade industrial waste heat are available in the form of hot water, flue gases, and low-temperature steam and other condensible vapors with temperatures ranging from 95°C to 260°C (200°F to 500°F). Although these waste-heat sources are normally considered to be of marginal value, they can be upgraded to serve as economically attractive feedstocks for the production of higher grade industrial process heat through the implementation of industrial heat pump systems. Utilizing waste heat in this manner provides an opportunity for the industrial sector to achieve substantial mid-term improvements in the utilization of fossil fuels, particularly natural gas.

Industrial utilization of all of today's heat pump cycles (both open and closed) are currently constrained by cost and technology limitations in mechanical vapor compressors. For example, the application of mechanical vapor recompression heat pumps to multi-effect evaporator processes (including waste steam upgrading) is well-recognized as a major energy-conserving option, but few installations have occurred because of the high investment and maintenance costs of today's mechanical compressors. Turbo-compressors, reciprocating, and rotary screw compressors all require complex design and manufacturing techniques which make their overall system installation costs prohibitive for application of vapor recompression to the large volumes of vapor being discharged from industrial evaporative and drying processes. Furthermore, the thermodynamic efficiency of the Rankine-cycle heat pump system could be enhanced by almost 40 percent if low-cost, high-compression-ratio mechanical vapor compression machines were readily available.

Sustantial government and private funds have been spent to introduce Rankine-cycle heat pump systems into a variety of industrial processes, but little money is being spent to develop cost effective, open-cycle mechanical vapor compression machines which are vital to the successful commercialization of all industrial heat pump cycles. As a result, GRI's industrial heat pump activities are directed toward developing cost-effective, open-cycle systems which take advantage of available hardware to offer minimum-complexity heat pumps with widespread applications.

GRI is allocating all of its current industrial heat pump funding to the development of three distinct types of open-cycle vapor compression heat pump systems:

o Open-Cycle Rotary Screw Steam Compressor

o Advanced Positive Displacement Epitrochoidal Rotary Vapor Compressor

o Open-Cycle Centrifugal Vapor Compressor

Each of the three heat pump projects is organized to address and overcome specific technology and end-use constraints currently limiting the commercialization of open-cycle heat pump systems. A fourth project involves the survey of U.S. industry and government and private sector R&D programs to identify the most promising industrial heat pump technology developments and specific process applications for vapor compression heat pumps.

<u>OPEN-CYCLE VAPOR-COMPRESSOR HEAT PUMP DEVELOPMENT</u>

The three vapor-compressor projects involve unique and separate vapor-compressor technology approaches. When integrated into a gas-fired heat pump system, each will have different applications in specific industrial process markets. The open-cycle rotary-screw steam-compression system is designed for processes requiring upgrading of clean and contaminated waste steam sources in the food, pulp and paper, and chemical industries. The advanced epitrochoidal rotary-motion vapor compressor is being developed to replace cost prohibitive complex turbo-compressors and inefficient reciprocating compressors. The open-cycle centrifugal vapor-compressor project involves development of a high-compression-ratio advanced centrifugal compressor which is driven by a gas turbine engine. It will have extensive application in evaporation and direct contact drying processes where the clean gas turbine waste heat can be utilized in a direct contact dryer. These three technical approaches to system development address different industrial heat recovery markets, and are currently in various stages of design, fabrication, field testing, and potential market introduction.

Open-Cycle Rotary Screw Steam Compressor

Thermo Electron Corporation is developing an open-cycle vapor compression steam heat pump to produce process steam. A flow schematic of the system is shown in Fig. 1. The system utilizes excess low-pressure steam, or that produced from an excess heat source with a waste heat boiler, and compresses this steam to the desired pressure level for process use. The compressor is driven by a gas-fired prime mover such as a gas turbine or gas engine. To enhance the system performance, the prime mover exhaust and cooling jacket heat are recovered to generate additional process steam or hot water. The system is analogous to a closed-cycle heat pump where mechanical work is used to upgrade energy from a low temperature level to a higher temperature level. Utilizing this concept, fuel consumption can be as low as 30 percent in comparison to a direct-fired boiler over the expected range of process conditions. This unique aspect of the heat pump system is the result of the fact that a major fraction of the energy in the steam is already available to the compressor as latent heat. As a result, only a small fraction of additional energy is required to raise its pressure and temperature to a useful level. Although significant energy savings can be realized by use of the steam heat pump system, such a concept must also be justified economically. Economic analyses have shown favorable simple payback periods of one to three years depending on the nature of the excess heat or steam source, steam flow rate, pressure ratios, and fuel cost.

In the development work to date, a proof-of-concept demonstration system has been built and is currently undergoing performance testing at Thermo Electron. An illustration of the proof-of-concept system is shown in Fig. 2. The system includes all the major components and is of adequate size to demonstrate the performance and operating characteristics of a full-size system. The system is based on a $1.04m^3/s$ (2200 CFM) screw compressor with a built-in pressure ratio of 3:1. The compressor is driven by a 373 kW (500 hp), 1200 RPM industrial gas engine through a 1:7.7 ratio gearbox. The system is designed for a nominal steam flow rate of 1.26 kg/s (10,000 lb/hr) at an inlet pressure of 207 kPa (30 psia) and an outlet pressure of 621 kPa (90 psia).

The system is currently being tested at Thermo Electron in a specially constructed laboratory to characterize the performance and operating characteristics under controlled conditions. After all the operational and control features of the system have been proven out, the unit will be installed in an industrial site for long duration testing. This will provide long-term working knowledge of the systems, leading the way to the commercialization of full-size systems.

Advanced Positive Displacement Epitrochoidal Rotary Vapor-Compressor Development

An advanced positive displacement rotary compressor concept that utilizes a unique epitrochoidal geometry shows excellent potential for open-cycle heat pump applications. Epitrochoidal and hypotrochoidal geometries have been used in rotary compression devices for many years but they have not been exploited commercially because of a lack of high speed and dependable valving and seal designs. Trochoid Power Corporation has developed a unique proprietary valving and sealing concept which it believes will enable the positive displacement epitrochoid rotary compressor to become commercially attractive.

The epitrochoid rotary compressor bridges the gap between reciprocating piston compressors and rotating turbo-compressors. Like the piston concept, it has the advantage of being a positive displacement device, but not the disadvantage of the unbalance inherent in reciprocating compressors. Like the turbo-compressor, it has the advantage of high rotary speeds, but not the disadvantage of a narrow duty cycle.

Due to its unique geometry, the epitrochoidal device possesses a higher surface-to-volume ratio than a reciprocating piston compressor. This permits greater heat dissipation which reduces the work of compression and thereby lowers the operating cost. Necessary heat exchanger surfaces for intercooling between compression stages are also reduced in size. Furthermore, the epitrochoidal geometry offers additional surface area for larger exhaust ports than is possible in a piston mechanism where the circular area of the piston is usually the limiting factor.

The Trochoid unit has a solid round powershaft with the rotor mounted eccentrically as shown in Figure 3. This provides greater strength and rigidity then is possible in a reciprocating device where the conventional crankshaft throws create inherent operating limitations due to crankshaft resiliency. By contrast, the epitrochoidal design is able to operate at high vibration-free speeds, thus increasing specific output. One further advantage of the Trochoid design is that the apex seals do not revolve, but rather are stationary and mounted in the outer housing. This is also true of the face and the rotor side seals as well.

It is GRI's intention to utilize the Trochoid unit in an open-cycle vapor compression heat pump system requiring smaller input and discharge capacities contrasted to those ideally covered by rotary screw machines which are more costly to manufacture.

A small single rotor (136 kg/h or 300 lbs/h) proof-of-concept unit has been engineered and is now in various phases of component fabrication and check-out in anticipation of an early 1982 testing program. This test unit is approximately 0.36 x 0.36 x 0.18 meters (14 x 14 x 7 inches) in size and is designed to operate up to 3450 rpm. Saturated steam at atmospheric pressure will be ducted to the compressor to evaluate its performance using steam as a compression medium up to compression ratios as high as 8:1. Seal, valving and rotor material, design and wear operating characteristics will be assessed during this developmental testing phase. Upon the successful completion of this proof-of-concept testing activity, a long-term endurance test unit will be fabricated and operated on steam to uncover any problems that may arise during uninterrupted running of the system.

Open-Cycle Centrifugal Vapor-Compressor Heat Pump Development

Present single-stage centrifugal compressors with their relatively low-pressure ratios can be used only in evaporators with low temperature differentials. Multistage compressors are required for evaporators with high temperature differentials but are expensive and therefore seldom used.

AiResearch Manufacturing Company is developing an advanced centrifugal compressor capable of pressure ratios up to at least 2.5:1. This compressor will be integrated into a gas turbine-driven centrifugal steam compression heat pump system designed for evaporator/direct contact drying processes. The exhaust gas from the gas turbine driver will be used as the heat source for the contact dryer.

The proposed mechanical vapor recompression (MVR) system will be installed at a powdered milk drying facility as shown schematically in Fig. 4. This schematic illustrates just one of several ways by which MVR can be adapted to a milk evaporator. The total energy input is the engine fuel, and a 47% reduction in fuel will be obtained for this application when the proposed system is installed. The compressor is driven by a gas turbine engine that is mechanically attached to the compressor through a gearbox. The compressor can increase the steam saturation temperature up to 13°C (45°F) at a design flow of 16.5 m^3/s (35000 cfm). The steam compressor and engine are illustrated in Fig. 5. It is estimated that, with this system, 3622 kJ (3434 Btu) of natural gas will be required to dry a pound of milk (0.45 kg), compared to 6884 kJ (6525 Btu) in a conventional system. In the conventional approach, about 70 percent of the input energy is typically used in the evaporation process.

The project schedule calls for equipment design to be complete in 1981, and the design and fabrication of a field installation at an industrial dry milk plant to be completed in 1982.

INDUSTRIAL APPLICATIONS

Industrial Survey

An industrial survey was undertaken with the broad objectives of gathering industrial plant data relevant to the applicability of vapor compression heat pumps in industry, and identifying the likely industrial applications of this technology. This survey included an examination of process waste heat streams and energy requirements. Information was gathered through contacts with corporate engineering groups and individual plant managers and engineers, review of existing energy studies and evaluations in the literature, and discussions with industrial consultants. All the major industrial process sectors were surveyed.

Within these sectors, energy consumption and numbers of establishments were examined on a 4-digit SIC level to select the individual industries to be included in the heat pump industrial survey. Typically about half of the industries on the 4-digit level were selected in any given industrial sector.

The industries with likely application opportunities for vapor compression heat pumps are identified in Tables 1-4 as a function of heat pump system type. In addition to listings of the industries on a 4-digit SIC level, each table suggests several of the more important application factors which should be considered in that particular sector.

Combined MVR/Drying Applications

The application described in the Centrifugal Vapor-Compressor section above is but one of several industrial processes where mechanical vapor recompression and drying processes occur in close proximity. Others include:

1. beet and cane sugar refining
2. brewery industry
3. cheese whey production
4. wet corn milling
5. textile finishing
6. many chemical processes

These industries offer a unique opportunity for gas-driven open-cycle heat pump systems. The latter three, in particular, represent a large opportunity for industrial energy savings and more efficient utilization of natural gas.

CONCLUSION

The development of advanced vapor compression heat pump systems is an important element of the Gas Research Institute's R&D program to improve the efficiency of natural gas use in industrial processes. By assessing the potential industrial applications of this technology and by supporting the design and development of advanced gas prime mover-driven compressor technology, it is expected that heat pumps will play an increasingly important role in the enhanced utilization of natural gas in U.S. industry.

ACKNOWLEDGEMENT

The authors would like to express their appreciation to Mr. Fred Becker, Thermo Electron Corporation, R. Hoffman, Trochoid Power Corporation, and T. Iles, AiResearch Manufacturing Company for their technical contributions to this paper.

Table 1. Temperature Boost Systems

SIC 20	2011	MEAT PACKING
FOOD	2013	SAUSAGES
	2016	POULTRY DRESSING
	2026	FLUID MILK
	2033	CANNED FRUIT AND VEGETABLES
	2041	FLOUR AND GRAIN MILL PRODUCTS
	2082	MALT LIQUORS
	2086	SOFT DRINKS

COMMON PROCESS NEEDS

- CLEAN-UP WATER
- BOILER FEEDWATER PREHEAT

COMMON WASTE HEAT SOURCES

- REFRIGERATION CONDENSER COOLING WATER
- PROCESS COOLING WATER

Table 2. Dehumidification Systems

SIC 20	2013	SAUSAGES
FOOD	2023	DRIED MILK
	2043	BREAKFAST CEREALS
	2046	WET CORN MILLING
	2082	MALT LIQUORS
SIC 22	2261, 2262,	
TEXTILES	2272	FINISHING PLANTS
SIC 26	2621	PAPER MILLS
PULP AND PAPER		
SIC 24	2421	SAWMILLS - KILNS
LUMBER &	2436	SOFTWOOD VENEER & PLYWOOD
WOOD		
SIC 28	ASSORTED AIR DRYING NEEDS AT 180-300°F	
CHEMICALS		

PROVIDES CLOSE CONTROL OF TEMPERATURE AND HUMIDITY

APPLICATION REQUIRES ACCEPTANCE AND
PROCESS INTEGRATION OF NEW DRYING SCHEDULES

Table 3. Low-Pressure Steam Production Systems

SIC 22	2061, 2062	FINISHING PLANTS
TEXTILES	2072	
SIC 26	2611	PULP MILLS
PULP & PAPER	2621	PAPER MILLS
SIC 28	VARIOUS STEAM SOURCES VIA VENTED STEAM AND	
CHEMICALS	WASTE HEAT EXCHANGERS	

REQUIRES HIGH TEMPERATURE (200°F) WATER SOURCE AND HIGH PRESSURE RATIO
COMPRESSOR, OR LOW PRESSURE VENT STEAM

PLANT STEAM BALANCE AND OTHER SOURCES OF LOW PRESSURE STEAM MAY LIMIT
APPLICATIONS

Table 4. Mechanical Vapor Recompression Systems

SIC 20   2033   CHEESE
FOOD   2023   CONDENSED AND EVAPORATED MILK
                2033   CANNED FRUIT AND VEGETABLES
                2046   WET CORN MILLING
                2062,63   SUGAR REFINING
                2082   MALT LIQUORS
                2085   LIQUORS

SIC 26   2611   PULP MILLS
PULP & PAPER

SIC 28, 29   CRUDE UNIT OVERHEAD VAPORS
CHEMICALS   SEPARATION OF CLOSE BOILING MIXTURES
                EVAPORATION AND DISTILLATION IN NUMEROUS CHEMICAL PRODUCTION
                PROCESSES

GROWING USE OF MVR IN EVAPORATORS AND DISTILLATION

STRONGLY INFLUENCED BY APPLICATION-SPECIFIC CONSIDERATIONS.

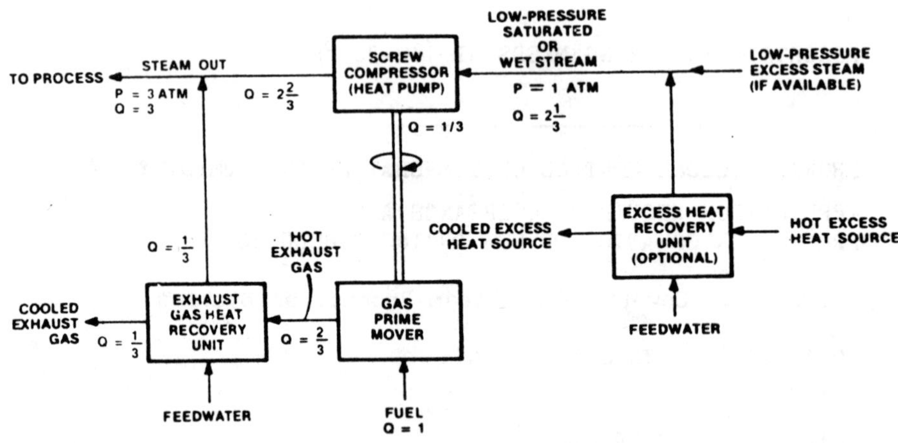

Fig. 1. Steam heat pump system flow schematic

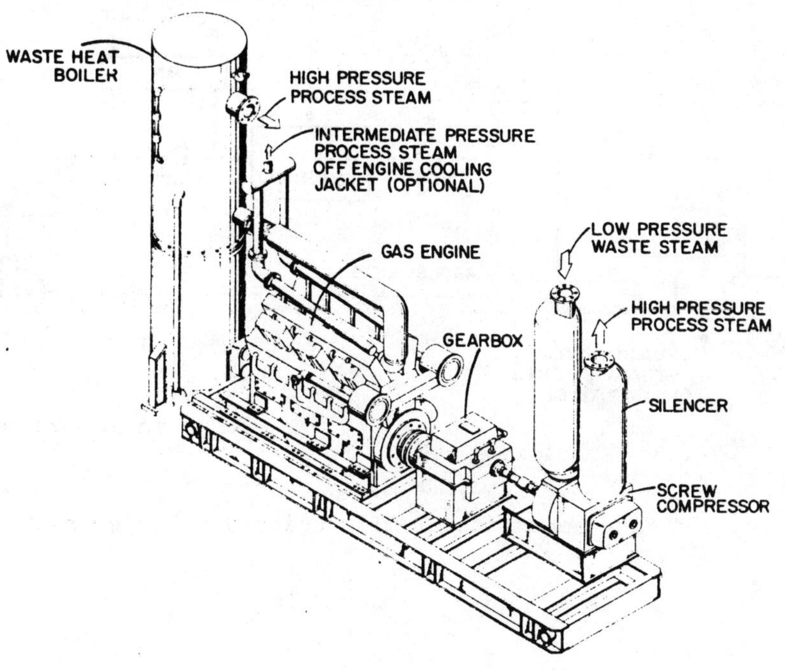

Fig. 2.   Screw compressor system configuration

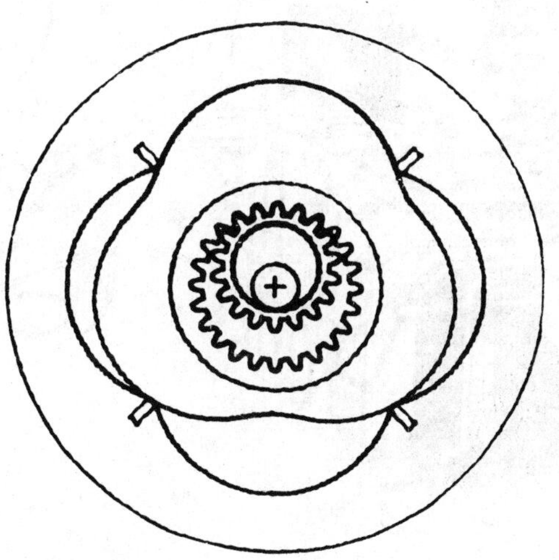

Fig. 3.   Epitrochoidal geometry

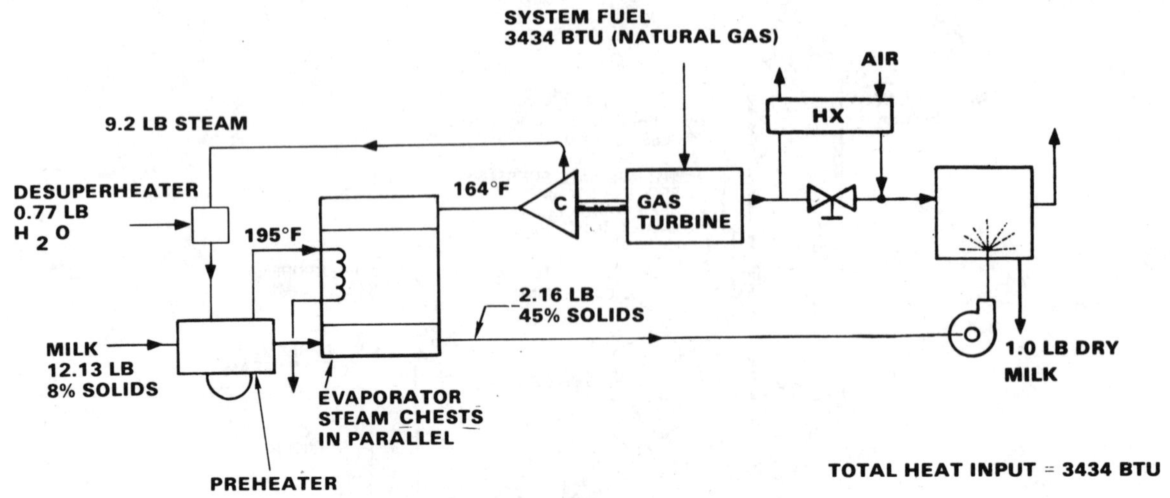

Fig. 4   Flow schematic of milk evaporator/dryer system

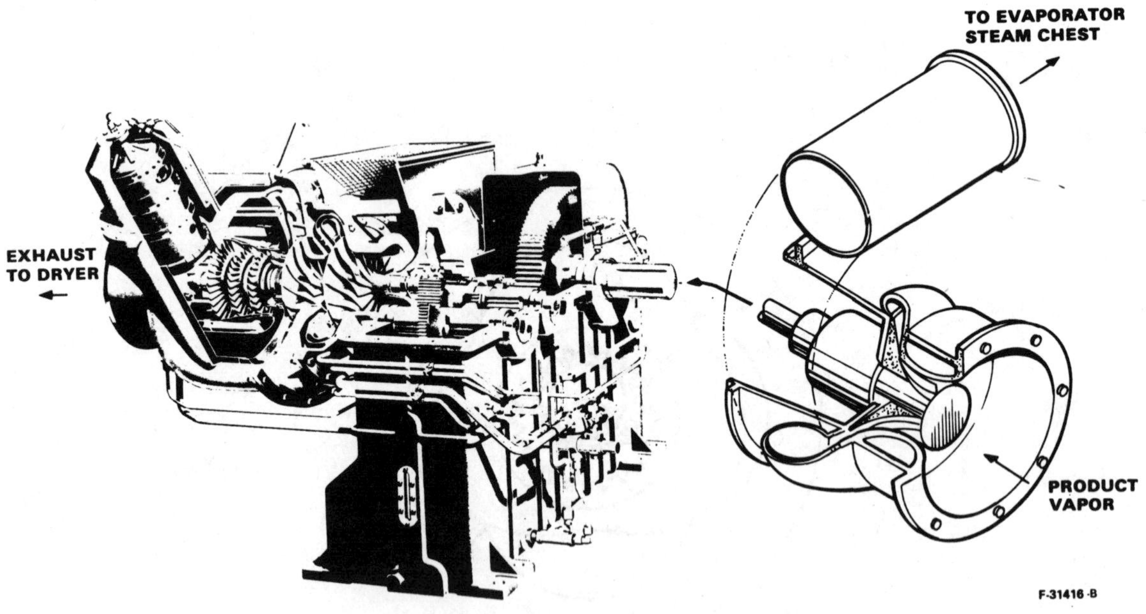

Fig. 5   Gas turbine-driven centrifugal compressor

PROBLEMS ENCOUNTERED AND RESULTS OBTAINED
IN THE APPLICATION OF COMPRESSORS
IN INDUSTRIAL PROCESSES FOR ENERGY SAVING

B. Degueurce and C. Tersiguel

Electricité De France, France

Summary

An extensive research program has been initiated by Electricité de France in order to achieve energy saving by an efficient use of compressors for heat pumps and steam compression.

Several research areas must be pursued, but some technical barriers to these applications have been removed.

Reciprocating compressors, using R 114, have been developed and tested with a condensing temperature of 120°C.

Steam compression tests have taken place using oil-injection free single screw, twin screw and lobe compressors.

Held at the University of Warwick, U.K.
Symposium organised and sponsored by
BHRA Fluid Engineering

1. INTRODUCTION

Since 1973, the increase in energy costs has fostered the development of many studies and projects at the "Direction des Etudes et Recherches" of ELECTRICITE de FRANCE.

Among the various equipment under study, heat pumps and steam compression are more and more frequently mentioned as potential energy saving media.

The aim of this paper is to describe how ELECTRICITE de FRANCE activities have been designed to improve the performances of compressors for heat pumps and steam compression.

2. COMPRESSOR TESTS

2.1 Objectives

Testing undertaken at EDF is designed to give an evaluation of the reliability and the performances of compressors to be used for industrial heat pumps or steam compression.

They allow manufacturers and users to assess the effects of the technical solutions adopted.

2.2 Test installations

EDF has three test rigs which allow compressors to be tested in operating conditions similar to those encountered in industry.

2.2.1 Principle

Each installation essentially comprises   a condenser and an evaporator (Fig 1.) The condenser is dimensioned to dissipate the energy carried in by the compressor and only receives the required flow. The condensed fluid vaporizes in contact with vapor or steam in the evaporator.

The operator controls suction pressure and temperature, as well as discharge pressure.

Two installations can handle steam whereas the third one is used with refrigerants.

The main characteristics of each installation are given in table 1 .

2.2.2 Measurements

The measurements taken give an assessment of the reliability and performance of the compressors undergoing test.

Pressures, temperatures, fluid flow and compressor power are measured for each test. The data collected is fed into a computer programme which calculates compressor efficiency.

2.3 Tests

Tests are undertaken on compressors which are to be used for industrial heat pumps or steam compression.

Twelve compressors have already been tested on the three rigs. Nine compressors should be tested ower the next two years.

2.3.1 Compressors for heat pumps

The introduction of the heat pump in industrial processes is a fairly recent
development although cooling systems have been in use for a long time.

We were aware, at EDF, that the availability of high temperature heat pumps
with a condensing temperature of 120°C could widen the scope for heat pumping in industrial processes.

Major improvements have been carried out in the design of reciprocating,
compressors for such use. At present, high temperatures (120°C) can be reached with a
refrigerant (R114), a polyalphaolefin oil, and reciprocating compressors (1) (2). In
these conditions and for an evaporating temperature of 70°C, the ratio of the practical
COP over the CARNOT COP is 0.47.

The practical coefficient of performance (COP) is defined as the ratio of the
power available at the condenser of the heat pump linked to the compressor, to the
compressor shaft power. The CARNOT theoritical coefficient of performance is the ratio
of absolute saturation temperature at discharge to the difference between saturation
temperatures at discharge and suction.

Investment cost is very often the stumbling block for the development of new
applications. EDF has tested a prototype compressor designed by a car manufacturer and
constructed from a mass produced engine. The mass production price of such a compressor
should be much lower than that of conventional reciprocating compressors.

Oil free compressors are a solution for condensing temperatures higher than
120°C. Low cost non-lubricated reciprocating compressors as well as oil-free single
screw compressors are to be developed.

2.3.2 Steam compression

Much emphasis is being placed on the use of steam compression systems in
evaporation and distillation processes. The curve in Fig 2 shows the state of developments under way or investment commitments which have already been made.

In order to achieve energy saving, economic compressors able to handle steam
must be designed and developed ; EDF has therefore undertaken tests on steam with
several types of compressors.

VOLUMETRIC COMPRESSORS

There are already installations using volumetric compressors for the compression of steam (7) but the performance characteristics of the compressors are
unknown.

Steam compressors of the lobe, single and twin screw type have been tested
at EDF. These compressors are oil free and equipped with water injection or cooling
devices allowing desuperheating of the compressed steam.

Tests on the lobe compressor designed by HIBON have shown the importance of
desuperheating on performance (3). Such a compressor which can be used industrially for
steam compression is now available on the market.

Tests on a single screw compressor designed by a french company (4) and a
twin screw modified for steam compression (8) have shown that this equipment could be
employed for suction pressures higher or lower than atmospheric pressure with coefficients of performance greater than those obtained with liquid ring or lobe compressors
in similar conditions, as shown in table 2.

The coefficient of performance for a single screw steam compressor is lower than for a twin screw, but its mass production price should not significantly exceed that of a liquid ring compressor and be below the price of non lubricated twin screw or piston compressors presently manufactured for steam compression.

CENTRIFUGAL COMPRESSORS

The type compressor used for steam compression depends on the required pressure ratio. In many applications, single stage centrifugal compressors have proven to be highly reliable and require very little maintenance.

The development of new high speed compressors with a higher pressure ratio can widen the use of such compressors.

A single stage centrifugal compressor, with a speed of 38 000 rpm and a pressure ratio of 1.7 has been tested for steam compression (5).

The results obtained are shown in table 2.

A two stage industrial compressor, designed by another french manufacturer with a pressure ratio of 4.5 will be tested soon.

3. CONCLUSIONS

The above mentioned studies have enabled ELECTRICITE de FRANCE to select and develop compressors using a variety of technical designs.

Major improvements have been carried out in the design and construction of reciprocating compressors for high temperature heat pumps.

Steam compressors of the lobe, single and twin screw as well as high speed centrifugal type have been tested and developed.

We shall be making further efforts in this field in order to offer industrial users a range of reliable high-performance compressors to suit their area of use from the technical as well as the economical point of view.

REFERENCES

1. <u>Tersiguel-Alcover</u> : "Testing of a COMEF reciprocating compressor for heat pumps at a condensation temperature of 120°C. EDF Report P33/4300/79-05 (in french).

2. <u>Tersiguel-Alcover</u> : "Testing of a SAMIFI BABCOCK reciprocating compressor for heat pump at a condensation temperature of 120°C. EDF Report P33/4300/79-45. (in french).

3. <u>Kleitz</u> : "Steam compression with a HIBON lobe compressor".EDF Report P32/D77-01 (in french).

4. <u>Degueurce - Pascal - Zimmern</u> : "Problems encountered and results obtained in direct steam compression using an oil injection free single screw compressor". Compressor Technology Conference - PURDUE - July 1980.

5. <u>Pascal</u> : "Testing a TECHNOFAN centrifugal compressor with steam". EDF Report P33/4300/80-46 (in french).

6   <u>Freytag - Viallier - Marchal</u> : "Original heat recuperation system by steam recompression. Application to continuous moisture separation of sheeted materials and extension to other types of applications". In : 9th UIE International Congress. Cannes - Session n° 6 - October 21 st, 1980. (in french).

7. <u>Solignac</u> : "Inventory of concentration units by evaporation and distillation with mechanical recompression of steams". EDF Report CR-HP/42/79-6 -January 1979 (in french).

8. <u>Tersiguel</u> : "Testing on ATLAS-COPCO twin screw compressor for steam". EDF Report P33/4300   (in french).

TEST INSTALLATION CHARACTERISTICS

FLUID	RIG N° 1	RIG N° 2	RIG N° 3
	REFRIGERANTS	STEAM OR ANY VAPOR	STEAM
Maximum flow - (m^3/h)	600	3 000	400
Maximum pressure - (bar)	30	80	2
Maximum temperature	130	250	120
Maximum electrical power - (kW)	250	500	20

MEASURED VALUES FOR STEAM COMPRESSORS

COMPRESSOR TYPE	SPEED RPM	POWER kW	EVAPORATING TEMPERATURE °C	TEMPERATURE DIFFERENCE °C	VOLUMETRIC EFFICIENCY	PRACTICAL COP	PRACT. COP / CARNOT COP	SOURCE
Liquid ring	740		100 120 140	10 10 10		7,8 12,8 14,8	0,20 0,30 0,44	Textiles Research Center Mulhouse 6)
Lobe	2 500 2 760 2 000	6 6 6	60 81 81	26 11 17	0,83 0,78 0,67	5,8 18,6 9	0,42 0,56 0,42	EDF (3)
Monoscrew	3 000 3 000 3 750	20 11 28	110 86 110	32 36 38	0,61 0,57 0,68	5,52 4,79 4,92	0,42 0,40 0,47	EDF (4)
Twin screw	8 730 8 730	104 96	103 106	30 25	0,82 0,78	8,49 9,79	0,61 0,60	EDF
Centrifugal	37 900	73	100	17		14,7	0,62	EDF (5)

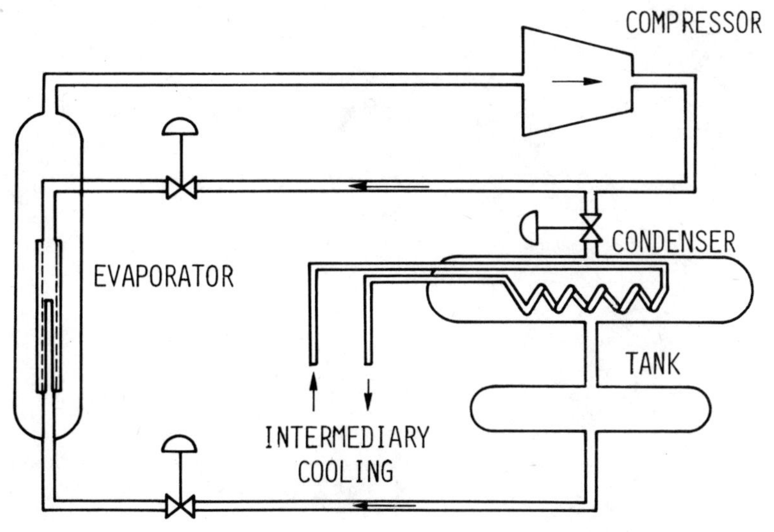

FIGURE 1 : TEST RIG PRINCIPLE

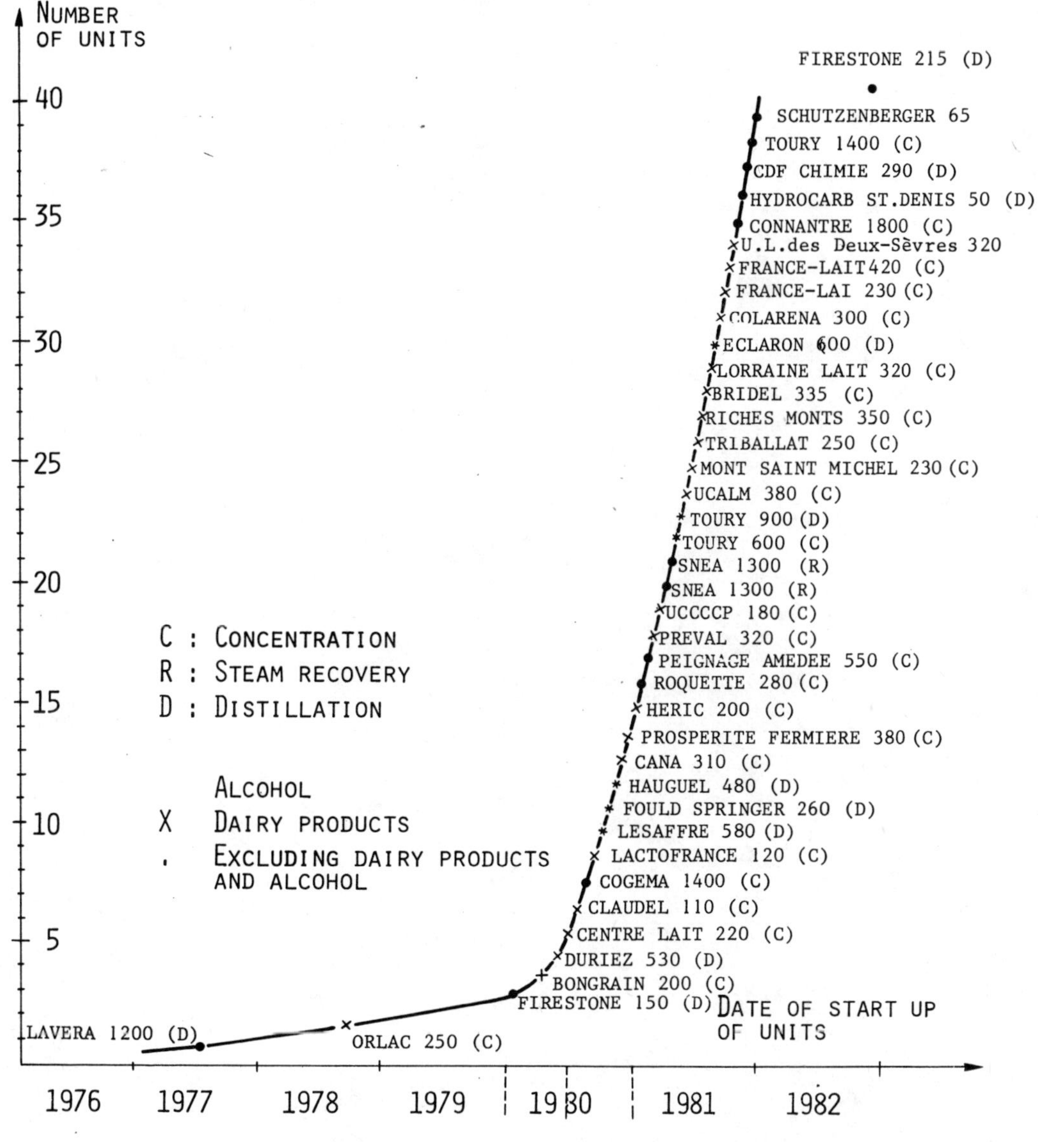

FIGURE 2 - DEVELOPMENT OF STEAM RECOMPRESSION IN INDUSTRY

THE EFFECTS OF CAPACITY MODULATION ON THE PERFORMANCE OF VAPOUR COMPRESSION HEAT PUMP SYSTEMS

S.A. Tassou, C.J. Marquand, D.R. Wilson

Polytechnic of Central London, U.K.

Summary

In this paper the effects of variable refrigerant flow rates on the performance of the individual components of capacity controlled heat pumps are examined. Special attention is given to the heat exchangers because these are the components which determine the efficiency of the heat transfer process between the heat source and the heat sink. Using experimental and analytical means, methods of obtaining optimum performance through control of the variable heat pump parameters are investigated. Control requirements are discussed and suitable control parameters for the optimisation of capacity modulated heat pumps using a microprocessor system are identified.

Held at the University of Warwick, U.K.
Symposium organised and sponsored by
BHRA Fluid Engineering

1.   INTRODUCTION

There are many industrial heating systems which could benefit from the use of heat pumps since they can reduce the primary energy requirements. The use of conventional heat pumps,   however, is constrained by the accuracy of the temperature control available, industrial process heating requiring much closer tolerances than are obtainable from on/off control. To achieve the required degree of accuracy, allowing at the same time for variations in both the heating load and the energy available in the low grade heat source, more sophisticated control techniques are needed. The availability today of advanced microcomputer technology at relatively low cost enables the introduction of improved control functions for the optimisation of industrial, commercial, and even domestic heating systems.

In all heat pump applications it is important both to match the supply of, and the demand for, heat as closely as possible in order to avoid the expense of supplementary heat or thermal storage, and to increase the control accuracy and efficiency by reducing or eliminating on/off cycling. This can be achieved by the introduction of capacity control to the heat pump. In general, capacity control gives improved efficiency due to the following effects (Ref. 1,2):

a)   Reduction of coil loading and thus lowering of the high-to-low side pressure differential which improves steady-state efficiency for loads less than the design load; and

b)   Reduction or elimination of on/off cycling losses for loads less than the design load.

Various methods of capacity control such as cylinder unloading, discharge gas by-pass or multiple compressors have been employed to match the system capacity to the load. The use of these methods, however, is restricted to relatively large air conditioning and refrigeration plants with no application in the residential area. The most promising method of employing capacity modulation in residential and small commercial heat pumps is through compressor speed control which can be either step wise or continuously variable. With step wise speed control there are still on/off effects when running continuously due to switching of contactors to effect motor speed changes. Only when the motor speed changes continuously with the load will reduced on/off cycling of the system reduce the on/off cycling of the individual components (Ref. 3).

The efficiency of a vapour compression heat pump depends on the individual component performance characteristics, but it is generally not determined through a simple cascading of the components. The coupling effects of combining the components into a system are important (Ref. 4). There is, in general, some optimum combination of sets of given component characteristics to meet a specified capacity.

In a variable speed heat pump, however, the component characteristics will change with variations in capacity. So, although the heat pump may operate at maximum efficiency at a given set of operating conditions, a change in conditions such as heat demand or energy available in the heat source will result in a decrease in efficiency. To maintain the efficiency, parameters such as heat load requirements, evaporator air flow rate and refrigerant supply to the evaporator should be correlated with the variable speed compressor operation. This correlation and thus the optimisation function can be performed using a microprocessor based control system (Ref. 5).

The development of control criteria for the microprocessor requires knowledge of the effects of capacity control on the performance characteristics of the individual components of the heat pump and their influence on the overall performance and efficiency. This knowledge can be attained by combined experimental and theoretical investigations. The results of such investigations are presented in the following sections.

2.   EXPERIMENTAL INVESTIGATIONS

2.1   Variable speed heat pump.

The experimental work has been carried out on an air-to-water heat pump designed for a maximum heat output of approximately 8 kW. The heat pump has been built of commercially available components. A conventional air cooled continuous fin

heat exchanger is used for the evaporator and a tube in shell heat exchanger for the
condenser. The compressor is a twin cylinder, open type, driven by a 3 kW shunt wound
motor. Speed control down to half the rated running speed is obtained by a thyristor
controller. The design also allows for a variable air flow rate across the evaporator
through the use of a variable speed axial fan. The degree of superheat at the
evaporator outlet is controlled by an electrically driven motorised expansion valve.

Test conditions for the evaporator are obtained in an air conditioning duct
designed to simulate climatic conditions in Western Europe. A range of heating loads
can be simulated using a water loading system. Any required water flow rate and
temperature at the condenser inlet can be obtained by mixing the hot water outlet
supply of the heat pump with mains cold water by means of a mixing valve and a
temperature controller.

The performance of the heat pump is monitored by a single-board microprocessor computer through comprehensive instrumentation. Apart from acting as a data
logging system the microprocessor controls the performance of the motorised expansion
valve. The software is currently being developed to cover the control of the whole
heat pump.

2.2   Performance characteristics.

The heating capacity of an air source heat pump increases with increased air
temperature. This is shown in Fig. 1 which presents the variation of the heating
capacity of the heat pump with air temperature for four compressor speeds. An
increase in the compressor speed causes an increase in the refrigerant mass flow rate
leading to increased system capacity.

The coefficient of performance (c.o.p.) of the heat pump at the four
compressor speeds is shown in Fig. 2. It can be seen that a reduction in the
compressor speed causes an increase in the c.o.p. which has been determined as the
ratio of the heat output to the total energy consumed by the heat pump. The
relatively low values of the c.o.p. are due to the fact that the heat pump has not yet
been optimised with respect to component sizing.

2.3   Effect of variable refrigerant flow rate on evaporator liquid supply control.

The thermostatic expansion valve is probably the most widely used refrigerant control at the present time. It regulates the flow of liquid refrigerant
entering the evaporator with response to the superheat of the refrigerant vapour
leaving the evaporator. Although the thermostatic valve is an efficient control
device when used in systems that operate near to their designed capacity, it is
unsuitable for use in capacity controlled heat pumps. This is because a valve chosen
for a particular set of conditions (pressure drop and capacity) will be undersized
at full load and oversized at partial load. An oversized thermostatic valve will not
control as well at full capacity as a properly sized valve and control will get worse
as the load decreases.

Fig. 3 shows the installed performance characteristics of an ordinary
thermostatic expansion valve as measured on the variable capacity heat pump. The
valve has been factory set for a 7°C superheat. It can be seen that the superheat at
the evaporator outlet does not remain constant but varies both with air temperature
and compressor speed. An increase in the compressor speed gives rise to increased
refrigerant superheat. This is mainly due to a higher pressure drop caused by a
higher refrigerant flow in the evaporator. The degree of superheat also increases as
the air temperature is lowered. This can be attributed to the pressure-temperature
relationship of the refrigerant. For R12 the change in pressure per degree of
temperature change decreases considerably as the temperature of the refrigerant
decreases (Ref. 6). Therefore the amount of superheat required to cause a given
increase in the pressure of the remote bulb of the valve is much greater at low
temperatures than at high temperatures. The result is that the amount of suction
superheat necessary to actuate the valve becomes excessive at low temperatures,
making much of the evaporator ineffective.

To overcome the deficiencies of the thermostatic expansion valve, a motorised
expansion valve has been designed for the variable speed heat pump. The operation
of the valve is controlled by the microprocessor, the aim being to maintain a constant

degree of superheat at the evaporator outlet. The control is performed in the
following way: The microprocessor measures the temperature and pressure of the
refrigerant leaving the evaporator using a thermocouple and a pressure transducer
respectively. The saturation temperature corresponding to the measured pressure is
then determined from a table stored in its memory. The difference between the
measured temperature and the saturation temperature gives the actual degree of superheat at the evaporator outlet. If its value is not equal to the prescribed superheat
the microprocessor, depending upon the error, either opens or closes the valve until
the two values agree with the desired accuracy.

The performance of the microprocessor controlled motorised expansion valve
is shown in Fig. 4. As can be seen, the accuracy of the superheat control is not
affected by variations in the air temperature or compressor speed.

3.   THEORETICAL INVESTIGATIONS

3.1   The heat pump model.

A mathematical model for determining the performance of variable speed heat
pump systems has been developed. The model is organised in three principal sections:
the condenser programme, the evaporator programme, and the compressor programme. The
heat exchanger programmes are based on their geometric characteristics and material
properties rather than manufacturers performance data. The compressor programme relies
on efficiency parameters that may be obtained from calibration tests. The heat pump
model is generalised so that it may be used to calculate performance and efficiency
over a wide range of operating conditions. It enables investigation of the performance interactions of the components and prediction of the effects of capacity
modulation on the heat pump performance.

3.2   Effect of evaporator parameters on heat pump performance optimisation.

A dry expansion evaporator coil in which complete vaporisation of the
refrigerant takes place can be divided into two distinct regions, a two-phase region
and a superheated region. The two-phase region consists of a mixture of liquid
refrigerant and its vapour, whilst the superheated region consists only of vapour.
The degree of superheat of the refrigerant vapour leaving the superheated region is an
important performance parameter because it is used as a signal for the operation of
the expansion valve, and it also ensures that no liquid refrigerant is carried to the
compressor. The overall heat transfer coefficient in the superheated region, however,
is much smaller than the overall heat transfer coefficient in the two phase region.
Therefore, high values of refrigerant superheat cause a reduction in the evaporator
capacity.

The effect of the degree of superheat on the compressor speed and the c.o.p.
of a variable speed heat pump for two levels of constant heat output is shown in
Fig. 5. It can be seen that for both output levels, an increase in the degree of
superheat causes a reduction in the c.o.p. This happens because as the superheat is
increased the evaporator capacity is lowered, forcing the compressor speed to
increase in order to keep the heat output constant. The increase in the refrigerant
flow rate caused by the increase in the compressor speed gives rise to a reduction in
the evaporating temperature and an increase in the condensing temperature. The higher
temperature and hence higher pressure differential increases the power consumption
of the compressor which in turn reduces the c.o.p.

As shown in Fig. 5, the degree of superheat has a greater effect on the
system c.o.p. at the lower heat output level. This is because at the lower output
level the same increase in the superheat causes a bigger percentage reduction in the
evaporator capacity and a larger percentage increase in the compressor power
consumption than at the higher output level. Therefore it is important, especially
at low capacities, to maintain the superheat at a low value. From Fig. 5, an optimum
value would be about 5°C. Reducing the evaporator superheat below this will have
little effect on the system efficiency and would create control problems due to the
oscillatory nature of the mixture-vapour transition point (Ref. 7).

The evaporator air flow rate is also an important parameter in the
performance of variable capacity heat pumps. Conventional heat pumps utilise constant
air flow rates across the evaporator which are optimised for fixed design conditions.

If constant air flows were used in capacity controlled systems, however, fan power would form a larger proportion of the total power consumption at low capacities than at high capacities, reducing the efficiency of the heat pump. Increased efficiency can be achieved by reducing fan power in addition to compressor power and refrigerant flow.

The heat pump model has been used to determine the effects of fan power on the performance of the variable speed heat pump. The results are presented in Fig. 6 which shows the variation of the optimum evaporator air volume flow rate with the heat pump output for two evaporator frontal areas. It can be seen that the value of the optimum evaporator air volume flow rate increases with increased heat power output. Also, at a constant heat output, an increase in the evaporator frontal area causes an increase in the optimum evaporator air volume flow rate.

3.3   Effect of condenser parameters on heat pump performance optimisation.

The heat transfer in the condenser can be improved by increasing the heat transfer coefficients of the two heat exchange fluids. In a capacity controlled heat pump, for given component characteristics and ambient conditions, the refrigerant mass flow rate is determined by the heat output requirements. The overall heat transfer coefficient therefore can be increased by increasing the water mass flow rate with the penalty of additional pumping power.

The effect of the condenser mass flow rate on the heat pump c.o.p. for three levels of heat power output is shown in Fig. 7. The results presented are for a constant water delivery temperature of 55°C. The delivery temperature is usually decided by the requirements of the heating process. Fig. 7 shows that for each output level there is an optimum water mass flow rate which increases with increased heat pump output. The gains in the c.o.p., however, are so small that, for constant delivery temperatures, the use of variable water flow rates with capacity controlled heat pumps is not justified.

The performance of a heat pump may be improved by subcooling the condensate leaving the condenser. Liquid subcooling reduces the losses due to throttling at the expansion valve and increases the evaporator capacity. It can be achieved by direct cooling using a cooling medium at a lower temperature or by a regenerative process in which the cold vapour leaving the evaporator cools the liquid leaving the condenser.

Fig. 8 shows the effect of condensate subcooling on the heat pump c.o.p. at two levels of heat power output and two ambient temperatures, when the cooling water entering the heat pump is used to subcool the refrigerant liquid leaving the condenser. It can be seen that the c.o.p. increases linearly with increased sub-cooling of the refrigerant liquid. For a heat pump output of 4.14 kW and an air temperature of 5°C an increase in the degree of subcooling from 0°C to 6°C produces a 2.7% increase in the c.o.p. For a heat pump capacity of 5.27 kW and an air temperature of 0°C the same increase in the degree of subcooling produces a 4.2% increase in the c.o.p. It can be seen therefore that the percentage increase in the c.o.p. due to condensate subcooling rises as the ambient temperature is lowered and the heat pump capacity increased.

Fig. 9 shows that when the regenerative method is used, refrigerant liquid subcooling causes a reduction in the c.o.p. This may be explained as follows: The condensate subcooling and the compressor suction vapour superheating which take place at the same time increase the enthalpy difference of the refrigerant across the heat exchangers. Although for constant output levels the refrigerant flow rate is slightly reduced, the reduction is not enough to prevent an increase of the condensing temperature caused by higher vapour superheat at the condenser inlet, and a lowering of the evaporating temperature caused by increased refrigeration effect in the evaporator. The higher temperature and thus higher pressure differential leads to increased compression power and hence a reduction in the c.o.p.

In the former method, condensate subcooling increases the c.o.p. because the heat extracted from subcooling the liquid refrigerant is rejected to the cooling water, slightly increasing system capacity. The reduced refrigerant flow rate that is needed to satisfy the required capacity prevents the increase in the temperature difference in the heat exchangers, giving rise to increased c.o.p.

4. CONCLUSIONS

From the results presented in this paper the following conclusions can be drawn:

1) Thermostatic expansion valves are not suitable for use in variable capacity heat pumps because the accuracy of the evaporator superheat control varies both with air temperature and system capacity.

2) Evaporator superheats above 5°C produce a reduction in the heat pump c.o.p. This reduction is considerable at high values of vapour superheat and increases as the heat pump capacity is reduced.

3) In microprocessor controlled heat pumps the control of the degree of superheat of the refrigerant vapour at the evaporator outlet can be effectively performed by a motorised expansion valve.

4) For each heat pump capacity there is a value of evaporator air flow rate which maximises the system c.o.p. This optimum value increases with increased heat pump output and evaporator frontal area.

5) Although at constant water delivery temperatures there is an optimum value of condenser water mass flow rate which increases with increased heat pump capacity, its effect on the system c.o.p. is so small that variable water mass flow rate through the condenser is not justified.

6) Subcooling the refrigerant liquid leaving the condenser using the cooling water entering the condenser increases the c.o.p. For the same amounts of subcooling the percentage increase in the c.o.p. rises as the heat pump capacity is increased.

5. REFERENCES

1. Tassou, S.A., Green, R.K., Wilson, D.R. and Searle, M.: "Energy conservation through the use of capacity control in heat pumps". Journal of the Institute of Energy, 54, 418, Mar. 1981, pp.30-35.

2. Muir, E.B.: "Capacity modulation for air conditioning and refrigeration systems". Air Conditioning, Heating and Refrigeration News, April 1979, pp.3-16.

3. Griffith, R.W.: "A study of the effects and economics of capacity modulation on seasonal energy efficiency ratios for air conditioners". ASHRAE Transactions, 86, 1, June 1980, pp.465-476.

4. Gluck, R. and Pollack, E.: "Design optimisation of air-conditioning systems". ASHRAE Transactions, 84, 2, 1978, pp.304-314.

5. Wilson, D.R., Green, R.K., Neale, D.F., Searle, M., Tassou, S.A. and Wang, Y.T.: "The minimisation of the power consumption of a thermodynamic heat pump by a microprocessor based control system". Commission of the European Communities, Luxembourg, Report No. EUR 7283 EN, April 1981.

6. American Society of Heating, Refrigerating and Air Condition Engineers: "ASHRAE handbook of equipment", New York, USA, 1978, pp.20.1-20.10.

7. Wedekind, G.L. and Stoecker, W.F.: "Transient response of the mixture-vapour transition point in horizontal evaporating flow". ASHRAE Transactions, 72, 2, 1966, pp.1-15.

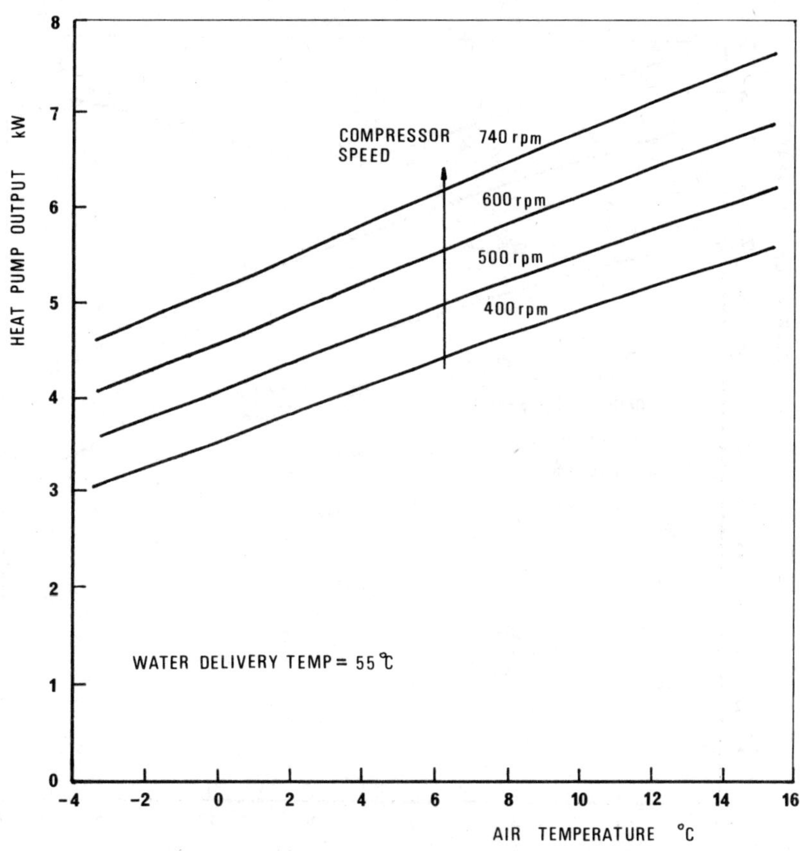

Figure 1   Variation of heat pump capacity with air
temperature

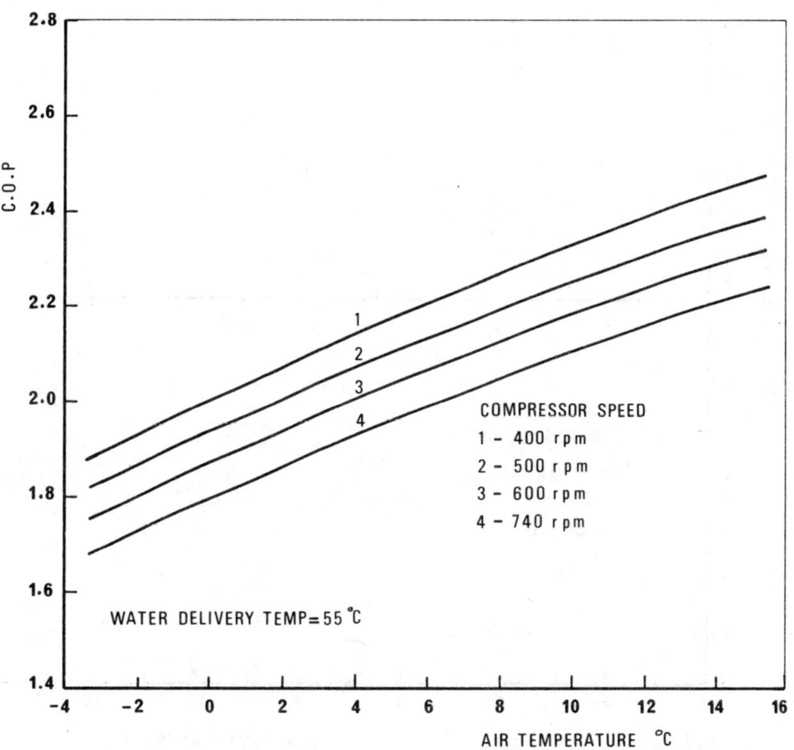

Figure 2   Variation of heat pump c.o.p. with air
temperature

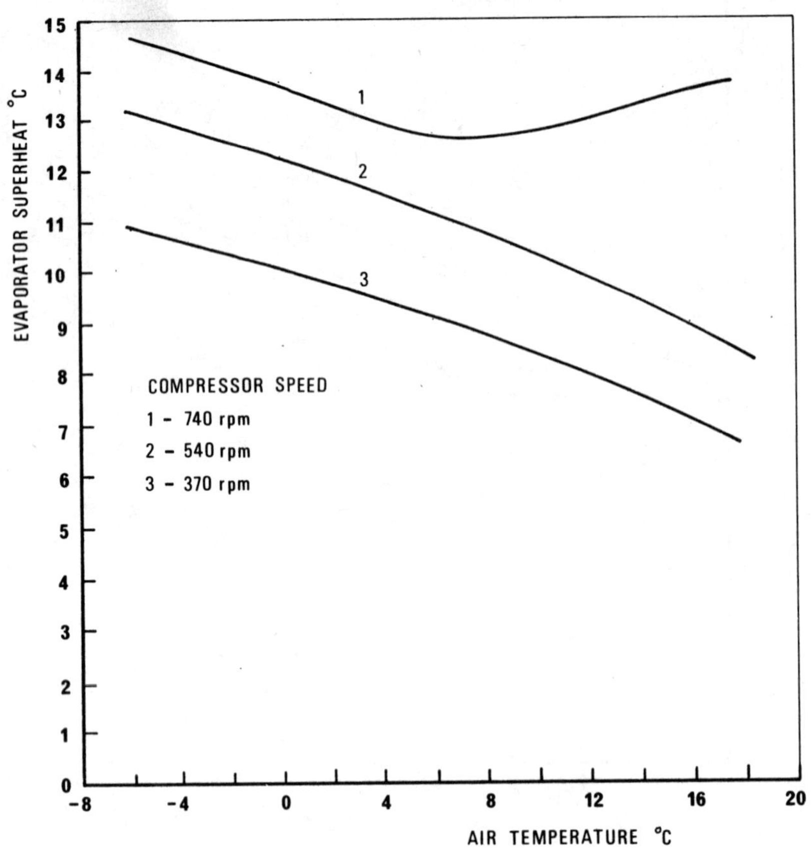

Figure 3   Installed performance characteristics of a thermostatic expansion valve

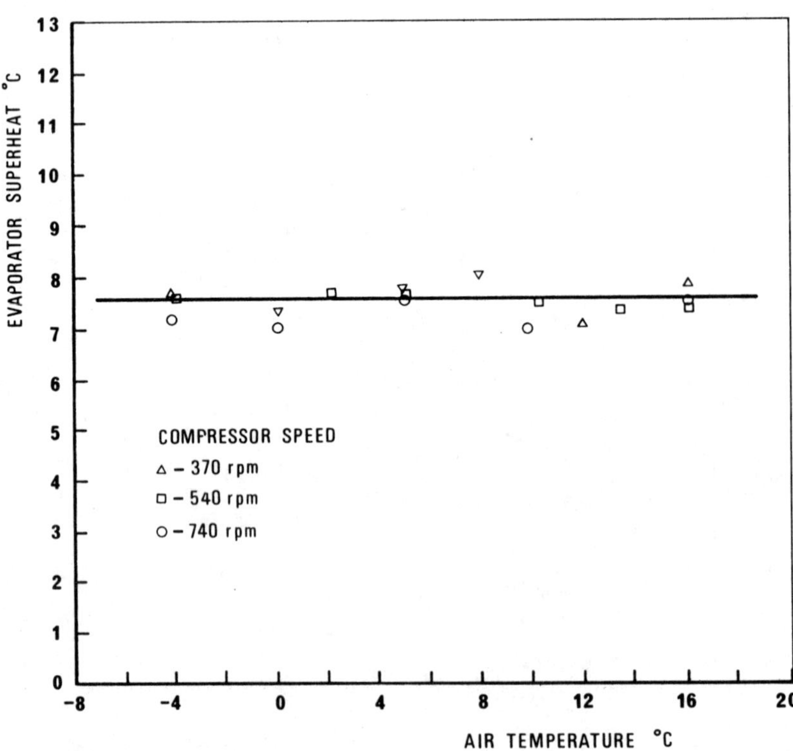

Figure 4   Performance of a microprocessor controlled motorised expansion valve

194

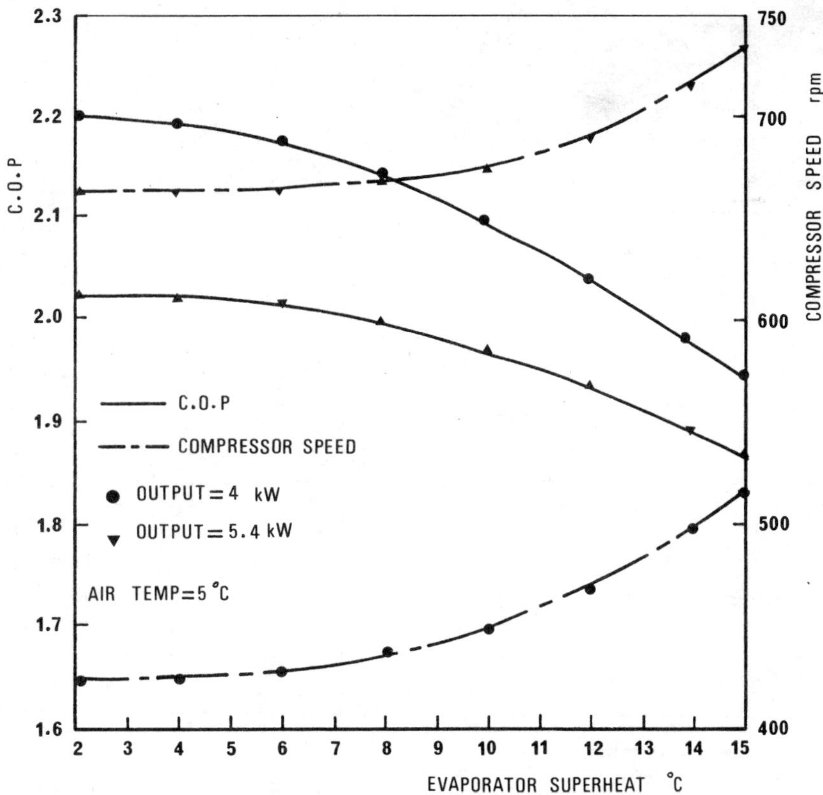

Figure 5   Effect of evaporator superheat on the performance of the variable capacity heat pump

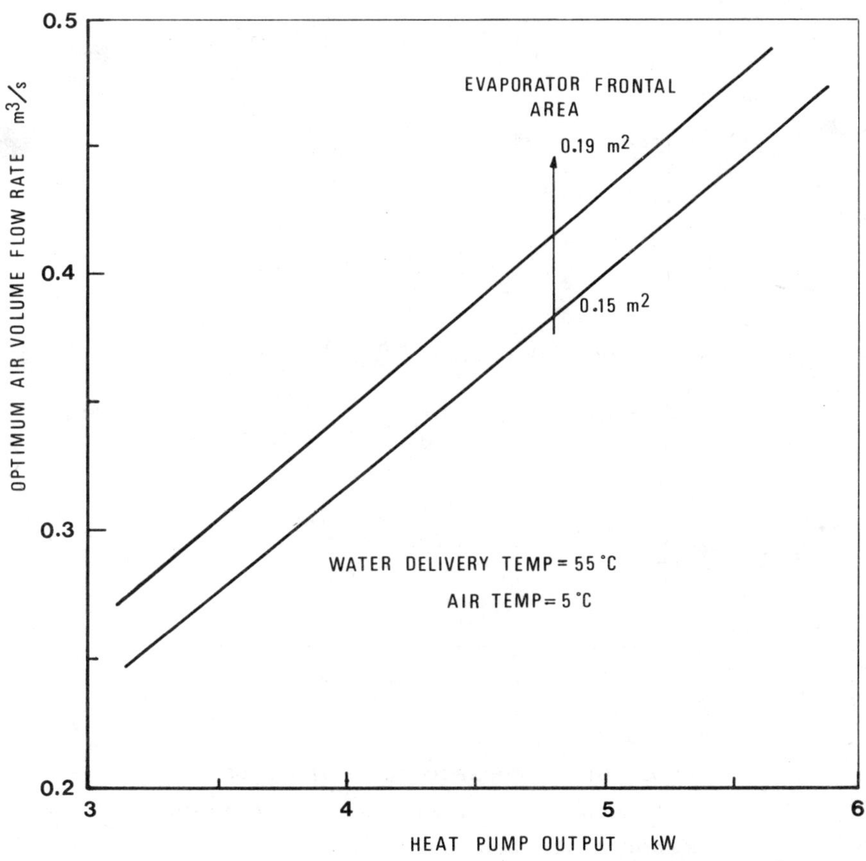

Figure 6   Variation of optimum evaporator air volume flow rate with heat pump output

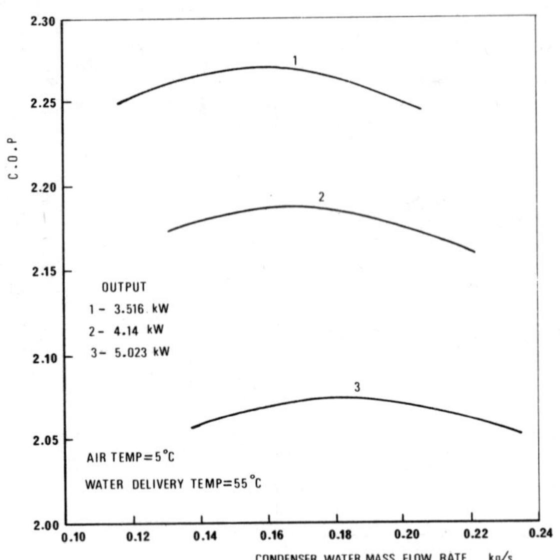

Figure 7

Effect of condenser water mass flow
rate on the heat pump c.o.p.

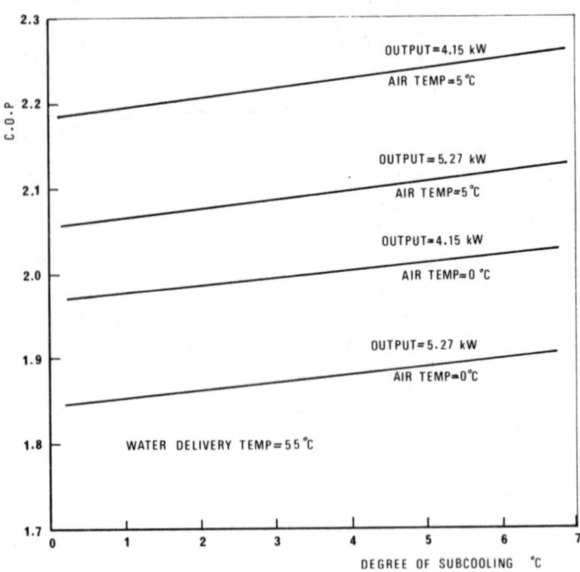

Figure 8

Effect of condensate subcooling on the
heat pump c.o.p. when the cooling water
is used as the subcooling medium

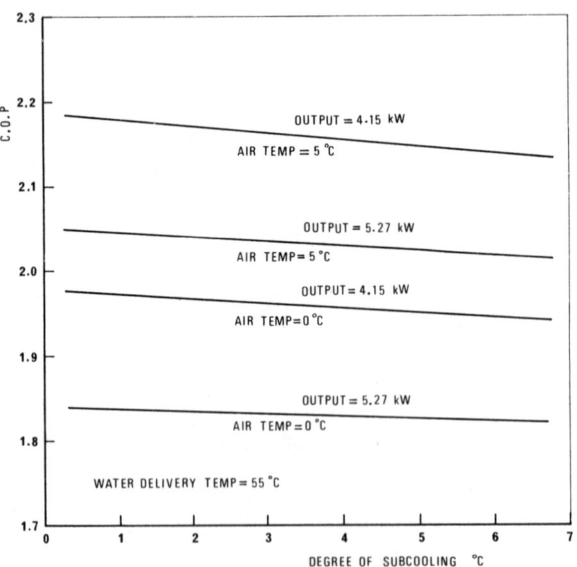

Figure 9

Effect of condensate subcooling on the
heat pump c.o.p. when the regenerative
method of subcooling is used

HEAT PUMPS – APPLICATION IN AN ENERGY INTEGRATED PROCESS

D. Boland and J.C. Hill

Imperial Chemical Industries PLC, U.K.

Summary

The paper reviews the effective use of heat in the process industries with particular reference to the quality and quantity of waste heat rejection. The scope for re-use of waste heat in the process industries is discussed. The relative merits of heat pumps and ORC's are examined against a background of the desirability for combined heat and power generation.

Held at the University of Warwick, U.K.
Symposium organised and sponsored by
BHRA Fluid Engineering

1.   Introduction

Heat pumps in the chemical industry seem to be a good idea yet there are
relatively few well proven applications. To understand why this is so,
we have to have some idea of how the industry operates.

In the chemical industry in general, and the large scale petrochemical
industry in particular, vast quantities of energy are used, typically
in the range $350^{\circ}C \longrightarrow 120^{\circ}C$, with rejection to atmosphere via air
cooled heat exchangers (fin-fans) or through cooling towers (via treated
cooling water systems). The key point in terms of heat pumps is that
these utilities dispose of substantial amounts of waste heat at around $110^{\circ}C$
with, say, a maximum temperature of about $120^{\circ}C$. The question to be
answered is how best to recover the heat for reuse.

2.   Pattern of Waste Energy Rejection : Correct Integration

The quantity and quality of surplus energy arise from independent effects:

1.   The quantity of waste heat is due in many instances to poorly
      integrated plant in which the energy recovery from the process
      could be greatly improved. In a typical process it is likely
      that for every 10 MWH of energy supplied at a high level than
      approximately 5 MWH are actually consumed by the process : the remain-
      ing 5 MWH are thus rejected as low grade waste heat. So for every
      5 MWH of energy rejected to the atmosphere the implication is that
      somewhere in the process 5 MWH of high value energy, usually a
      fossil fuel, has been consumed to no avail (See Fig 1)

      Of course, it is rarely possible to design a process in which
      there are no waste energy streams. For this to occur hot and cold
      streams are required to balance perfectly with heat input at the
      highest level only (see Fig 2). For most processes this state of
      perfect balance does not exist and external cooling is required in
      addition to heating (see Fig 1). However, it is possible to reduce
      this rejection of energy towards a minimum value which is
      determined absolutely by the thermodynamics of the process. This
      concept has been fully reported in reference 1 and in other
      publications.

2.   The quality of the waste energy is largely determined by the
      nature of the processes. In other words, it turns out that most
      chemicals, are produced by process routes which broadly speaking,
      involve unit operations in the $150^{\circ}C$-$350^{\circ}C$ range. For example,
      typical reaction stages require energy at $180^{\circ}C$ - $200^{\circ}C$ as do many
      separation processes. Energy at this level is usually, and convenient-
      ly, supplied in the form of steam. For process temperatures around
      $250^{\circ}C$ - $300^{\circ}C$ extremely high pressure steam is needed, which in
      practical terms is unacceptable. Direct fired heaters or
      indirect, hot-oil circuits provide the necessary alternative supply
      of energy.

      The demand for process energy below $150^{\circ}C$ is surprisingly limited;
      energy of this quality is conventionally used to cover feed preheat
      requirements, or to preheat boiler feed water for steam raising.
      But demand is usually relatively small and significant waste heat
      is available. Certainly, proper thermal integration of a process
      will reduce the energy being rejected but some surplus will undoubtedly
      be present.

Of course for a large site, there are practical limitations to just
how much integration can be achieved between hot and cold streams
and this effect may restrict the approach to minimum rejection of
heat. For example, when a centralised combined heat and power
station is employed it can be difficult to achieve maximum use of
condensate and optimum preheat boiler feed water; waste heat
is dispersed in small quantities across the total large site, and
recovery costs to a central location are prohibitive.

The percentage of the total energy usage (Fig 3) which is dissipated
to waste heat is variable but significant. On a typical
petrochemical plant the energy rejected at a low level can be between
40% and 80% of the energy input to the process. After thermal
integration this figure, as a percentage, will reduce but the major
benefit is that the absolute quantity of energy input will be reduced
by 10% - 60%. (Derived from various studies carried out in ICI).

The pattern of waste heat rejection is not confined to air coolers
and cooling waters. Large quantities of waste heat arise in flue
gas streams from both steam raising plant and direct fired process
heaters. Typically the flue gas temperature to atmosphere should
be 170°C - 180°C. Lowering the stack temperature further may result
in acid condensation with subsequent corrosion of waste heat recovery
units in the stack. However the reduction in energy usage to be gained
from further reducing flue gas temperatures (say to 110°C) is
significant and represents a 3.5% reduction of the fuel fed to the
burners. The possibility of corrosion resistant equipment therefore
arises (see later).

3.   Considerations arising from Use of Combined Heat and Power

Another area of concern is the pattern of energy usage, principally
the ratio between heat and power generation. On a large chemical
manufacturing site the ratio of heat to power is, in our experience,
around 3:1 (a figure of 2.76:1 has been quoted for the whole of the
chemical industry in reference 2).

A conventional steam raising heat and power station will typically provide
a heat to power ratio of 3.66:1. Thus for sites supplied by conventional
combined heat and power stations (CHP) there will always be an imbalance.
The deficiency in electricity production has to be met by using far more
expensive electricity from the national grid. The possibility of a more
enlightened design of the overall CHP system (ie the heat engine design)
arises. (See Later).

4.   Economic Constraints

In the current economic climate capital spending is severely restricted
and any investment must be assured of a rapid return. This in effect
means that modification projects, including those to save energy, should
preferably achieve payback times measured in _months_ rather than _years_.
Our experience indicates that, in a climate where only small amounts of
money is available, being selective in this way is not unreasonable. In
fact, by imposing a severe financial criterior for judging prospective
projects it appears that the engineer is encouraged to look beyond the
more obvious solutions - towards the best short pay-back projects.

It is undesirable that the engineer faces a complex problem. He must
take account of the methods now available for thermally integrating
process streams towards minimum energy consumption. However, he must
recognise that these techniques require support by achieving the best
heat/power interface. In addition, waste energy, the ratio of heat to

power and the economic constraints currently pertaining to the industry
all have implications. The engineer must consider all these aspects in
deciding what type of hardware is most cost effective.

5.   <u>Improving the Heat/Power Interface</u>

In a previous paper, ref 3, the relative merits of the common types of
heat engine and heat pump cycles were discussed in relation to their
efficiency and reliability. This paper is intended to throw into
perspective the relative scope of the various devices for redressing the
imbalance at the heat/power interface for process plant.

5.1   Heat Pumps - Theoretical Considerations

The principle of heat cascading has already been discussed in
reference 1. The diagram in figure 4 represents a process in which
energy is supplied from an external high temperature source (Q1) and
slowly degraded in quality by process interchange until residual
energy is rejected at low temperature to the surroundings (Q2).
Most processes exhibit a pinch temperature in this energy cascade
diagram. Above this temperature the net effect of all the process
stream enthalpy changes is to provide a heat sink maintained in
balance by an external energy supply (Q1). Conversely below this
temperature the process is a net heat source maintained in balance
by external cooling (Q2). It follows therefore that for minimum
energy consumptions there must be no heat transferred across the
pinch temperature. If energy were transferred across the pinch
then the net heat sink above the pinch would become larger thus
requiring Q1 to increase, simultaneously the net heat source below
the pinch would increase and Q2 would increase to maintain the
balance.

From this analysis it is clear that there is only one position in
a totally integrated process where a heat pump can actually save
energy (Fig 5). Only if the heat pump takes energy from the net
heat sink above the pinch can any reduction be made in the values
of Q1 and Q2. A heat pump operating wholly above or wholly below
the pinch will simply result in the work input to the compressor
being rejected to external cooling (Figs 6 (a) and (b)). Thus,
although the heat pump appears to recover and reuse heat effectively
on the micro-scale, in terms of the macro system the net effect is a
decrease in system efficiency. For a complex chemical plant, it is
desirable to identify the pinch and verify that a heat pump which
looks to make good sense on a local scale is equally viable in terms
of the whole system.

While the ideal situation discussed above involves a fully
integrated thermal process onto which heat pumps can be effectively
superimposed, this is only likely to arise on a greenfield site
where the macro system has been correctly integrated. For an existing
system, designed by traditional methods, complete integration by
retrofit is likely to be uneconomic; implementation of the more
attractive retrofits will only provide partial integration. In this
situation retrofit installation of a heat pump at some point other
than across the pinch may save energy but careful analysis is
required of the implications on the macro system. For example, it
should not be automatically assumed that the installation of a heat
pump to a distillation column necessarily saves energy within the
total system.

To summarise, there will be instances where a heat pump, or
more preferably direct mechanical vapour recompression using
suitable process fluids, can provide a real energy saving but
the important message is that the entire process system must be
examined in order to determine the real energy benefit.

5.2   Heat Pumps - Practical Considerations

As stated previously, most waste heat in the chemical industry is
available in the range $100^{\circ}C$ - $120^{\circ}C$ and the majority of processes
require energy at around $180^{\circ}C$ - $200^{\circ}C$. This situation requires the
working fluid in the heat pump to be raised, on average from $100^{\circ}C$
to $200^{\circ}C$, assuming temperature differentials of $10^{\circ}C$ in both evaporator
and condenser. This operational requirement severely restricts the
usefulness of conventional designs of heat pump. Firstly the
operating temperature is too high for the practically proven,
established working fluids (principally refrigerant fluorcarbons)
and secondly the very high temperature lift required low coefficients
of performance (COP), typically around 2.0 2.5. This situation
tends to eliminate heat pumps as a viable way of converting waste heat
into useful process grade heat within the process industries.

The only apparent outlet for heat pumps is in within a single unit
operation involving change of phase, for example, distillation,
evaporation or drying. Of these, only distillation plays a major
part in the petrochemical industry in terms of energy consumed.
Further, the higher temperatures involved indicate that working
fluids such as toluene or steam are necessary to ensure a reliable
system. However, despite being able to overcome any technical
problems involved, the economics of heat pumps in the chemical
industry are unfavourable. Recent evaluation studies undertaken
within ICI indicate that for the cheapest form of heat pump (one
with an electric drive) payback is of the order of 3 years. The
heat and power imbalance on large chemical sites with combined
heat and power stations is however further exacerbated by the
introduction of electrically driven heat pumps which consume power
while reducing the demand for heat. It would therefore appear
that alternative drivers for heat pumps will provide more
attractive propositions if the pay-back times reduce significantly.
Ideally heat pumps driven by steam turbines utilising waste steam
as both motive force and low grade heat source would provide
the most effective way of recovering waste heat.

Future developments to improve the viability of heat pumps are
discussed in Section 6.

5.3   Alternative Uses of Low, Grade Energy

Even if heat pumps are unlikely to be viable in economic terms
in the foreseeable future, there are still considerable savings
to be made in more efficient use of the waste heat currently
being discarded. A major contribution is likely to be made
by systems which generate power from low grade heat. Not only
do such systems generate the highest value energy utilised in the
chemical industry, electricity, but also they will help to redress
the heat and power imbalance which can be a very costly on in terms
of imported electricity charges.

Schemes of this nature all involve the use of turboexpanders coupled
to generating equipment. There is currently a fair level of
interest in organic rankine cycles to perform this type of duty;
however, the power output of these units is generally less than 1MW
and for large scale operations this represents a marginal contribution.
The installation of such a system does not currently appear to be
economic on a chemical manufacturing site even for larger machines
generating around 5 MW. In fact, the economics of scale tend to favour
smaller units in this case where advantage can be taken of
relatively large production volumes and standardised components.

In spite of the drawbacks, this method of utilising waste heat is
currently nearer to economic viability within the chemical industry
than that of heat pumps.

There is also some scope for power generation within processes
themselves usually those involving large volumes of off-gas or
large pressure let-downs, again by using turboexpanders.

The developments necessary to bring this type of scheme into
economic viability are further discussed in section 6.

6.   <u>Recent Studies and Future Development Requirements</u>

A recent study on an ICI plant indicated that the achieveable savings
utilising heat pump technology (in this particular case direct mechanical
vapour recompression) amounted to about 14% of the total savings
achievable by retrofit projects. The total energy saving package was
formed from a number of independently definable modification projects
which varied in payback time from around 7 months to just over two years.
The project containing the MVR proposal was at the upper end of this
time scale thus demonstrating that there are as yet significant economies
to be made without resorting to 'complex' technical proposals. The short
pay back projects in this instance all involved improvements in energy
recovery by low risk modifications to the plant, utilising established
technology.

The main developments required in improving the viability of turboexpander
schemes and heat pump schemes lie in the same area. Firstly the
development of compact heat exchangers which not only give large heat
transfer surface per unit cost but also allow closer temperature driving
forces to be utilised. Secondly, the development of turbines and
compressors in standard unit sizes of a capacity suitable for most
chemical industry applications. Current work in both rankine cycles
and heat pumps is aimed at marketing machines of 1MW or less capacity or
power requirement. As stated previously larger machines will be
necessary, around 5MW, if they are to become a viable proposition
within the chemical industry. Off the shelf bolt-on packages are the
most promising way in which systems of the aforementioned types will ever
make any impression on the existing plant market.

Further areas for development of energy saving technology lie in the areas
of integrating gas turbines into process plant. The advantage of
this form of heat engine is that the heat to power ratio is biased in
favour of power production thus alleviating the present imbalance.

The waste heat in flue gas amounts to a considerable quantity of
available energy. In order to reduce stack temperatures further
the development of ceramic heat transfer surfaces capable of resisting
acid condensation is required. Such a development would also be
beneficial in improving the efficiency of oil fired turbines for
combined heat and power generation.

References

1.   The Preliminary Design of Networks for Heat Exchange by Systematic
      Methods D Boland and B Linnhoff
      The Chemical Engineer, April 1979, Pg 222-228

2.   Energy Conservation in the Chemical and Process Industries
      C D Grant   I Chem E / Godwin 1979

3.   An Assessment of the Future Applicability of Heat Engines and Heat
      Pumps in the Process Industry
      D Boland, J C Hill and D W Townsend Paper presented at the
      Symposium Heat Pumps - Energy Savers for the Process Industry,
      Salford University 7th-8th April 1981.

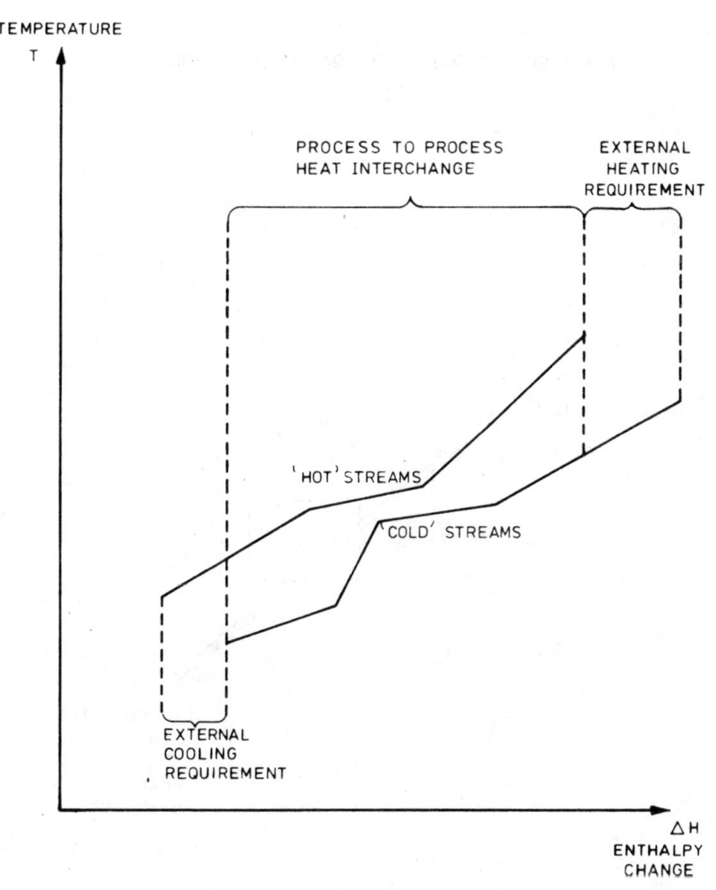

COMPOSITE CURVES FOR TYPICAL PROCESS

Fig I

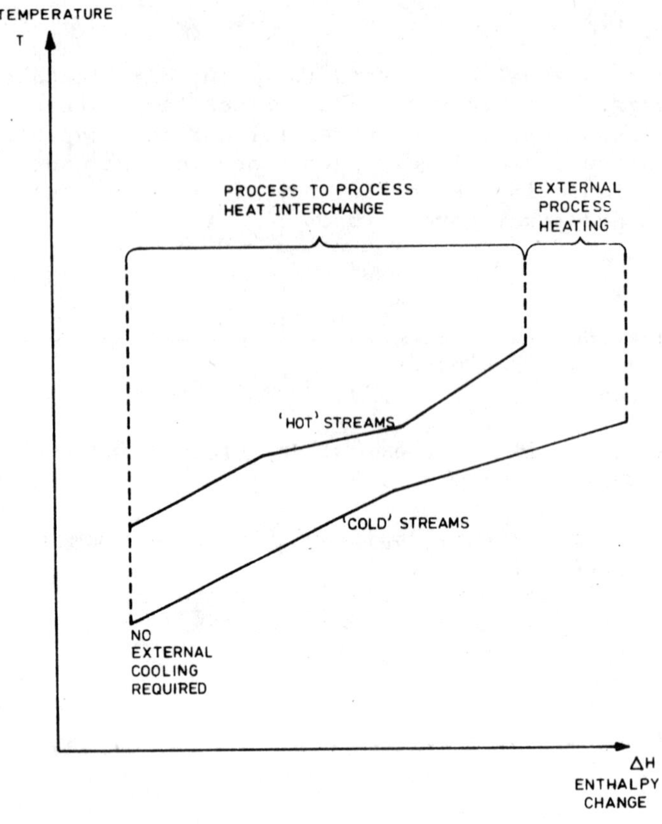

COMPOSITE CURVES FOR BALANCED PROCESS
Fig 2

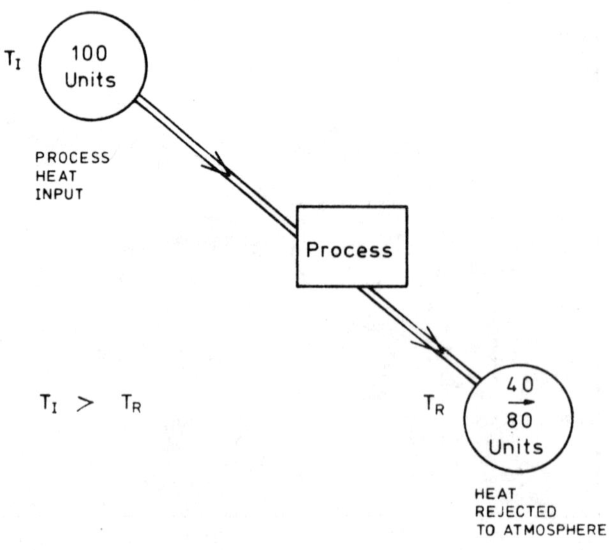

HEAT REJECTED TO ATMOSPHERE
Fig 3

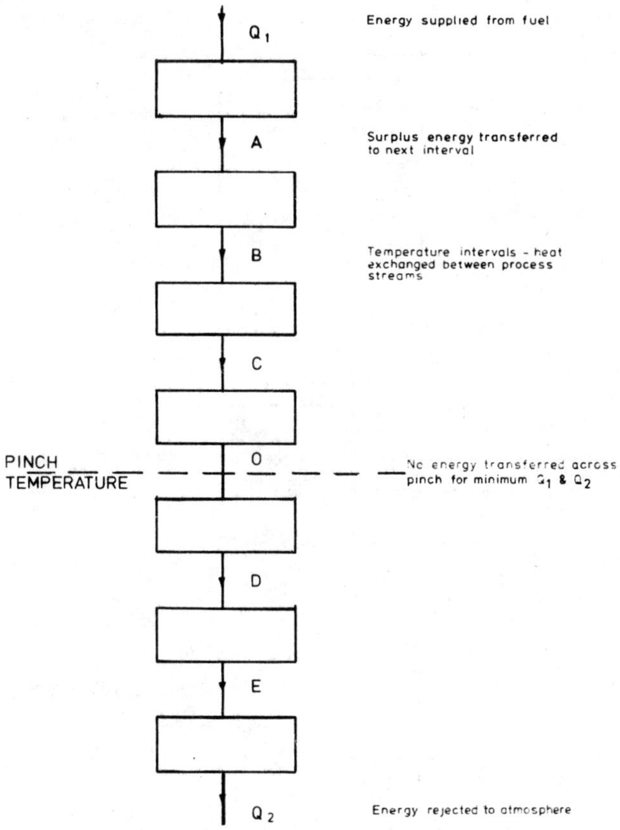

ENERGY CASCADE DIAGRAM
Fig. 4

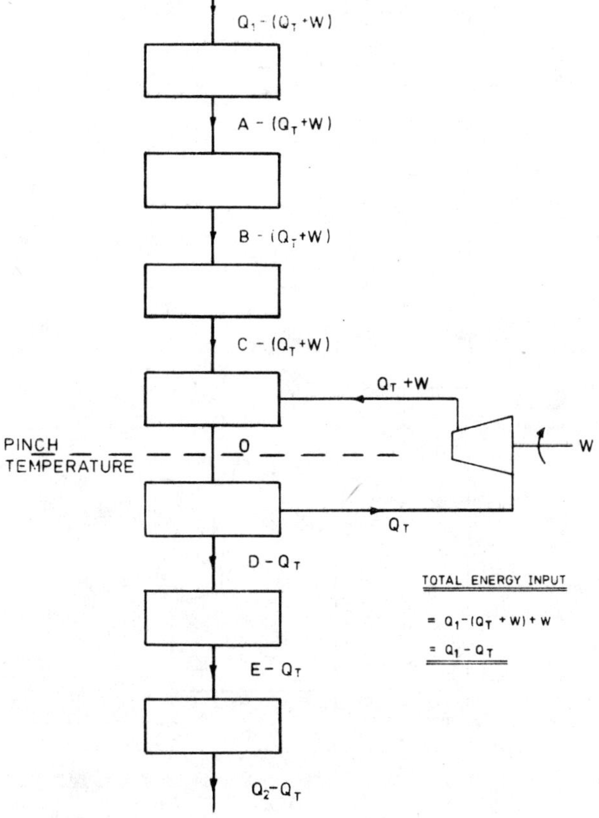

HEAT PUMP OPERATING ACROSS PINCH
Fig. 5

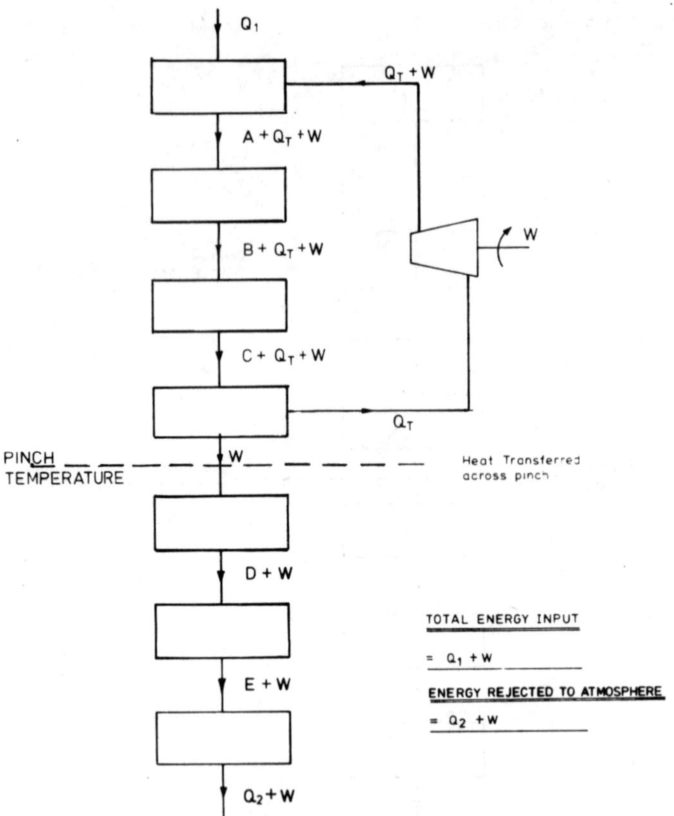

HEAT PUMP OPERATING ABOVE THE PINCH
Fig. 6 (a)

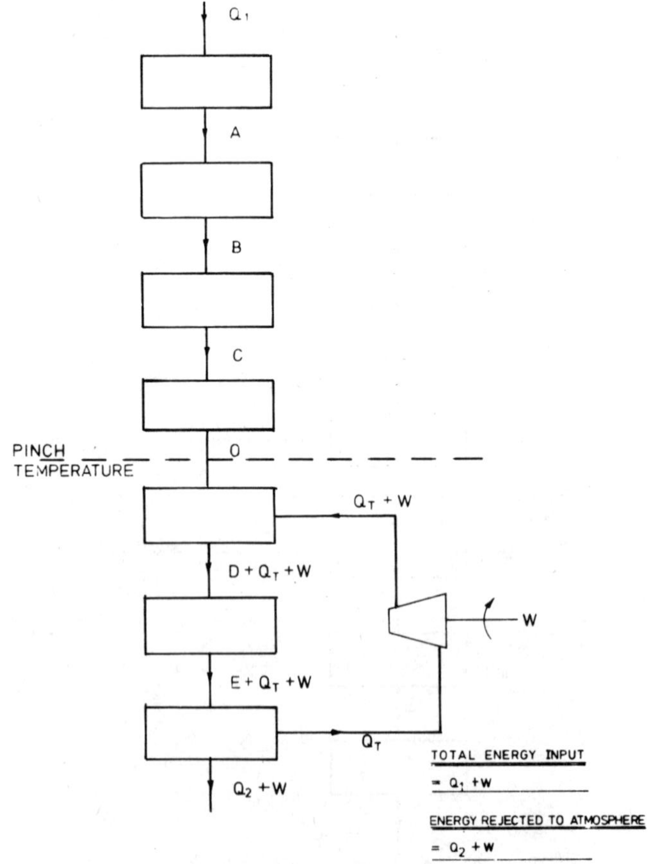

HEAT PUMP OPERATING BELOW PINCH
Fig. 6 (b)

HEAT PUMPS FOR THE PRODUCTION OF PROCESS STEAM

P. A. Kew

International Research & Development Co. Ltd., U.K.

Summary

The potential market for heat pumps in industry will increase greatly when systems are available capable of supplying process energy in the form of steam. Currently prototypes are under test with condensing temperatures of 120°C and producing low pressure steam. One such prototype is discussed in this paper. In order to attain higher temperatures significant advances must be made in working fluid, lubrication and compressor technology.

Held at the University of Warwick, U.K.
Symposium organised and sponsored by
BHRA Fluid Engineering

1.   INTRODUCTION

Although the concept of the heat pump has been known for over a century it is only
recently that rising fuel prices have led to a growing interest in heat pumps as
energy saving devices in process industries. At present few industrial heat pumps are
operational in the UK but as their usefulness and reliability are demonstrated in
existing applications the demand for them is likely to increase.

Improvements in the performance and operating range of commercially available
industrial heat pumps will increase the number of potential applications. One significant advance in this direction would be an increase in maximum condensing temperatures
to a level where steam could be produced by the heat pump.

Several problems are inherent in heat pumps with condensing temperatures above 100°C.
The majority of working fluids and lubricants used in low temperature heat pumps,
refrigeration and air conditioning units are unsuitable for use at the high temperatures involved, the limit of condensing temperatures using halocarbon refrigerants
with conventional compressors and synthetic lubricants is unlikely to rise much above
130°C.

Improvements in compressor, working fluid and lubricant technology are required before
condensing temperatures above 120°C can be attained with the long term reliability
necessary in an industrial application. The stability of working fluids in contact
with the lubricant and compressor materials presents the major obstacle.

Alternative cycles (especially absorption cycles) are being developed for high
temperature use, but their evolution is lagging behind vapour compression units and it
will be necessary to demonstrate that they are well developed technically before their
market potential can be fully evaluated.

2.   CLASSIFICATION

The classification of heat pumps and refrigeration machines by cycle is well known.
The majority of proposed industrial heat pumps are of the vapour compression type.
Absorption, chemical and gas cycles are all being considered for heat pump use, however, unless otherwise stated the heat pumps discussed in this paper are of the vapour
compression type operating on the reversed Rankine cycle.

Since different problems are involved in the design of heat pumps at different
operating temperatures it is also worthwhile to classify heat pumps in terms of their
condensing temperatures. Heat pumps condensing below 100°C can be referred to as low
temperature machines, between 100 and 130°C as high temperature machines and above
130°C very high temperature machines. The three ranges are illustrated in the staged
heat pump system proposed by Salford University (Fig.1)(1). This arrangement demonstrates
the way in which the three classes of heat pump differ and the ranges over which they
operate. The low and high temperature machines are conventional vapour compression heat
pumps, Using halocarbon refrigerants chosen to give optimum performance and stability
at the operating temperatures. The very high temperature unit is of less conventional
design, in this case an absorption cycle using water and a salt is proposed, however as
yet unused fluids might make the use of a vapour compression cycle feasible. It should
be noted that, even when the technical problems of the high and very high temperature
units are eliminated, it is unlikely that a device with a temperature lift of much more
than 80°C will be economically viable.

3.   MARKET POTENTIAL

It is very difficult to assess the potential market for industrial heat pumps
accurately, there are data regarding energy usage (2) and also data concerning the
effluents discharged by sections of industry available, the difficulty arises when
linking the two sets of data to assess the extent of coherent effluent supplies
(sources) and energy demands (sinks) where heat pumps would be viable. One cannot
even look at single processes which require heat and produce effluent since it will
often be the case that one process will produce effluent while another provides the

sink. Similarly in an integrated plant high temperature effluent from one process
will provide heat to a lower temperature process. An example of the latter being the
petrochemical industry where operations exist which use and reject heat over a large
temperature range, but, by using an integrated plant approach, it is possible to
gradually downgrade heat through a series of processes, the relative demands are such
that there is in fact a surplus of steam at temperatures up to 130°C in many petrochemical plants (3). The only market for heat pumps in the petrochemical field is thus
in the very high temperature range.
In other industries the situation is slightly different and the potential market for
heat pumps of different condensing temperatures is probably closely proportional to the
amount of energy used at the relevant temperatures. Table 1 shows the energy demand
by UK industries (excluding Petrochemicals) for energy at various temperatures (4).
It should be noted that much of the comparatively low grade energy used(80°C)is supplied via steam coils to the process, thus favouring steam producing heat pumps.

TABLE 1 - Energy Use for Low Temperature Process Heating (4)

Supply Temperature $^{\circ}$C	80	80-120	120	200
Heat Supply Medium	Hot Water/Steam			Direct Fired
Energy used kWh x 10^9	26.7	73.5	54.2	23

In addition the industrial and commercial space heating requirement is 210 x 10^9kWh/
year.

It can be seen from Table 1 that while currently available heat pumps have a potential
market of 26.7 x 10^9kWh/year plus the space heating duty high temperature heat pumps
would extend this potential market by 127 x 10^9kWh/year while very high temperature
heat pumps would add a further 23 x 10^9kWh/year to the market. Estimates of the market
penetration which will be achieved differ markedly, varying from 20% (4) to 2.5% (2).
In either case the adoption of heat pumps capable of producing steam at up to 120°C
would increase the number of potential applications (excluding space heating) by a
factor of 4, even if one neglects the operational advantages of producing steam which
can be used to supplement an existing steam main supplying a plant rather than supply
a single process.

4.   DESCRIPTION OF A STEAM PRODUCING HEAT PUMP

International Research & Development Co Ltd (IRD) has designed, constructed and
conducted preliminary tests on a gas engine driven, high temperature industrial heat
pump which is capable of producing steam at 110°C while drawing energy from a source
of liquid effluent at $60-80^{\circ}$C (5)(6).

A circuit diagram of the prototype installation is shown in Fig.2. The specification
of the unit is given in Table 2, together with the performance measured during
preliminary trials.

An outline feasibility and design study (5) recommended that the prototype would have
a 75 kW drive provided by a reciprocating spark ignition gas engine powering a wet
reciprocating compressor. The heat pump was designed to give a thermal output of
356 kW with an evaporating temperature of 60°C and a condensing temperature of 120°C
when removing heat from a water source at 80°C and delivering saturated steam at 110°C.
The type of compressor and engine chosen was dictated by the size of the unit, a screw
or centrifugal compressor might be utilised in a larger system. At higher shaft
powers (above 500 kW) a dual fuel engine could prove viable in some installations.
The main problem in the design of the heat pump was the selection of working fluid and
compressor lubricant and the design of the engine cooling water supply.

The ideal working fluid would be one which had a critical temperature well in excess
of 120°C, operates through a cycle which is tolerable in a typical compressor, the

main limitations of which are summarised in Table 3 and Fig.3, these lead to the constraint that the vapour pressure must be below 22 bar absolute at 120°C, and several other desirable properties summarized below.

TABLE 2 - <u>Performance of IRD High Temperature Heat Pump</u>

	Actual	Design
Gas power to engine	221 kW	242 kW
Steam power from system	333 kW	356 kW
Steam flowrate	0.141 kg/s	0.151 kg/s
Primary energy ratio (PER)	1.50	1.47
Engine RPM	1080 r/min	1100 r/min
Shaft power from engine	68 kW	75 kW
Heat pump coefficient of performance (COP)	3.12	3.19
Heat recovery from engine and exhaust	121 kW	117 kW
Heat lost from engine and exhaust	32 kW	50 kW
Percentage energy recovery from engine & exhaust	86%	79%
Refrigerant flowrate	2.54 kg/s	3.01 kg/s
Condensing temperature	120°C	120°C
Compressor discharge temperature	126°C	125°C
Evaporating temperature	60°C	60°C
Steam outlet temperature	110°C	110°C
Effluent inlet temperature	80°C	80°C
Effluent outlet temperature	65°C	65°C
Effluent flowrate	2.3 kg/s	2.6 kg/s
Heat removed from effluent	145 kW	164 kW
Compressor oil differential pressure	1.5 Bar	1 to 2 Bar

TABLE 3 - <u>Limits of Operation of Typical
Reciprocating Heat Pump Compressor</u>

Maximum evaporating temperature	66°C
Minimum evaporating temperature	-24°C
Maximum discharge temperature	140°C
Maximum discharge pressure	22 Bar abs
Maximum pressure difference	18.5 Bar
Maximum pressure ratio	10
Minimum pressure ratio	1.5
Maximum speed	1750 RPM

The fluid must have good thermal stability at all temperatures encountered in the
cycle. Inspection of the pressure-enthalpy diagram for the reversed Rankine Cycle
(Fig.4) shows that other desirable properties are a small increase in temperature
during compression at constant entropy, high latent heat and low specific volume at
the compressor suction condition. The first of these properties ensures a low
discharge temperature while the latter two result in a low compressor swept volume
being required. Depending upon the application flammability and toxicity will have
varying importance attached to them.

A compromise between these ideal properties is obviously necessary however the cycle
would not work successfully with the fluid in a supercritical state therefore working
fluids with low critical temperatures were immediately dismissed, these include R12,
R22 and R502 three of the most commonly used refrigerants. Some of the potential
working fluids remaining are listed in Table 4.

TABLE 4 – <u>Suitable Refrigerants with Respect to Critical Temperature</u>

ASHRAE Ref. No.	Refrigerant Name	Chemical Formula	Critical Temp $^{\circ}$C	Critical Pressure MPa
717	Ammonia	NH_3	133	11.3
600a	Isobutane	C_4H_{10}	134	3.5
40	Methyl chloride	CH_3CI	143	6.6
114	Dichlorotetrafluoroethane	$CCIF_2CCIF_2$	146	3.3
600	Butane	C_4H_{10}	152	3.8
764	Sulphur dioxide	SO_2	157	8.0
630	Methylamine	CH_3NH_2	157	7.8
21	Dichlorofluoromethane	$CHCI_2F$	178	5.1
631	Ethyl amine	$C_2H_5NH_2$	183	5.6
160	Ethyl chloride	C_2H_5CI	187	5.2
11	Trichlorofluoromethane	CCI_3F	198	4.4
611	Methyl formate	$C_2H_4O_2$	214	4.2
113	Trichlorotrifluoroethane	CCI_2FCCIF_2	214	3.4
1130	Dichloroethylene	$CHCI=CHCI$	243	5.5
30	Methylene chloride	CH_2CI_2	249	4.6
1120	Trichlorethylene	$CHCI=CCI_2$	271	5.0
610	Ethyl ether	$C_4H_{10}O$	272	2.6
718	Water	H_2O	374	22.0

Fluids with discharge pressures in excess of 21 bar abs and highly flammable fluids
were then eliminated. The fluids which were considered to be worthy of further study
were Refrigerants 11, 21, 113 and 114 and water, of these R21 is highly toxic and water
has a very high discharge temperature (412°C) and an extremely high volumetric flowrate.

Stability of the working fluid in contact with the compressor components and lubricant
is of crucial importance if the heat pump is to be reliable. Table 5 gives the
maximum cycle temperatures to be encountered and the maximum recommended temperature
for continuous exposure for the three potential fluids.

The slope of the R114 of saturated vapour isentropes is such that isentropic
compression will result in liquid droplets forming, it is therefore necessary to use

an intercooler to superheat the compressor suction vapour. Ideal compression from the suction conditions will result in saturated vapour leaving the compressor, the 5°C superheat predicted is due to the anticipated inefficiencies of compression.

TABLE 5 - Stability of Heat Pump Candidate Working Fluids

Working Fluid Ref.No.	Maximum Cycle Temperature $^{\circ}$C	Maximum Recommended Operating Temperature $^{\circ}$C
11	141	107
113	133	107
114	125	121

It is worth remembering that the maximum recommended operating temperatures are mainly derived from sealed tube decomposition data and the results of these tests may differ markedly from in service decomposition rates due to differences in lubricants and materials present (7).

Of the three fluids R114 has the lowest volumetric flowrate at the condition specified and is also operating at only 4°C above its recommended maximum temperature, for this reason it was chosen as the prototype working fluid. A similar conclusion has been drawn by other workers (8) but both R113 and R11 should also be considered as candidate fluids for high temperature heat pumps, especially if their stability can be enhanced with highly refined symthetic oils (9).

Because the compressor suction pressure of a heat pump operating with R114 at an evaporating temperature of 60°C is relatively high (5.7 bar abs) there is a tendency for the refrigerant to dissolve in the compressor lubricating oil. This fact, together with the high temperature at which the oil operates leads to significant problems with a dilution and hence viscosity reduction of the oil and frothing as the refrigerant leaves the solution. Fig.5 shows the relationship between viscosity and temperature and pressure for two oils in the presence of R114 (9). Synthetic oils are now produced for high temperature heat pump use which do much to overcome this problem and also reduce the decomposition of the lubricant-refrigerant mixture. Tests with the IRD prototype have shown that frothing is significant when the heat pump is running at its design conditions but causes no operational problems, however during transient conditions when the suction pressure is reduced rapidly this frothing can lead to excessive oil carry over and hence cause the sump oil level to be lowered to an unacceptable level. Longer term tests are required to assess the stability of the oil and R114 combinations during continuous operation.

When using an engine to drive a heat pump compressor it is highly desirable to recover the heat rejected in the engine cooling jacket and exhaust and add this to the useful heat supplied by the heat pump condenser. In the case of a low temperature heat pump where energy is supplied to a fluid stream by sensible heat it is normal to put the condenser, engine jacket and exhaust boiler in series so that, for a given system outlet temperature the condenser leaving temperature and hence condensing temperatures are lowered. When steam is produced approximately 95% of the heat is supplied as latent heat at the saturation temperature of the boiling water, it is therefore necessary to recover heat from the engine and exhaust boiler at or above this temperature. Use of a two phase engine cooling system results in the production of steam either in the engine cooling jacket passages or in the pipework leading from the engine.

Water is supplied to the engine at or slightly below its saturation temperature corresponding to the coolant pressure. As the coolant passes through the waterways in the engine cylinder head heat is transferred to the fluid, however, rather than resulting in a temperature rise, the majority of the heat is absorbed as latent heat of evaporation as the water changes phase. Since there will inevitably be a pressure

drop in the pipework between the engine and the steam separator the saturation
temperature will drop and a proportion of the water will 'flash' to steam after
leaving the engine.

It is essential that all engine coolant passages have water or very wet steam flowing
through them at all times. If any portion of the cooling system dries out the heat
transfer coefficient will fall resulting in a dramatic rise in the cylinder head
temperature in the region of the dry passage. Water may be supplied to the engine
by means of a pump (or pumps) or through natural convection. If a pumped system is
used a large excess of water must be supplied to the engine and the coolant supply
and outlet manifolds must be carefully designed to ensure a good flow distribution in
the engine. A properly designed free convective system will automatically maintain
the coolant passages in a wet condition, the upward leg will be almost full of steam
while the downward leg supplies water. With a 1.2 metre head the static pressure
change in the upward leg is approximately 12 kN/m^2 which corresponds to an elevation
in saturation temperature of approximately 2.5^{o}C thus if the flowrate is sufficiently
high a substantial proportion of the phase change will occur in the upward leg of the
convective system.

Figures 6, 7 and 8 show alternative cooling arrangements to provide water to the
engine ebullient cooling system and exhaust boiler. Fig.6 illustrates a free convective
system to the engine, the exhaust boiler either being incorporated in the free convective
loop or supplied separately by a pump. A single pump may be used to provide the water
supply to the engine and exhaust boiler, as shown in Fig.6, but this necessitates use of
a very large excess of water to ensure complete wetting of all heat transfer surfaces.
Use of these pumps, supplying the exhaust manifold, cylinder block and heads and exhaust
boiler maintains a more even flow distribution thus permitting a lower water circulation
rate, the arrangement shown in Fig.8 utilises this configuration.

Steam raising from the exhausts of gas turbines and large diesel engines (greater than
1500 kW) in marine use is established practice, however it is rare to recover heat in
the form of steam in smaller installations, and no suitable exhaust boiler was
commercially available and units are unlikely to be available in the sizes required
for heat pump use unless they are specifically designed for this purpose. The
boiler used for the IRD prototype was designed for the use and is of finned tube in a
cylindrical shell with the exhaust gases flowing over the fins while water boils in
the tubes, it is illustrated in Fig.9.

The heat pump evaporator is of the shell and tube direct expansion type with refrigerant evaporating in the tubes, the flowrate being controlled by a thermostatic
expansion valve and as such differs little from a low temperature heat pump evaporator.

Since water is boiling in the condenser it is necessary to incorporate a steam
separator in the condenser shell, this also acts as a steam separator for the very
wet steam returning from the engine and exhaust boiler. Because of this and the non-
availability of heat exchangers designed for use with R114, the evaporator and
condenser were purpose designed. The condenser is mounted above the evaporator to
minimise the floor space required.

Figs.10 and 11 show the assembled prototype heat pump. An industrial unit would
require a smaller floor area, the prototype was built in such a way as to allow ample
access to the engine and heat exchangers during testing.

It was concluded from the design study and test programme on the high temperature heat
pump described above that it is feasible to build and run a gas engine driven high
temperature heat pump substantially using commercially available components, the novel
aspect of the design being the integration of these components in a high temperature
heat pump. The payback period of a high temperature system would be similar to that
for a low temperature unit operating over the same temperature difference.

   FURTHER DEVELOPMENTS

Most of the problems encountered in the design and operation of high temperature heat
pump systems are similar to those which would be expected with low temperature units
and development is   required in the fields of system control and optimisation over
a large range of operating conditions, compressor and heat exchanger design and
reliability. Areas which particularly require further research in the high
temperature case are the stability of conventional working fluids and investigation of
potential new working fluids and compressor lubricants.

Before steam producing heat pumps can be marketed commercially it is necessary to
demonstrate the reliability of prototype machines in industrial processes.

6.   CONCLUSIONS

Existing technology can be adapted for use in heat pumps producing steam at
temperatures up to $110\text{--}115^\circ C$. The main obstacle to the introduction of these machines
is the lack of operating experience acquired with them.

It will require considerably more development effort to extend the maximum operating
temperature of heat pumps above $120^\circ C$, mainly since the working fluids which have been
used in low temperature heat pumps and refrigeration units are unsuitable.

REFERENCES

1. Holland, F.A.:   "What is a Heat Pump?". Heat Pumps Energy Savers for the Process
                     Industries. Institution of Chemical Engineers. North Western
                     Branch, Symposium Papers No.3, 1981.

2. Currie, W.M. et al: "Heat Pumps in the UK". ETSU Note N-2/81.

3. Townsend, D.W. et al: "The Future of Heat Engines. Heat Pumps in the Process
                     Industries". Heat Pumps Energy Saves for the Process Industries
                     Institution of Chemical Engineers, North Western Branch,
                     Symposium Papers No.3, 1981.

4. Masters, J., Pearson,J., Read, M.A.: "Opportunities for Gas Engine Driven Heat
                     Pumps in Industry and Commerce". British Gas Corporation,
                     Midland Research Station Communication 1129.

5. Macmichael, D.B.A. and Reay, D.A.: "Feasibility and Design Study of a Gas
                     Engine Driven High Temperature Industrial Heat Pump". IRD
                     Research Report 78/19. Under EEC Contract 175-77 EEUK.

6. Kew, P.A.:   "Design Construction and Testing of a Gas Engine Driven High
                     Temperature Industrial Heat Pump". IRD Research Report 81/29
                     under EEC Contract 369-78-3 EEUK.

7. Gainer, G.C., Luck, R.M., Moreland, W.C.: "Thermal Stability and Materials
                     Compatibility Tests on Five High Temperature Working Fluid
                     Candidates". Westinghouse Electric Corporation. Topica;
                     Report 1978.

8. Degueurce, B.   "Etude et Mise au Point D'un Compresseur a Pistons Lubrifies
                     Pouvant Equirer une Pompe a Chaleur a Temperature de
                     Condensation de $120^{o}C$". Electric de France P.33/4300/78.11
                     March 1978 (in French).

9. Duffield, J.S., Hodgett, D.L. "The choice of Working Fluid for Heat Pumps
                     Condensing at up to $120^{o}C$". XV International Congress of
                     Refrigeration, Venezia 1979.

10. Stirling, R., Drakesmith, F.G.: "Refrigerant 114: Viscosity and Solubility
                     Measurements in Hydrocarbon and Synthetic Lubricants". XV
                     International Congress of Refrigeration, Venezia, 1979.

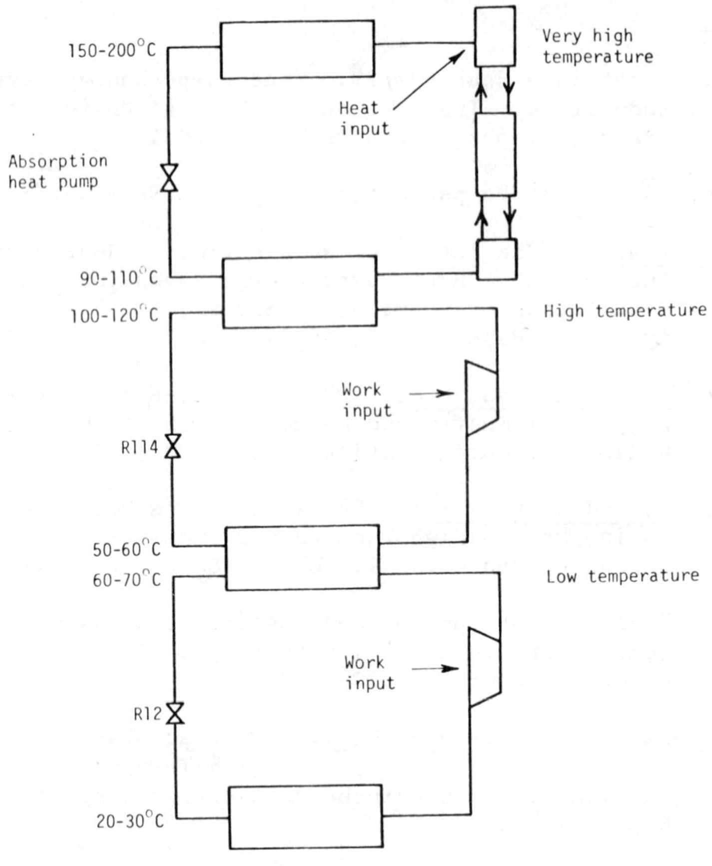

FIG.1 STAGED HEAT PUMP SHOWING CLASSIFICATION BY TEMPERATURE

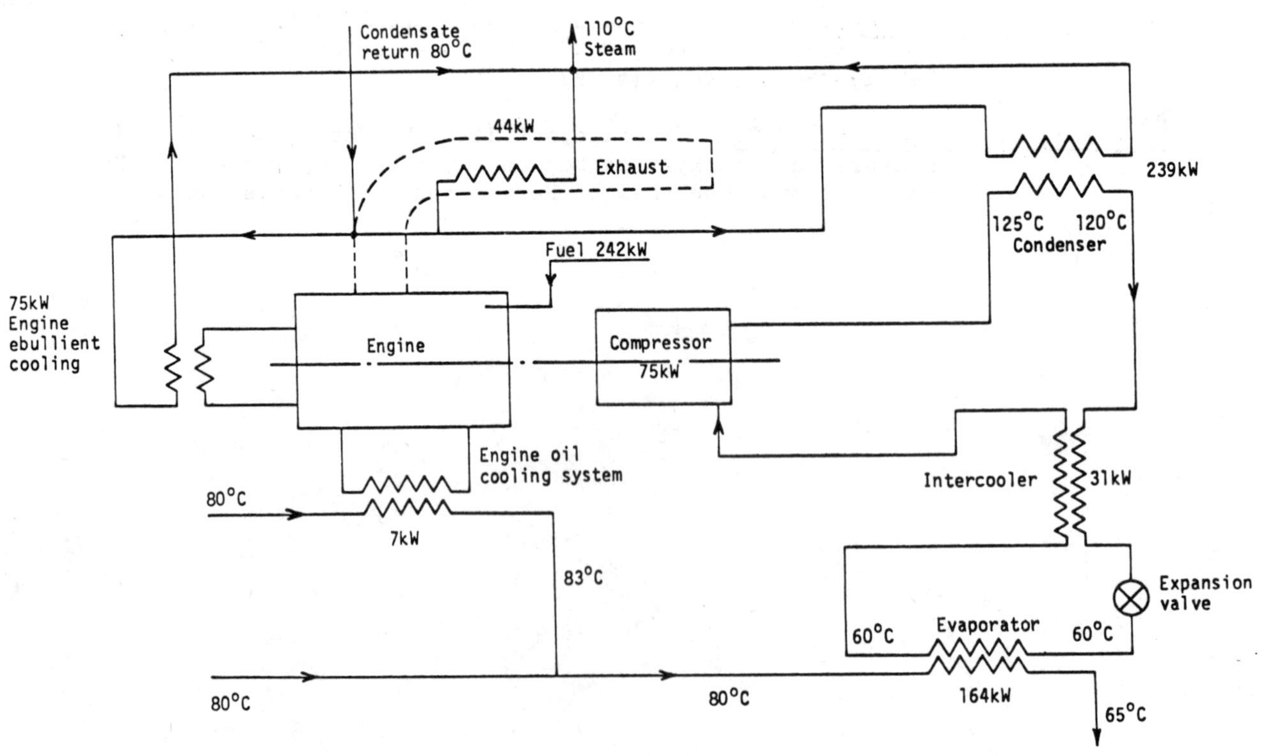

FIG.2 SCHEMATIC OF THE IRD GAS ENGINE DRIVEN HIGH TEMPERATURE HEAT PUMP

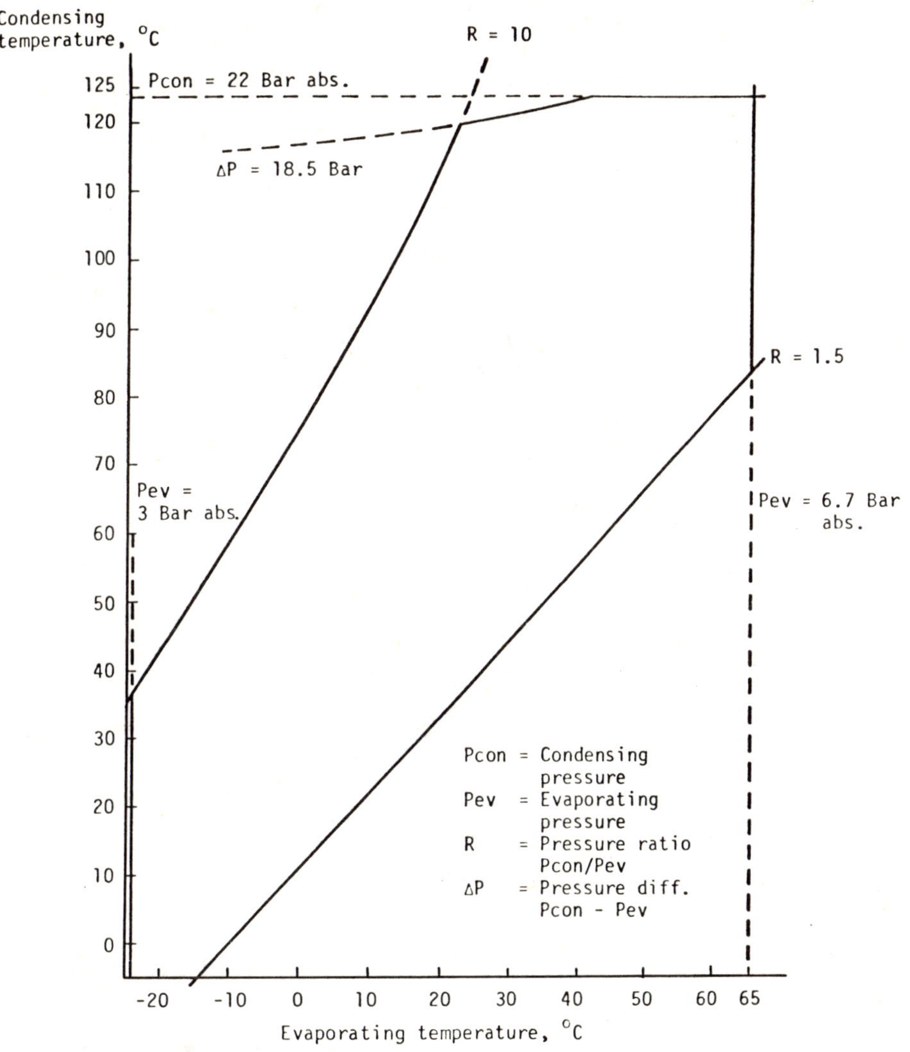

FIG.3 FIELD OF OPERATION OF RECIPROCATING COMPRESSOR

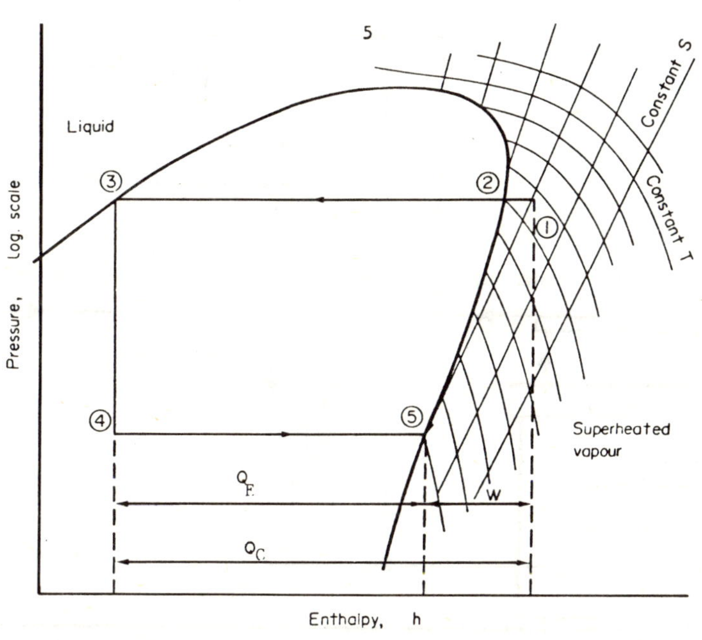

FIG.4 THE IDEAL VAPOUR COMPRESSION CYCLE ON A PRESSURE-
ENTHALPY DIAGRAM

217

FIG.10 HIGH TEMPERATURE HEAT PUMP - HEAT EXCHANGERS

FIG.11 HIGH TEMPERATURE HEAT PUMP - ENGINE AND COMPRESSOR

INDUSTRIAL APPLICATIONS OF A RANKINE POWERED HEAT PUMP FOR THE GENERATION OF PROCESS STEAM

W. F. Koebbeman

Mechanical Technology Incorporated, U.S.A.

Summary

This paper describes a Rankine-powered heat pump (RPHP) being developed for the generation of industrial process steam. The RPHP system utilizes waste heat to provide the latent heat of vaporization for the process steam and, as a source of energy, to boost that steam to the required pressure. Key components are the organic Rankine cycle (ORC) turbine power unit (TPU) and the centrifugal steam compressor with integral speed increaser. The TPU which drives the steam compressor derives its power through a classical Rankine cycle utilizing Trichlorotrifluoroethane (Refrigerant 113) as a working fluid.

Also included in the paper is an application study conducted to determine the system's technical and economical viability in the industrial arena. The food, paper, chemical, and petroleum industries were explored, with chemical and petroleum processing emerging as the best market areas. The methodology utilized to select the design basis for the development unit from the application study data is described.

Held at the University of Warwick, U.K.
Symposium organised and sponsored by
BHRA Fluid Engineering

<u>NOMENCLATURE</u>

RPHP = Rankine-powered heat pump

ORC = organic Rankine cycle

TPU = turbine power unit

DCF = discounted cash flow

Pressure = All pressures are given as gauge pressure.

1. INTRODUCTION

The work described in this paper was performed under contract with the United States Department of Energy. The objective of the project was to develop a system for the generation of useful process steam from industrial waste heat. The system utilizes waste heat (which is too low in temperature to generate useful steam through direct heat exchange) by heat pumping to the required temperature. The application study described here was performed to determine the potential for energy savings through this technology and the size of system that should be developed to best meet the needs of industry. A development unit is presently under construction.

2. THE RANKINE-POWERED HEAT PUMP SYSTEM

2.1 System operation.

The Rankine-powered heat pump (RPHP) converts low temperature heat (100-200°C) to usable process steam. In the RPHP system, the resource temperature, normally industrial waste heat, is raised through the addition of mechanical work. An organic Rankine cycle (ORC) turbine extracts energy for this mechanical work from the heat source to drive the heat pump compressor. Fig. 1 illustrates the operation of a RPHP system. The heat stream first passes through the water evaporator to generate low pressure steam, which is then compressed by a centrifugal compressor to the desired pressure. Steam is discharged at superheated conditions and may be desuperheated at the user's option. The highest temperature is utilized in the water evaporators, minimizing the steam compressor pressure ratio and also the power required by the Rankine drive. After the water evaporator, the waste heat passes through the refrigerant evaporator and into the refrigerant preheater. There, the Rankine cycle working fluid (Refrigerant-113) is heated from condenser conditions to saturated liquid at evaporator conditions. The refrigerant evaporator vaporizes the working fluid, which is then expanded through the turbine to produce the required compressor power. The turbine exhausts into the condenser, where the refrigerant vapors are condensed to liquid and returned to the hotwell to repeat the cycle.

2.2 System description.

The RPHP system consists of two separate modules: the heat exchanger module and the power module. The heat exchanger module consists of three shell and tube heat exchangers: the water evaporator, the refrigerant evaporator and the refrigerant preheater. All three units are connected in series on the tube side, which contains the waste heat stream. Waste heat enters the tube side of the water evaporator and evaporates fresh water on the shell side. Then, vapor separation is achieved by gravity in the open volume above the tube bundle. Vapor generated in the water evaporator is piped to the compressor inlet.

The power module consists of the steam compressor and the turbine power unit. The steam compressor is a centrifugal unit, driven through an integral speed increaser. Pressure ratios of 1.8:1 can be achieved with a single stage. For pressure ratios to 3.6, a twin pinion model is available, which offers two stages from a single driver. The turbine power unit (TPU) shown in Fig. 2 is a mechanical-drive turbine designed for use in an organic Rankine cycle system employing Refrigerant-113. The turbine is designed to accommodate the range of applications shown in Fig. 3. The applications envelope shown was developed out of the applications study described in the following sections of this paper.

The drive turbine is an axial-flow type with an exhaust diffuser. High turbine efficiency is maintained over the range of potential turbine design conditions by changing the turbine flow path components. Changes which can be readily accommodated within the TPU include: number of turbine stages (either one or two), blade and nozzle height, use of partial arc admission, and turbine operating speed. In addition, the turbine casing can accommodate up to three separate inlets to allow for partial arc admission at off-design conditions.

The TPU utilizes the process fluid for lubrication and cooling of the output shaft and all bearings, thus eliminating the use of a second fluid in the system and the potential for system contamination. The liquid used in the seals and bearings is supplied by the system condensate pump. A filter is incorporated in the supply line.

The output shaft seal is a double mechanical face seal. The radial bearings are five pad tilting shoe and the axial bearings are self-equalizing tilting pad, which have been specially designed for operation in liquid refrigerant.

The main shaft and casing contain provisions for adding the main system feed pump if required by the application.

The system controls are designed for unattended operation. A programmable controller provides automatic start-up and monitors critical system parameters during normal running. Automatic shutdowns are preprogrammed for specific upset conditions.

3.0 INDUSTRIAL APPLICATION SURVEY AND EVALUATION

The objective of this survey and evaluation was to determine that significant numbers of RPHP applications do exist and to identify a "best" application to be used as a design basis for the development unit. The methodology for identifying and evaluating viable heat pump applications is shown in Fig. 4. The procedure is a multilevel screening process designed to filter out the best applications for the application studies. Six sites were selected as candidates for on-site plant visits based on Level 1 and Level 2 threshold requirements. Based on the information obtained during the site visits, a single site was selected for the design basis for a development unit.

3.1 Preliminary data collection and screening.

Screening criteria for the applications were grouped into three levels to expedite the screening procedure by eliminating unpromising sites at the earliest possible date. This procedure allowed concentration of effort on those sites which showed the most potential.

Level 1 criteria apply to information that is easily obtained from secondary sources.

Level 1 Criteria

Heat source type	– condensing or liquid stream
Heat source size	– >21 x 10^6 kJ/hr
Heat source temperature	– >85°C
Heat source duty cycle	– >4000 hr/yr
Steam usage	– >11,340 kg/hr of steam below 10.3 bar

Level 2 criteria apply to information which requires contact (by questionnaire and telephone) with plant personnel familiar with the process. Special material requirements, steam usage (flows and pressure levels), steam value and fuel cost criteria are involved at this level.

Level 2 Criteria

Quantity of replaceable steam	– >4536 kJ/hr
Value of replaced steam	– >$1.90/$10^6$ kJ
Estimated discounted cash flow (DCF)	– >15%

Level 3 criteria evaluate the data obtained during visits to plant locations for details affecting installation and further data refinement. Six plants were visited, then graded and ranked.

Level 3 Criteria

Adequate space for installation
Low risk to process and production
Management enthusiasm about project
DCF based on: heat source
                   steam usage
                   installation data

Companies were initially contacted by telephone to gauge interest level and available waste heat sources. If a potential application existed, a follow-up letter was mailed.

Contact was made with 61 companies for Level 1 data; of these, 13 returned data sufficient for a Level 2 screening. Table 1 shows the breakdown by industry of candidates
for the various levels of screening. Ten of the largest food processing companies were
contacted but no waste streams that met the Level 1 criteria were found. (The lack of
applications in the food industry is due to the relatively low process temperatures involved and small waste heat streams available.) The results of the Level 2 application
data collection and screening are shown in Table 2. A variety of heat source streams
were received that varied in size from 22 to 158 x 10^6 kJ/hr and from 100 to 195°C in
temperature. Average steam value is \$3.88/10^6 kJ; 10 out of 13 applications require a
two-stage steam compressor. A single RPHP turbine power unit can cover the entire range
of applications with only turbine wheel and nozzle modifications to maintain good efficiency. DCF analyses were run based on this preliminary data; they varied from 20%
to 45%.

3.2 In-plant data collection and screening.

As a result of the Level 1 and Level 2 screening, six companies were selected for in-
plant visits. Application K was eliminated from consideration when the plant identified
a direct heating use for their waste heat source. The Level 3 screening of the other
five candidates is shown in Table 3. Based on all criteria, Application E emerged as
the design basis for the development unit.

3.3 Application engineering study.

A case study was conducted on Application E using refined data collected during the
plant visit.

The design basis heat pump concept incorporates the use of an organic Rankine cycle to
drive a heat pump compressor (see Fig. 1). The waste heat stream (contaminated steam)
enters the water evaporators at 116°C (saturated vapor) and leaves at 116°C (saturated
liquid) giving up 18.4 x 10^6 kg/hr to generate 8437 kg/hr of steam at 1.5 bar. This
steam is then compressed to 2.8 bar.

The waste heat stream enters the refrigerant evaporator/preheater assembly at 116°C and
leaves at 86°C giving up 11.5 x 10^6 kJ/hr to generate refrigerant vapor at 102°C. The
vapor is expanded through a turbine and condensed at 40°C, generating 365 kW. The condensate is then pumped back to the evaporator to complete the regrigerant cycle. 5300
lit/min of plant cooling water are required to condense the refrigerant exhaust vapors.

The combination of steam system heat input of 18.4 x 10^6 kJ/hr and steam compressor work
of 1.3 x 10^6 kJ/hr brings an energy addition of 19.7 x 10^6 kJ/hr to the 1.72 bar plant
steam manifold. Steam at the compressor discharge is superheated to about 188°C. If
saturated steam is required, an additional 363 kg/hr of steam could be generated by de-
superheat spray.

3.3.1 Design basis system performance.

Overall Energy Balance (in millions of kJ/hr)

Heat extracted from process by regrigerant system	11.5	11.5
Heat rejected by regrigerant condenser		−10.2
Net energy delivered by refrigerant turbine		1.3
Heat extracted from process by RPHP steam vaporizer	+18.4	18.4
Total heat extracted from process		
Total recovered energy delivered to steam manifold	29.9	19.7

Compressor Design Parameters

Working Fluid	Steam
Inlet Temperature	111°C
Inlet Pressure	1.5 bar
Pressure Ratio	1.89
Stage Efficiency	79%
Mass Flow	8210 kg/hr
Discharge Temperature	168°C
Discharge Pressure	2.8 bar
Delivered Steam Energy	19.7 x 10^6 kJ/hr

Turbine Design Parameters

Working Fluid	R-113
Inlet Temperature	102°C
Inlet Pressure	3.6 bar
Efficiency	82%
Condenser Temperature	40°C
Shaft Power Out	365 kW
Rankine Cycle Efficiency	11%

12,790 kg/hr is the combined vented steam from 11 continuous and 3 batch evaporators and represents the maximum operating condition for the system. The system will deliver 19.6×10^6 kJ/hr at this condition. However, the plant cannot use this amount of steam continuously due to seasonal variations in steam usage. The yearly average steam output for the system will be 15.4×10^6 kJ/hr, and this value was used in the economic study.

3.3.2 Economic evaluation.

Based on the computer cost model, production equipment costs are estimated to be $870,000. This model is derived from historical cost data and has been verified by a detailed cost study. Installation costs, although highly site specific, will generally fall within a range of 50% to 100% of equipment costs. This analysis assumes 100% of equipment costs or $870,000 for installation. The total installed cost, therefore, is estimated at $1.74 million.

An economic evaluation indicated an energy rate of $5.48/$10^6$ kJ based on energy cost projections made by site personnel. Operating hours are based on 97% utilization. Other input assumptions are as follows:

- Energy and electrical rates will rise 10% per year.

- Annual operation and maintenance costs other than labor are based on 1% of equipment costs plus parasitic power requirements and cooling water costs. Operating labor is estimated at one hour per shift.

- The system is expected to have a useful life of 15 years and could be depreciated over a period of 12 years.

- Local property taxes and insurance will be 2% of the initial investment; the combined federal and state income tax rate will be 51%; the investment tax credit for energy conservation equipment will be 20%.

- In conservation projects such as this, the installed equipment generates savings by reducing the fuel bills. It is assumed these savings are reinvested at a conservative 6% interest rate.

- The initial investment includes the cost of the equipment plus 100% for site preparation, construction, tie-in, and start-up.

A basic financial investment analysis was made to evaluate this investment decision. The analysis showed that, for this investment, the DCF is 37% and the break-even point, 2.9 years. This analysis is based on the annual average output of 15.4×10^6 kJ/hr although the system has a maximum capacity of 19.6×10^6 kJ/hr. The 15.4×10^6 kJ/hr takes into account seasonal variations in the waste steam and clean steam usage.

4.0   CONCLUSIONS

A total of 61 companies were contacted in the chemical, petroleum, food and pulp and paper industries. From these, 13 economic applications were obtained from plant personnel. A good response was obtained from the chemical and petroleum industries (10 applications). Because they are the top energy users in the United States, have large waste heat streams, utilize low pressure steam, and have high annual operating times, the chemical and petroleum industries were identified as the principal markets for the RPHP. Three responses were obtained from the pulp and paper industry. Return on investment tends to be lower in the pulp and paper industry because waste heat stream temperatures are lower. The pulp and paper industry represents a secondary market. No responses were obtained from the 10 food companies contacted. There appears to be no potential for this type of heat pump in the food industry due to the lower temperatures and magnitudes of waste heat streams in this industry.

The following conclusions are based on the results of the site-specific application studies that were conducted.

- Substantial quantities of waste heat are available in the chemical and petroleum industries at temperature levels required for economic operation of the heat pump.

- Evaluation of replaceable quantities of low-pressure steam requires careful consideration of the plant steam balance. The replaced steam must eventually result in a boiler turndown to be a true savings. In some cases, because of back-pressure turbine power requirements, adding low-pressure steam through waste heat will only result in steam being vented or condensed at the turbine exhaust; however, the replacement of the back-pressure turbine by an electric motor can be an economically attractive alternative.

- Direct heat exchange possibilities should be exhausted prior to consideration of the Rankine-powered heat pump because of the relative economics. For the same heat savings, direct heat exchange will normally result in lower capital cost and, hence, better return on investment.

- At present steam values, which average $\$3.89/10^6$kJ, the RPHP was found to be economically attractive in all 13 of the applications which met Level 1 criteria.

- Of the applications, 3 out of 13 required a single-stage compressor; the rest required two stages.

- A single RPHP turbine power unit can cover the expected range of applications with only blade and nozzle modifications.

A nylon processing application was selected as the design basis for the development system based on the three-level screening process of applications. The principal factors involved in this selection were:

- Interested and enthusiastic personnel in plant engineering and management.

- An application which should yield a DCF of over 30% in follow-on production configurations.

- High value placed on low-pressure steam.

- Low risk to process.

- Clean, noncorrosive waste stream.

- Adequate space available for installation of equipment.

<u>TABLE 1</u>

<u>HIGH COP CANDIDATES BY INDUSTRY</u>

	Inquiry Lists Level 1	Tech-Econo Studies Level 2	Plant Visits Level 3	Application Engineering Study Design Basis
Chemical Processing	35	6	3	1
Refining	12	4	3	−
Pulp and Paper	4	3	−	−
Food Processing	10	−	−	−
Total	61	13	6	1

TABLE 2

RESULTS OF LEVEL 2 APPLICATION DATA COLLECTION AND SCREENING

Plant	Heat Source Stream Vapor/Liquid	Heat Available Over 65.6°C 10^6kJ/hr	Source Temp (°C)	Required Steam Pressure bar	Steam Value $/$10^6$kJ	Annual Operating Hours	Compressor Pressure Ratio	Rankine Cycle kW	Delivered Steam Energy 10^6kJ/hr	Installed Cost $ x 10^6	DCF (%)	Cash Flow B.E.P. (Yr)
A	Boiler Blowdown	37	182	8.3	4.74	7880 90%	2.66	531	17.1	1.5	37	2.9
B	Pentaerythritol Unit Corrosive Steam	85	108	3.1	4.44	8160 93%	3.33	671(2)	19.6(2)	2.0	30	3.7
C	PCC Unit Vapor Mixture	79	157	10.3	4.74	7880 90%	4.28	1040	23.2	2.6	33	3.2
D	Alkylation Unit #2	35	113	2.6	3.98	7880 90%	1.41	321	21.1	1.2	45	2.3
E	Contaminated	23	116	1.7	3.70	8470 97%	2.00	349	15.0	1.2	35	3.0
F	Pentaerythritol Unit 1 psig, Corrosive Steam	32	102	1.4	4.44	6000 68%	2.20	410	17.7	1.6	20	5.0
G	Blowdown + Clean Steam Liquid Vapor	38	106	10.3	4.36	7880 90%	2.20	410	15.7	1.4	34	3.1
H	T/M Pulping Process Dirty Steam	42	100	3.4	3.79	7880 90%	5.19	820	18.0	2.3	23	4.4
I	T/M Pulping Dirty Steam	42	100	2.1	2.84	7880 90%	3.60	744	21.1	2.2	22	4.7
J	3 Streams Vapor Mix	22	195	2.8	3.79	7000 80%	1.63	191	12.8	0.9	34	3.1
K	ETOH Water Cycle	158	145	5.2	3.79	8000 91%	2.42	1107(2)	40.5(2)	2.6	40	2.6
L	Condensing Hydrocarbons	32	159	10.3	2.84	8400 96%	2.32	530	18.8	1.9	22	4.4
M	Coagulator	23	121	4.1	3.08	8470 97%	2.50	448	16.4	1.4	28	3.9

TABLE 3

LEVEL 3 SCREENING

Plant	Space for Installation (10 – Good Space Available)	Risk to Process (10 – Very Low Risk)	Enthusiasm of Personnel (10 – Best)	DCF (40% – 10)	Average
E	10	10	10	9	10
L	8	8	9	6	8
C	4	4	8	6	6
B	0	10	5	8	6
A	5	10	0	9	6

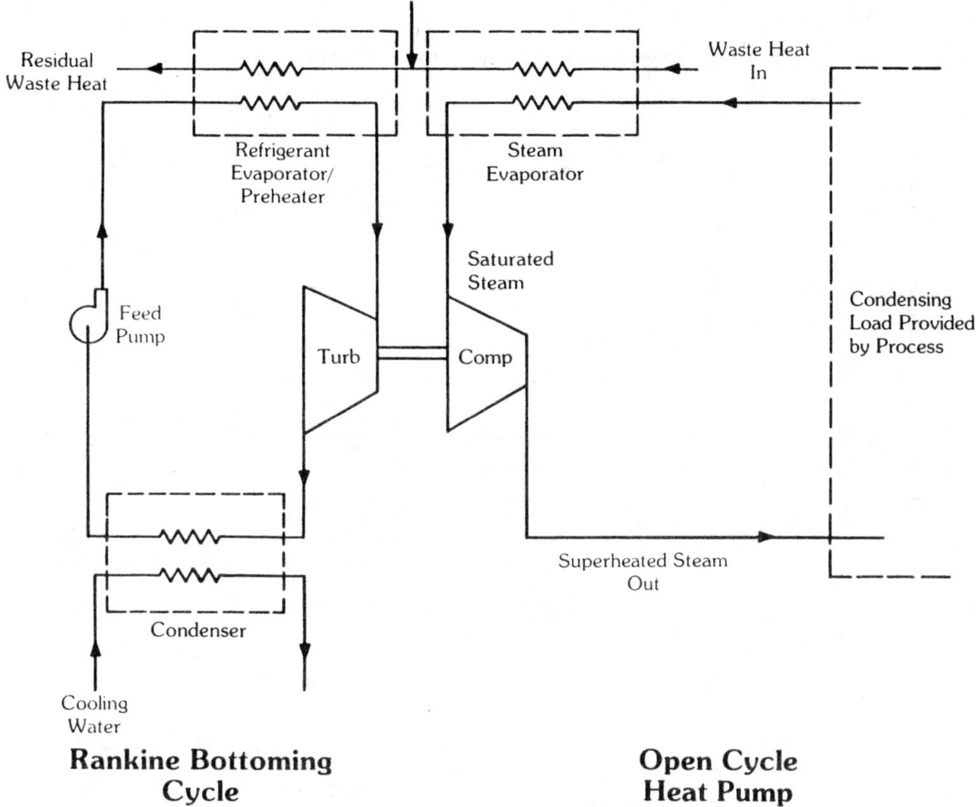

Figure 1.   Rankine-powered heat pump schematic

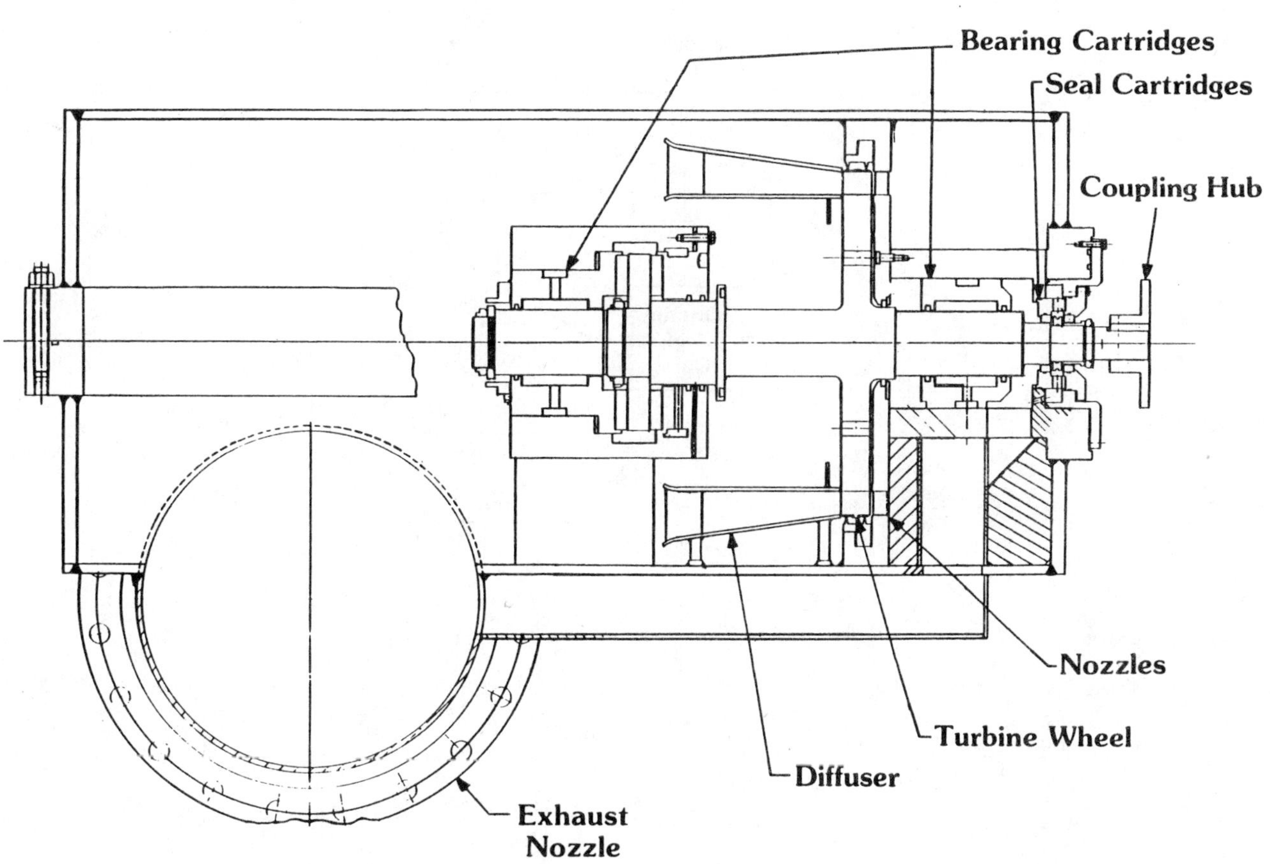

Figure 2.   Turbine power unit

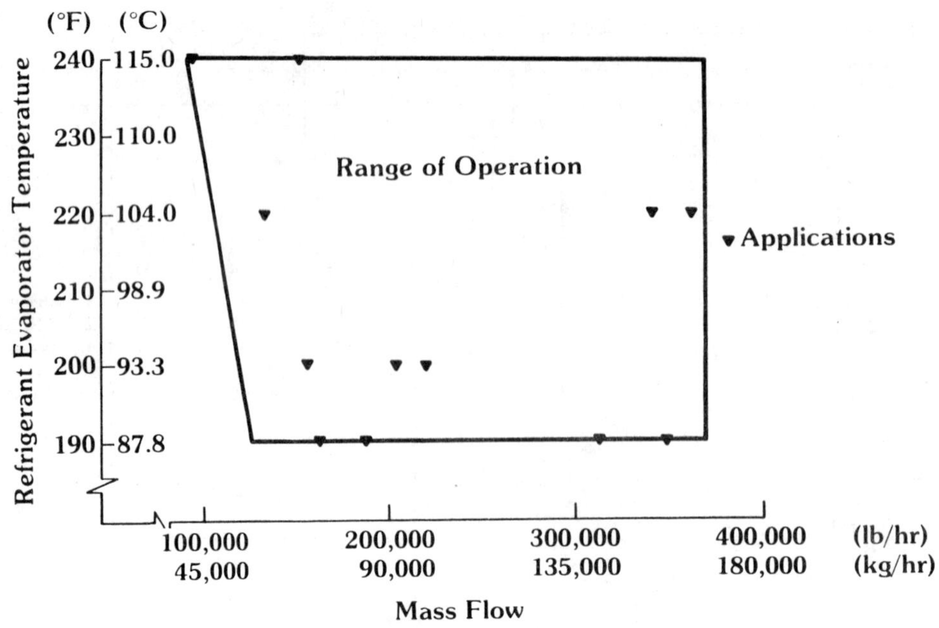

Figure 3. Turbine application range

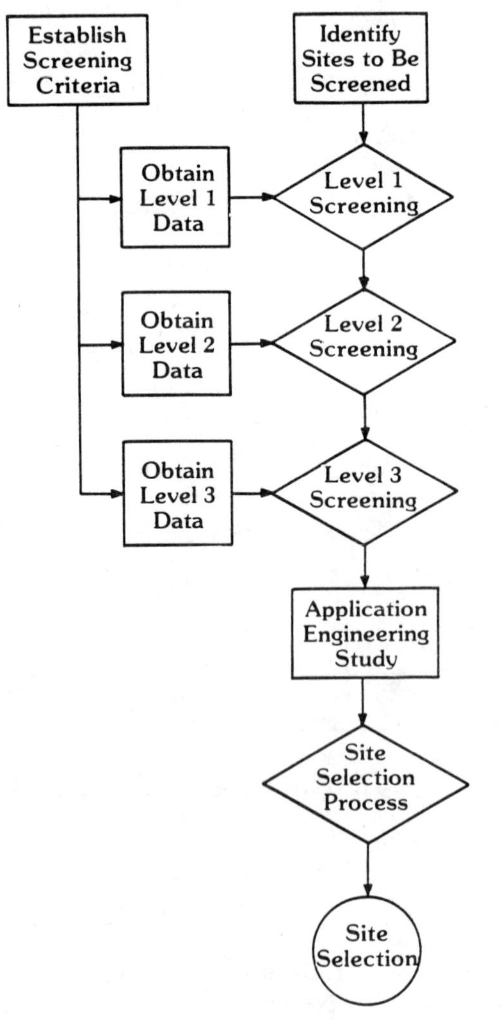

Figure 4. Methodology for high COP heat pump application evaluation

INDUSTRIAL APPLICATION OF HEAT PUMPS

DEVELOPMENT AND DESIGN OF AN ELECTRIC HEAT PUMP
FOR THE GENERATION OF PROCESS STEAM

W.C. Moreland and R.W. Wolfe

Westinghouse Electric Corporation, U.S.A.

Summary

This paper gives a functional description and the design choices of a high temperature heat pump being developed under a U.S. Department of Energy program. This specific design is for waste heat recovery from the effluent of a thermomechanical pulping line in a specific paper mill. In addition to this description, the costs of the major equipment items are presented and an overall system cost is estimated. This cost estimate is based on quotes for all the components in the system. On the assumption that the installation cost will be equal to the equipment cost, calculation of payback time (using energy prices at this mill) is less than one and a half years. The next milestone in the development program is verification of the technology by fabrication and testing of a prototypical unit.

Held at the University of Warwick, U.K.
Symposium organised and sponsored by
BHRA Fluid Engineering

1.0 INTRODUCTION

The Westinghouse R&D Center, under contract to the Department of Energy, is developing a high temperature heat pump which will extend the technology more than 56°C (100°F) above that of available commercial units. This is a two-phase program in which the first phase covers the development, design, fabrication, and laboratory testing of a prototypical unit. The second phase would cover the actual installation and demonstration of this heat pump at an industrial partner's plant.

Under this contract the critical aspects of the basic technology required for extending the high temperature capability of heat pumps have previously been identified and investigated (Ref. 1). On the basis of this work, several candidate working fluids for the new heat pump were chosen and extensive testing was performed to determine their thermal stability and compatibility with various alloys and elastomeric materials (Ref. 2).

This technical effort has been supplemented by a number of application studies (Ref. 3) which indicate that such electrically driven heat pumps can be cost effective in reducing fossil fuel consumption in particular applications. These studies have concentrated on the pulp and paper industry. This industry is one of the major consumers of fossil energy and has both process energy requirements and waste heat sources within the thermal capability of the high temperature heat pump.

2.0 SUMMARY OF PREVIOUS WORK

The critical aspects of the basic technology requirements depend primarily on the characteristics of the working fluid. There were eight criteria that were used to evaluate potential working fluids. They were:

1. Thermodynamic performance
2. Chemical stability
3. Corrosivity
4. Cost and availability
5. Volume flow rate
6. Head requirements (for compression)
7. Hazard
   a. Explosive and flammable limits; b. Toxicity; c. Carcinogenicity.
8. Availability of thermodynamic properties

Before these criteria could be applied to all potential candidates, however, an initial screening was required to reduce the number of candidates to a tractable number. The screening procedure was as follows:

First, it was decided that the working fluid had to have a critical temperature greater than 205°C (400°F). Since the heat pump will be operating over a conventional reversed Rankine cycle, we wanted to operate well below the dome of the saturation curve at the highest application temperature. A minimum critical temperature of 205°C (400°F) will insure that the top of the dome is at least 30°C (50°F) above our maximum delivery temperature.

Next, we decided that the normal boiling point of the working fluid had to be below our lowest source temperature. This insures that a positive pressure exists throughout the entire system during operation. This is certainly a necessity if rotating or sliding seals are required and air is to be excluded from the unit.

An autoignition temperature above the highest operating temperature is required to eliminate the possibility of spontaneous ignition in case of leakage.

The thermal stability of the working fluid obviously has to be excellent up to and beyond the maximum operating temperature, particularly if significant superheating occurs on compression.

Of the more than 1000 compounds considered, approximately 100 passed the initial screening. Ten of these had to be rejected because insufficient data were available on various thermodynamic properties. Of the remaining group, 11 fluids were selected as being potentially suitable for this application. These were then compared in detail on the basis of the remaining seven criteria.

This comparison narrowed the field to four: methanol (CH_3OH), thiophene (C_4H_4S), n-Hexane (C_6H_{14}), and Fluorinol 85 (a mixture of 85 mole percent trifluoroethanol and 15 mole percent water).

Of these, methanol was the preferred candidate on the basis of associated equipment cost, assuming that it demonstrated acceptable chemical stability and compatibility with the materials to be used in the heat pump system. Sealed tube tests at elevated temperatures have been underway for over a year and have shown that all of these candidates have satisfactory chemical compatibility based on visual evidence. Further analytical chemistry techniques have been used to quantitatively measure the thermal stability of methanol. Minor amounts of dimethylether and water have been detected but these are not expected to result in any system problems. The dimethylether will be purged as a noncondensible during normal system operation. We will need system operating experience to determine if a molecular sieve will be required to remove the minor amount of water generated.

The primary advantage in raising the delivery temperature of the industrial heat pump substantially higher than presently available units is to produce process steam. This is the form in which most of the heat in the pulp and paper industry is utilized. From the viewpoint of a heat pump, the generation of steam is a desirable way to deliver heat since a latent heat sink remains at a constant temperature. This minimizes the mean temperature difference in the working fluid condenser/steam boiler. The smaller this temperature difference, the smaller the "lift" or temperature rise the heat pump must produce. This in turn leads to a higher Coefficient of Performance (COP).

For the same reasons, a latent heat source has been found to be virtually essential to supply heat to a large capacity industrial heat pump.

In a paper/pulp plant the largest consumer of process steam is the paper/pulp drier. In paper making, for example, the process steam condenses in rotating drums and the latent heat liberated evaporates the water from the web of paper coming from the fourdrinier. Obviously it would be advantageous if the water evaporated from the web could be condensed in the evaporator of the heat pump to serve as a heat source. If all or most of the water vapor from the web could be condensed in the evaporator it would be possible to operate a drier-heat pump combination as a symbiotic unit whose only energy input would be electricity. Unfortunately, we found that even in the most modern driers there is so much air introduced that the temperature of the effluent stream is prohibitively low. Moreover, if heat is removed, the temperature falls even further.

There are two types of equipment in some paper plants that provide an attractive heat pump source. These are continuous digesters and thermomechanical pulping (TMP) units. They are attractive in that they produce a steady flow of steam at a pressure above atmospheric and contain relatively small amounts of noncondensibles. They are less than ideal, however, in that they contain both particulates and corrosive chemical constituents. Accordingly, it is necessary to provide a filter/washer to remove the particulates and to use a high quality stainless steel for the surfaces in contact with this steam.

This paper describes the design of a heat pump applied to a specific TMP unit in a particular paper mill in the northwestern part of the United States. The application is generic in nature, however, and with moderate design changes could be applied to other TMP units, or alternatively, to continuous digesters.

3.0 GENERAL SYSTEM DESCRIPTION

The high temperature heat pump as it is presently designed will serve as a waste heat recovery system for a TMP unit in a paper mill. In this application low pressure waste steam from a TMP unit is the thermal source for the heat pump. The output of the heat pump is 5.4 bars (63 psig) steam with 11°C (20°F) superheat, which can be used in the paper drier and other mill processes. The energy added to the output steam is approximately 4-1/2 times the electrical energy required to drive the compressors.

3.1 Functional Description

The schematic of the heat pump system is depicted in Fig. 1. Since methanol is used as the working fluid (refrigerant) in the heat pump, each system component in the following discussion will be identified by its function in the methanol loop. For example, the methanol condenser/steam boiler will be referred to as the condenser. The thermodynamic states of the methanol can be followed on the pressure-enthalpy diagram in Fig. 2. The thermodynamic states at different points in the heat pump system are identified by the eight locations indicated by Δ's on both the schematic and pressure-enthalpy diagram.

Heat is added to the system when low pressure waste steam enters the tubes of the methanol evaporator (1) and condenses. The heat of condensation is transferred to the methanol liquid surrounding the tubes and the methanol evaporates. The evaporated methanol enters a superheater (2) where its temperature is raised three degrees above its saturation temperature with heat from hot methanol condensate leaving the condenser at state point 7. This superheating is required to prevent partial condensation of the vapor when it accelerates into the compressor inlet at point 1 on the pressure-enthalpy diagram. The superheated vapor entering the first stage compressor (3) is then compressed to approximately 2-1/2 times inlet pressure (point 2). The compression process (with an anticipated isentropic efficiency of 75 percent) adds a significant amount of superheat, which, if left unchanged, would decrease the efficiency of the second stage of compression and cause undesirably high discharge temperatures. Correcting this situation requires the addition of an intercooler (4) where liquid methanol of lower enthalpy but higher pressure is sprayed into the superheated vapor from the first stage discharge. Evaporation of the methanol liquid spray brings the methanol vapor to within 3°C (5°F) of the saturation temperature. Any overspray is separated from the vapor in the knockout drum and returned to the evaporator. The intercooler discharge corresponds to point 3 on the enthalpy-pressure diagram. The vapor with three degrees of superheat enters the second stage compressor (5) where it is compressed to approximately 2-1/2 times the interstage pressure. The exit condition corresponds to point 4 on the enthalpy-pressure diagram. As before, the methanol vapor is superheated during the compression process. The superheated methanol is used to superheat the steam exiting the condenser (6). This is accomplished in the desuperheater (7). The vapor exiting from the desuperheater corresponds to point 5 on the enthalpy-pressure diagram. The exiting vapor enters the tubes of the condenser where it condenses to the saturated liquid state at the discharge pressure. The saturated methanol is then subcooled by 3°C (5°F) in a partitioned section of the condenser to an enthalpy corresponding to point 7 on the diagram. (Although a separate subcooler is not required for this application because the feedwater temperature is so close to the boiling temperature, if the feedwater is available at a

lower temperature in another application a separate subcooler as shown in Fig. 1 would
improve the COP of the heat pump.) After its exit from the condenser at point 7
on the enthalpy-pressure diagram the methanol flow is split into two streams. Roughly
10 percent of the liquid flow goes to the intercooler (4) as discussed previously,
while the remainder of the methanol is admitted to the evaporator (1) through a float
controlled expansion valve (9). The methanol enters the evaporator at a condition
corresponding to point 8 on the enthalpy-pressure diagram. With a methanol flow in
the condenser of approximately $9.3 \, kg \, s^{-1}$ (74,000 lbm/h) this thermodynamic cycle has a
net output of $8.79 \times 10^6 \, J \, s^{-1}$ (30×10^6 Btu/h).

In addition to the basic heat pump system there are several subsystems which
are required for operation of the high temperature heat pump. These subsystems are
the hot gas bypass system, the methanol storage system, the nitrogen injection system
and the nitrogen purge system. While most of these subsystems are not required during
normal operation, they serve important functions in protecting the equipment and
providing desired levels of safety when the system is either not running or not
operating normally.

The hot gas bypass system has two main functions: a) to prevent the
compressor from surging under abnormal system conditions such as loss of source heat
and b) provide some measure of capacity control. Surging is prevented by modulation
of the hot gas bypass valve (10) to lower the discharge head when the pressure ratio
across the compressors increases and brings them close to their surge line. The
bypassed gas is throttled and returned to the evaporator. The gas enters the
evaporator through a special distributor plate to prevent violent bubbling which
could cause liquid carry-over into the compressors. In addition to protecting the
compressors, the hot gas bypass system can be used to maintain operation even when the
heat source is reduced in capacity or cut off. Although this mode of operation
reduces the COP of the system significantly, it does provide flexibility in handling
capacity variations for short periods of time.

The methanol storage tank (11) serves to maintain a reservoir of liquid
methanol for make-up of losses by the compressor seals and purge devices. It is
large enough to hold the total system's capacity in case it is necessary to remove the
methanol from the heat pump for servicing. This capability is provided by a separate
methanol pump (12), a heat exchanger (13) for cooling the methanol before returning it
to storage, and appropriate valves.

Because methanol is potentially explosive when mixed with air and because
methanol with water in it is corrosive, the methanol and the heat pump system must be
kept dry and separated from the atmosphere at all times. The methanol is kept dry
while in storage by the use of a dry nitrogen blanket which is maintained over the
liquid methanol surface. The dry nitrogen is supplied from a liquid nitrogen storage
tank (14) through pressure reducing valves. A nitrogen injection system is also used
to keep the methanol side of the heat pump system dry and above atmospheric pressure
in case of prolonged shutdown. As with the methanol storage tank, nitrogen is
automatically injected into the heat pump system if the system is not operating and
the pressure drops below a fixed set point. This prevents air from leaking into the
system at any time.

While the injection of nitrogen has obvious benefits in protecting the heat
pump and providing safer operation, it does present a problem if present when the heat
pump is started. Nitrogen is a noncondensible gas at system temperatures and its
presence in any significant amount in the heat pump system, in particular in the
methanol condenser, would have a serious effect on the operation of the system's heat
transfer and performance. As a result, a system for purging the nitrogen from the
heat pump is required. Actually two systems are required, one for the high purge rate
at start-up (15) and one for a small continuous purge during normal operation (16).
Both of these systems will condense methanol from the nitrogen-methanol mixture,
returning the major portion of the methanol to the heat pump system. The remaining
vapor, having a reduced methanol content, will be vented to atmosphere through the
storage tank vent.

Throughout its operation the high temperature heat pump system's temperature,
pressure, and liquid levels will be monitored. At the same time, the surrounding
environment will be monitored for the presence of methanol leakage. Should significant
levels of methanol be detected, alarms will be sounded and the system automatically
shut down.

3.2 System Performance Specifications

An abbreviated version of the heat pump system performance specifications are listed below:

a) Heat Source – The TMP effluent used as a heat source is to be supplied to the heat pump evaporator at a pressure of 1.7 bars (10 psig) and at a flow rate of 4.114 kg s^{-1} (32,600 lbm/h). It is expected to be compatible with 316 stainless steel and contain not more than 1 percent noncondensibles. The heat pump will extract 78 percent of the latent heat in the steam. The remaining 22 percent will provide a safety margin for possible variations in the TMP process and in normal operation this excess will be vented to atmosphere. Fouling rates will be predetermined by test and a cleaning schedule and procedure will be developed prior to installation of the heat pump at a demonstration site. (For performance verification tests at a laboratory test site, clean steam will be used as the thermal source.)

b) Feedwater – Boiler feedwater at a temperature of 152°C (305°F) and a flow rate of 4.114 kg s^{-1} (32,600 lbm/h) is to be supplied at a pressure greater than 5.5 bars (65 psig). The quality of the feedwater (oxygen content, pH, and dissolved solids) shall be compatible with low carbon steel. As presently configured, the boiler blowdown is manually operated and a suitable service schedule and blowdown frequency will have to be determined after the exact quality of the feedwater is known.

c) Thermal Capacity – The output of the heat pump will be 5.4 bars (63 psig) steam with 11°C (20°F) superheat. The steam flow rate will be 4.114 kg s^{-1} (32,600 lbm/h). The total capacity of the unit is 8.79 x 10^6 J s^{-1} (30 x 10^6 Btu/h).

d) Electrical Power Consumption – The total electrical power input to the heat pump including auxiliary power for ancillaries and controls shall not be more than 1850 kW. (This is based on the compressor manufacturer's rated power requirements for this service with the inclusion of additional losses for system pressure drop, a rated compressor motor efficiency of 96 percent and an allowance of 5 kW for auxiliary electrical power.)

e) Overall Performance – On the basis of 152°C (305°F) feedwater and steam output of 4.114 kg s^{-1} (32,600 lbm/h) at 50 psig with 20°F superheat, the COPH is

$$\text{COPH} = \frac{\text{Steam rate (kg s}^{-1}) \text{ x Enthalpy Difference (J/kg)}}{\text{Power Input (kW) x 1000 J s}^{-1}/\text{kW}}$$

$$= \frac{4.114 \text{ kg s}^{-1} (2774.2 \times 10^3 - 639.3 \times 10^3)}{1850 \times 1000} = \frac{8.79 \times 10^6 \text{ J s}^{-1}}{1.85 \times 10^6 \text{ J s}^{-1}} = 4.75$$

3.3 Major Components

Figure 3 shows the detailed layout of the heat pump design with the major components identified. Given below is a general description of these components as well as a list of their costs.

a) Hardware Description:

1. Methanol Evaporator: This is a kettle drum type heat exchanger which contains the bulk of the methanol in the system in liquid form on the shell side. The shell is 2 m in diameter and approximately 7 m long and is made of low carbon steel. With a design temperature difference of 8°C (15°F) it is anticipated that the boiling process will be relatively quiesant. Nonetheless, a disengagement space is provided as well as a demister. The relatively high temperature bypass gas flow is introduced at the bottom of the shell below a sparging plate which is provided with small (approximately 2 mm) holes and extends the full length of the shell. Calculations have been made which show that this arrangement will assure that all of the superheat in the bypass gas will be removed as it bubbles up through the liquid methanol. The tube bundle is a two pass arrangement made of stainless steel unfinned tubing. There is a tube sheet at both ends of the bundle rather than "U" tubes in order to facilitate the mechanical cleaning of the inside of the tubes if that proves to be necessary.

2. Methanol Superheater: This is a small capacity heat exchanger whose
configuration had been altered from that shown in Fig. 2. It is now simply a 0.6 m
diameter cylinder 0.8 m long with single pass tubes carrying the methanol vapor. The
methanol liquid is on the shell side. It is all of low carbon steel construction.
Because of the lower pressure on the vapor side it would normally be expected that the
vapor would be on the shell side. However, because of the small size of the exchanger
it is convenient to mount it directly on a 0.6 m diameter flange at the top of the
evaporator with the axis of the tubes pointing vertically upward.

3 and 5. Dual Compressors with Lube Oil System and Control Package: It was
found for the capacities of interest that reciprocating and rotary screw compressors
were more expensive than centrifugal compressors. Furthermore, they had the disadvantage of requiring contact between lubricant and working fluid which was unacceptable
from a chemical compatibility viewpoint. Even so, it was found that the cost of the
standard heavy duty chemical industry centrifugal compressors was such that, when
designed into a heat pump system, they jepardized its economic viability. We were
fortunate in finding, however, a small manufacturer with a unique high impeller tip
speed design. This unit is very compatible with the compression of methanol which is
a high head fluid which requires a high impeller tip speed. They build five frame
sizes with an integral gearbox and in addition, for applications requiring two stages,
they build a compact unit wherein two compressors are driven from a single gearbox.
This provides additional economies of cost and space as can be seen in Fig. 3. The
compressor, gearbox, lube oil system, control package, and drive motor are mounted and
shipped on a single foundation which requires minimal installation effort. Due to its
relatively light weight construction, large forces and moments due to the thermal
expansion of attached piping must be avoided and it can be seen from Fig. 3 that
generous pipe loops are provided.

4. Intercooler and Knockout Drum: Because the heat transfer in the intercooler is achieved by direct contact between superheated methanol gas and subcooled
methanol liquid, a heat exchanger as such is not required. Rather, the liquid is simply
sprayed into the gas through a nozzle mounted in the piping. To assure that no liquid
droplets are carried over into the second stage compressor a large tank called a knockout drum is provided. This reduces the gas velocity to very low values so that liquid
droplets fall out of suspension. In addition, the drum is fitted with demister pads
to eliminate even fine spray and thereby assure that the gas entering the second stage
compressor can be accurately controlled to have 3°C (5°F) of superheat.

6. Methanol Condenser: The configuration of the methanol condenser is very
similar to the methanol evaporator. In this case, however, the methanol is condensed
in the tubes and boiler feedwater is boiled on the shell side. Because both fluids
are clean no stainless steel is required. Further, there is a single tube sheet and
the two pass arrangement is provided by "U" tubes because no need for mechanical
cleaning of the inside of the tubes is anticipated.

7. Methanol Desuperheater: Since dry and saturated steam is delivered by
the methanol condenser it is desirable to provide a modest amount of superheat to
minimize condensation in the steam piping. Accordingly, a simple single pass gas to
gas heat exchanger is provided in which the steam is on the shell side because of its
lower pressure.

Other components whose costs are itemized below are auxiliary equipment
for which detailed discussion is not justified-

b) Hardware Costs:

1.	Methanol Evaporator	$157,000
2.	Methanol Superheater	22,400
3 & 5.	Dual Compressors with Lube Oil	
	System and Control Package	308,340
4.	Intercooler and Knockout Drum	13,000
6.	Methanol Condenser	94,700
7.	Methanol Desuperheater	14,900
8.	Methanol Expansion Valve	980
9.	Hot Gas Bypass Valve	1,170
10.	Methanol Storage Tank and Vents	8,200
11.	Methanol Pump	3,500

	12.	Cooldown Heat Exchanger	4,500
	13.	Nitrogen Storage Tank*	–
14 & 15.		Nitrogen Purge Systems	3,300
	16.	Compressors' Drive Motor	70,000
	17.	Misc. Valves, Traps, and Hardware	18,650
	18.	Instrumentation and Controls	22,665
	19.	Methanol (6000 gal)	3,000

$ 746,305

> *The nitrogen tank is normally a rental
> item. The rental cost and nitrogen
> consumption is estimated at $2800/yr.

3.4 Installation Costs

Each application site for a heat pump has to be evaluated on an individual basis to obtain an accurate estimate of installation cost. This is due to the fact that in some situations extensive piping lengths are involved or significant rearrangement of facilities must be made to provide electric power or physical space for the heat pump. As a means of estimating the approximate cost of installation, a rough rule-of-thumb has been suggested which equates installation cost to equipment cost. In the one application evaluated in detail to date, this rule proved valid to within 5 percent.

3.5 Payback

Simple payback is frequently used as a preliminary economic measure to evaluate the cost effectiveness of energy conservation systems. The payback period is defined as the first cost divided by the net annual savings. The annual savings in this application is the difference in annual cost between supplying the heat from an oil-fired boiler and from the electrically driven heat pump. The calculation of payback is shown below using present costs of fuel and electricity at the proposed plant site:

$$\text{Annual Fuel Cost} = \frac{\text{Fuel Price } (\$/10^9 \text{ J}) \times \text{Capacity } (\text{J s}^{-1}) \times \text{Operating Time } (\text{s/yr})}{\text{Boiler Efficiency}}$$

$$= \frac{(\$30/\text{bbl}) \times (8.79 \times 10^6 \text{ J s}^{-1}) \times (7900 \text{ h/yr}) \times (3600 \text{ s/h})}{(6.645 \times 10^9 \text{ J/bbl}) \, (0.88)}$$

$$= \$1,282,500/y$$

$$\text{Annual Electric Cost} = \frac{\text{Electric Rate } (\$/10^9 \text{ J}) \times \text{Capacity } (\text{J s}^{-1}) \times \text{Operating Time} (\text{h/yr})}{\text{COP}}$$

$$= \frac{(\$0.008/\text{kWh}) \times (8.79 \times 10^6 \text{ J s}^{-1}) \times (7900 \text{ h/yr})}{(1000 \text{ J/kW s}) \times 4.75}$$

$$= \$117,000/\text{yr}$$

$$\text{Annual Savings} = \$1,282,000/\text{yr} - \$117,000/\text{yr} = \$1,165,000/\text{yr}$$

The first cost or capital cost is determined from the equipment cost plus installation cost. Using the rule-of-thumb discussed previously we have

$$\begin{aligned}
\text{Capital Cost} &= 2 \times \text{Equipment Cost } (\$) \\
&= 2 \times (\$746,300) \\
&= \$1,493,000
\end{aligned}$$

$$\text{Payback} = \frac{\text{Capital Cost}}{\text{Annual Savings}} = \frac{\$1,493,000}{\$1,165,000/\text{yr}}$$

$$= 1.3 \text{ yr}$$

    It should be noted that the electric cost used in these calculations (0.8¢/
kWh) is very low but is the present rate being paid by the mill, which is located in
the northwestern part of the United States. Even if this rate were four times as high,
the payback would still be less than two years.

<u>REFERENCES</u>

1. Snyder, P. H. and Moreland, W. C., "The Selection of Working Fluid Candidates for
    Use in High Temperature Industrial Heat Pumps," Westinghouse Research Report No.
    79-9E9-SOTEM-R1.

2. Gainer, G. C., Luck, R. M., and Moreland, W. C., "Thermal Stability and Materials
    Compatibility Tests of Five High Temperature Working Fluid Candidates,"
    Westinghouse Research Report No. 79-9E9-SOTEM-M1.

3. Wolfe, R. W. and Moreland, W. C., "High Temperature Electric Heat Pumps for Waste
    Heat Recovery in Industry," Westinghouse Research Report No. 81-9E9-SOTEM-R1.

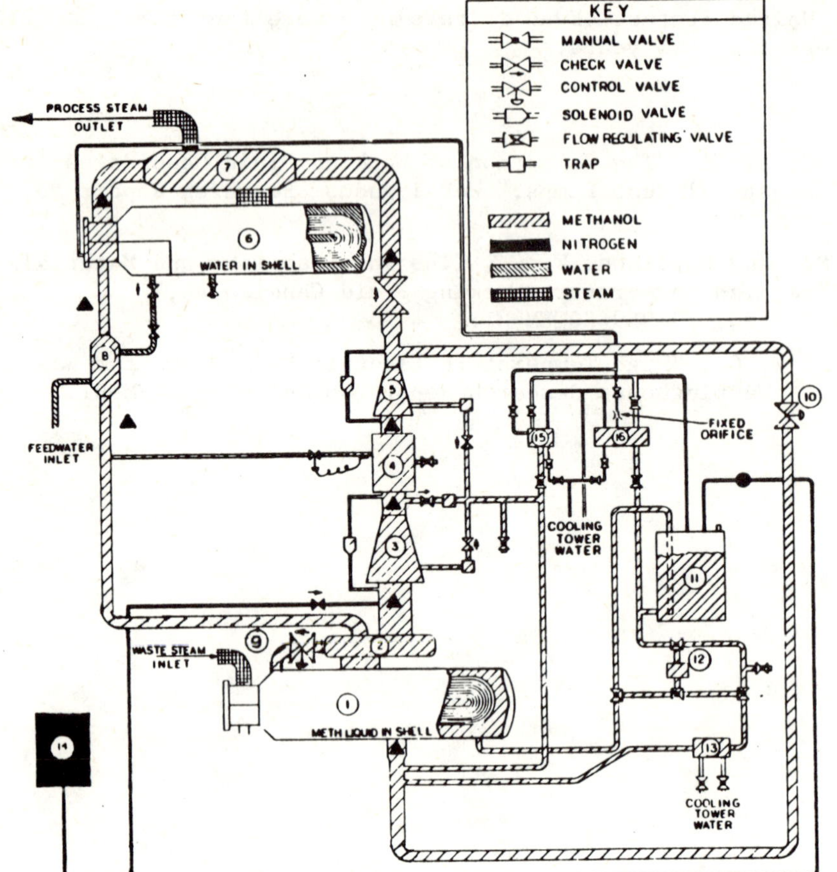

Fig. 1   Schematic of Heat Pump System

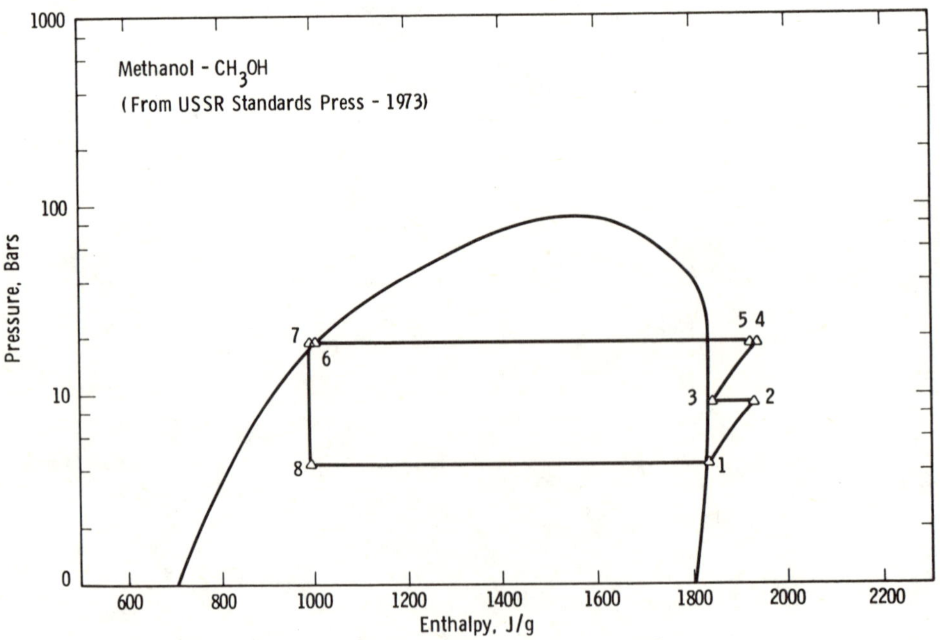

Fig. 2   Pressure-Enthalpy Diagram Indicating Thermodynamic States of Heat Pump Cycle

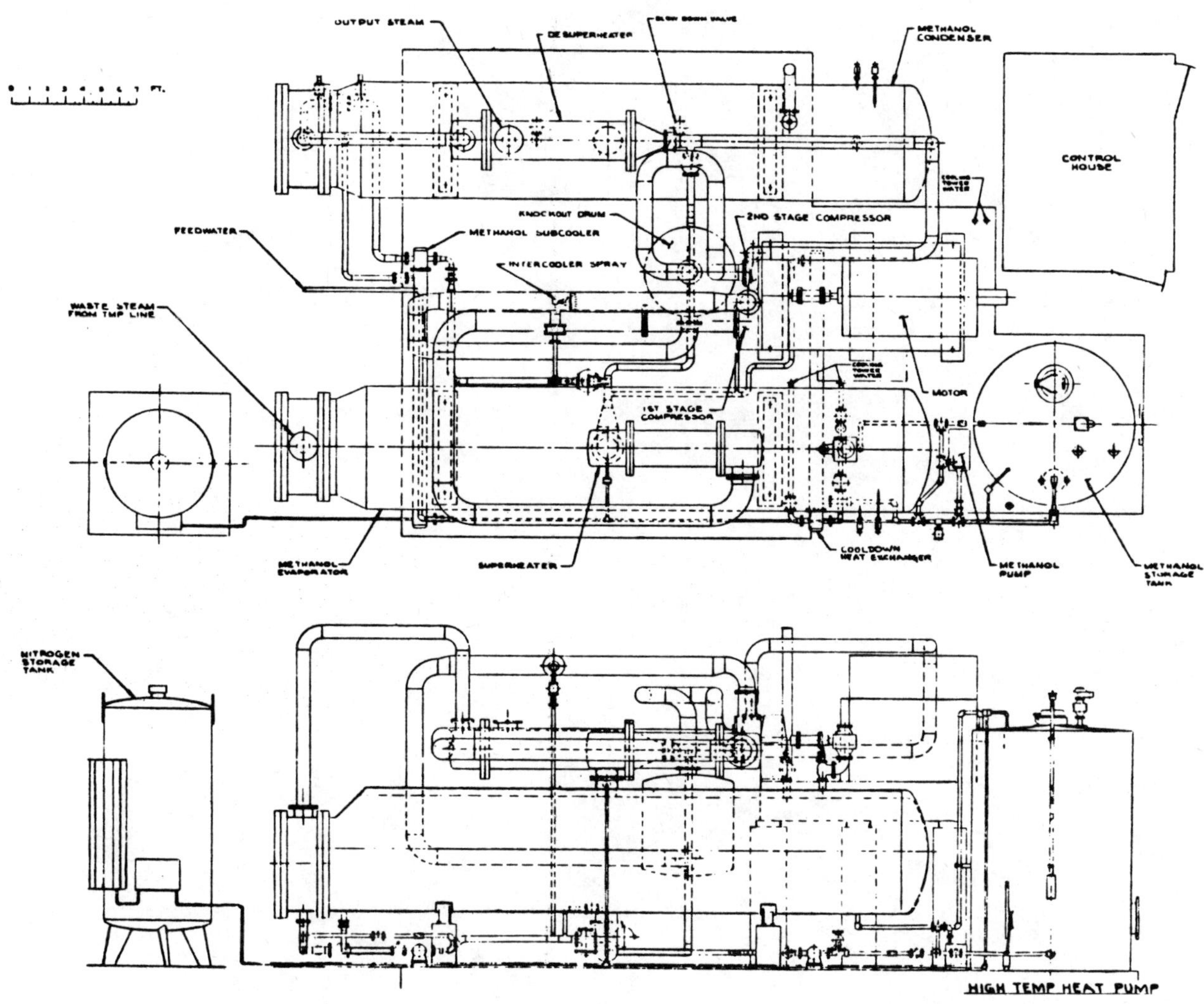

Fig. 3   Layout of Heat Pump System

AN OVERVIEW OF WORK ON INDUSTRIAL AND DOMESTIC HEAT PUMPS IN THE ENERGY R+D PROGRAMME OF THE EUROPEAN COMMUNITY

P. Zegers and J.A. Knobbout

Commission of the European Communities, Belgium

Summary

In the framework of its Energy R and D programmes the Commission of European Communities is carrying out R and D on heat pumps. Work in these programmes was focussed on heat sources, improvement of electric and ICE driven compressor heat pumps, absorption heat pumps and heat pumps for use in industry.

In the first Energy R and D programme (1975-1979) 21 heat pump projects were carried out at a total cost of 3 Mio ECU (1 ECU = 0.6 t). The utilization of soil and sewers as heat pump heat sources has been investigated. Both horizontal and vertical soil heat exchangers have been studied. In different ways the C.O.P. of conventional compressor heat pumps has been improved (heat exchangers, use of microprocessors for part load operation, fluid mixtures). Absorption heat pumps were found to be a promising option.

In the second R and D programme (1979-1983) 25 heat pump projects have been selected. A large part of this programme is now devoted to work on absorption heat pumps in particular to work on the development of a cheap and reliable fluid circulation pump and new working fluids. The work on improvement of compressor-heat pumps is being continued. Research on industrial heat pumps, with drying applications in the paper, milk and food industry, has been considerably expanded. Also the possibilities for heat pump operation above 120°C are being explored. (e.g. Brayton cycle heat pump, heat transformer).

Held at the University of Warwick, U.K.
Symposium organised and sponsored by
BHRA Fluid Engineering

1. CEC ENERGY R AND D.

Heat pump research is executed in the framework of the Energy R and D Programme of the European Communities. This programme is subdivided in five sub-programmes: energy conservation, solar, geothermal, hydrogen and energy modelling.
The European Community became actively involved in Energy R and D in 1975 when a four-year programme was approved by the Council of Ministers. A second four-year programme started in 1979. The allocation for the two successive programmes was 59 and 105 Mio ECU respectively (1 ECU = 0.6 ₤). The distribution of this money over the different sub-programmes is given in Table 1.

Heat pump R and D is mainly carried out in the Energy Conservation R and D sub-programme. In the first programme (1975-1979) 21 heat pump projects were carried out at a total cost of 3 Mio ECU. In the second programme (1979-1983) 25 projects have been selected of which the cost amounts to 7 Mio ECU. The CEC contribution is normally 50 %.

Table 1.　　　　Budget for the CEC Energy R and D Programme

	1975-1979	1979-1983
Energy Conservation	11.4 Mio ECU	27 Mio ECU
Solar Energy	17.5	46
Hydrogen	13.2	8
Geothermal energy	13	18
Energy Modelling	3.9	6
T o t a l	59	105

2. PRESENT STATE OF HEAT PUMPS.

- At present about 40.000 and 3.000.000 heat pumps are in operation in the European Community and the USA respectively. The type of heat pump most frequently used is an electrically driven air source heat pump.

- The main potential for energy saving with heat pumps lies in the heating of commercial and domestic buildings. Heat pumps may also be used for the production of low temperature process heat (50-150° C) in industry but the energy saving potential is an order of magnitude smaller.

- Large scale introduction of heat pumps for domestic and commercial heating could bring energy savings of 10-15 % of the total final energy consumption. But the expensive heat pumps will be installed only when cheaper energy saving devices such as insulation, draft prevention and double windows have been installed.

- Only 2 % of the building stock is renewed per year. Heat pumps should therefore be designed for retrofitting in existing houses and be compatible with existing radiators.

- Heat pumps for apartments and district heating at present are mainly large internal combustion engine driven heat pumps.

- Production of process heat with heat pumps is presently limited to temperatures
  between 50 and 150° C. Applications may be found in the textile, paper and
  food industry. About 3,5 % of the total final energy consumption is used for
  process heat in this temperature range. If we assume a 50 % higher energy
  conversion efficiency with heat pumps then we may estimate the energy saving
  potential for heat pumps in industry to be 1,2 %. If heat pumps can be developed
  for temperatures up to 400° C then the energy saving potential may increase to
  3,5 %.

3. FIRST PROGRAMME 1975-1979.

With the above data in mind the first heat pump programme of the CEC was largely
designed to develop heat pumps for the domestic sector. It focussed on efficiency
improvement and cost reduction of heat pumps producing heat at 50-100° C.

The work executed in the first programme had four focal points :

- Heat sources

- Compressor heat pump improvements

- Large heat pump systems

- Advanced heat pumps.

3.1. Heat sources.

Heat extraction from soil can be realized with horizontal or vertical tubes with
a circulating fluid (e.g. glycol-water).

Results of a series of experimental and computer simulation studies can be
summarized as follows :

- The optimum depth of horizontal tubes is as close to the soil surface as
  practically possible (+/- 0,5 m).

- The required soil surface : 2 x heated surface for horizontal tubes

  1/2 x heated surface for vertical tubes (10 m)

- Vertical tubes require heat recovery for deeper layers in the summer.

A feasibility study showed that with the available waste water in Germany heat
pumps could provide 5 % of the heat requirements for heating of buildings and save
2,5 % as compared to conventional heating systems.

3.2. Improvement of compressor heat pumps.

In order to improve the efficiency and reduce the cost, R and D was executed in
the following areas :

Partload operation : As compared to the usual on-off operation, micro processor
based control of the rotation speed of the compressor and of the expansion valve
lead to considerable energy saving.

Fluid mixtures : The COP can be considerably improved when instead of a single
fluid, a fluid mixture is used to have a temperature variation during the vaporization and condensation stages, parallel to that of the outside fluid with which
heat exchange takes place. Savings up to 50 % as compared to conventional heat
pumps have been shown possible.

Lubrification of compressor : In several experiments problems arose with the
lubrification of compressors , due to the fact that most oils dissolve in the
refrigerant. This problem will be further studied in the second programme as it
influences the C.O.P. of the heat pump.

With a better regulation system for the expansion valve an improved performance
can be obtained. The presently used thermostatic expansion valve was not satisfactory. Motorized expansion valves gave a 10 % improvement but they are more
expensive.

The problem of defrosting of the evaporator of heat pumps with air as heat source
can be satisfactory resolved.

It is expected that with realization of the improvements studied in this programme
electrically driven heat pumps may obtain a seasonally averaged efficiency based
on primary fuels of 80-100 %, this is comparable with very good conventional
heating systems. Thus the energy saving potential with electrical heat pumps is
limited.

3.3. Large internal combustion engine driven compressor heat pumps.

When heat pumps are driven by internal combustion engines and the waste heat of
these engines is recovered, considerable higher efficiencies may be obtained
(150-180 %). The draw back is that one is limited to the use oil or gas.
Experiments with a 468 kW gas engine driven heat pump using air as heat source
heating 64 apartments with floor heating gave satisfactory results. The pay back
time was estimated to be 8 years. An industrial gas engine driven heat pump
producing steam of 110° C, which up to now is the upper limit of compressor heat
pumps, is expected to have a pay back time of 2,3 years. A feasibility study for
a 10 MW diesel engine driven heat pump for district heating, gave a pay back time
of 3 years. This heat pump is now being constructed in the framework of the CEC
demonstration programme.

3.4. Advanced heat pumps.

Whereas electrically and ICE driven compressor heat pumps are already in
operation, more advanced systems such as absorption heat pumps and Organic Rankine
Cycle engine driven compressor heat pumps, are still in the Laboratory stage.
ORC driven heat pumps of 10 kW are being developed with the ORC turbine and the
vapor compression of the heat pump on the same axis. This avoids transmission
losses, provides oil free lubrification due to gas bearings, silent operation and
flexibility in fuel use. The heat pump may be fired with oil, gas, coal, wood or
even waste heat .COP values of 1,20 have been obtained.

Another advanced concept is the absorption heat pump. It has no rotating parts,
limited maintenance, is likely to be cheap and can be easily extrapolated to larger
sizes for industrial use. The efficiency is estimated to be around 130 % and it
may use any type of fuel. From the different projects executed in the first programme results that future R and D should develop cheap and reliable fluid circulation
pumps, new working fluids and a capacity control system.

4. SECOND PROGRAMME 1979-1983.

From results of the first programme it is expected that electric compressor
heat pumps with energy conversion efficiencies based on primary fuels up to 100 %
will be replaced in the middle and long term by absorption and ICE driven heat
pumps which are expected to have energy conversion efficiencies of around 130 %
and 150 % respectively. This is reflected in the content of the second R and D
programme.

- A considerable part of the heat pump work is now focussed on absorption heat
  pumps and in particular on the development of a fluid circulation pump and
  working fluid pairs.

- work <u>to improve the efficiency of mainly ICE driven compressor heat pumps</u>
  is basically a continuation of the first programme with research on variable
  speed compressors, micro processor regulated expansion valves, fluid mixtures
  and soil as a heat source for heat pumps. A study of the effect of compressor
  oil dissolved in the refrigerant on the heat pump performance was added. The
  suitability of ICE driven heat pumps for single houses is being tested.

- Finally the development of <u>industrial heat pumps</u> is emphasized. In particular
  the possibilities for heat pump operation above 120° C are being explored.

4.1. <u>Absorption heat pumps.</u>

R and D in the first programme showed the need to get a more detailed knowledge on fluid pairs for absorption heat pumps. It was decided to study alcohols
and amines as solutions and different halides and nitrates as solvents. Vapor
pressure curves in the whole concentration range, viscosity and the solution limit
will be determined.

Experiments have started with thiocyanates of Li and Na in a methyl amine
solution and Li and Na halides and thiocyanates in methanol. This work is carried
out in close collaboration with R and D on working fluid pairs for industrial heat
pumps described below.

Another subject of investigation is the development of a fluid circulation pump.
In the first programme it was established that the only suitable pump available on
the market, a membrane pump, had serious problems with shaftsealing and was moreover very expansive. Of the different pumps investigated the internal gear pump
turned out to be the most interesting one. This pump will be further developed.
Experiments will be carried out with E 181-R 22 fluid pairs, at 40-60° C, with a
pressure difference of 23 bar and a mass transport of 800 kg/h. This pump runs
quietly, has a small volume and is cheap. It is expected that the problem of
shaftsealing can be solved. As a long term solution a hermetic unit which includes
both the pump and the driving motor is being developed. A submerged motor will be
used for this purpose. Also complete absorption heat pump systems are being
developed, R and D is carried out on a 10 kW methanol/mixed bromide heat pump. The
aim of this project is the development of a heat pump which is suitable for retrofitting in buildings and which therefore produces a heat output at 60-70° C in
order to be compatible with conventional radiator systems.

Also work on a absorption - resorption heat pump is being carried out.

4.2. <u>Compressor heat pump improvements.</u>

R and D on the improvement of compressor heat pumps is centered around the
reduction of the compressor rotation speed in part load operation, improvement of
evaporator and condensors, use of fluid mixtures, electronically regulated expansion valves and the influence of oil dissolving in the refrigerant.

It is well known that compressor oil dissolves in the refrigerant and that this
impairs the operation of the heat pump and its performance. This was also found
to be the case in several projects in the first programme. It was therefore
decided to investigate this problem systematically. Experimental work is carried
out with a rotating vane compressor of 7-10 kW with R 12 as a refrigerant.
Oil-refrigerant ratios of 1-20 % will be studied. The influence of the oil on the
performance of the evaporator the condensor and the overall heat pump operation
will be investigated.

One solution to this problem is the development of an oil which does not interact with halogenated refrigerants. This however is out of the scope of this programme. A second possibility is the development of a heat pump which can operate without oil. To this end a screw compressor has been developed where the oil circuit has been suppressed. A prototype of 22 kW, with a throughput of 2070 l/min was found to have an efficiency of 74 %. An electrical heat pump of 30 kWel with this type of screw compressor will be developed. The working fluid will be R 22.

Also the optimization of two electrically driven, heat pumps with air as a heat source is being studied. Here a two speed compressor, variable speed ventilators and micro processor control are used.

Internal combustion engine driven compressor heat pumps are more suitable for the regulation of the heat output as the speed of the engine can be easily varied and in addition one or several cylinders may be switched off. In a project with a 170 kW ICE heat pump the speed regulation is controlled with a microprocessor. In another experiment a small 20 kW ICE driven heat pump is being tested. It is often assumed that this type of heat pump is not suitable for single houses as they are very complicated and require careful maintenance. This 20 kW heat pump is installed in 6 houses in order to test the validity of this assumption. Switching off and on, which causes wear, is reduced by adapting heat supply to the demand by varying the speed of the motor and shortcircuiting of cylinders with a micro processor. Results of to now are satisfactory.

Working fluid mixtures developed in the first programme will be tested in four air/water and two water/water compressor heat pumps.

Finally several heat sources are being tested such as an energy roof, vertical soil heat exchangers with normal and concentric tubes, and heat pipes.

4.3. Industrial heat pumps.

In the temperature range 50-150° C, several heat pump applications in industry are being investigated.

The performance of a compressor heat pump of 356 kW driven by a gas motor of 242 kW, which was developed in the first programme is now being tested. The heat pump has a six cylinder wet reciprocating compressor and uses R 114 as a working fluid. The heat source is hot water of 80° C and the steam outlet temperature is 110° C. Applications foreseen are in sulphuric acid manufacture, paint pigment and paper drying and food processing.

Another application of ICE driven air source heat pumps is believed to be feasible in grain drying. The heat pump produces both chilled air of 4-5° C (heat source), which can be used for refrigerated storage of undried grains, and hot air which is used for drying. In this way the grain can be stored for 3-4 months when chilled to 4-5° C. A dryer can then be utilized over a period 2-3 times longer than is typical with current practice. The construction of a 5 tonne batch grain dryer and a 20 tonne grain store has started. A 1 kW refrigerator with a 1 kW ventilation fan will be assembled as a heat pump driven by an air cooled ICE engine. If tests are satisfactory a 100 tonne store will be built in order to attempt to dry this quantity over a 4 month period.

A start has been made with the development of absorption heat pumps for industrial applications. A 10 kW prototype is being built with a lithium bromide/water fluid pair. The heat pump will transform heat at 20-50° C into heat at 40-80° C. The results of the prototype will be used for the design of a heatpump with a heat output of 300 kW.

In order to increase the energy saving potential of heat pumps in industry it is essential that high temperature heat pumps should be developed.

A Brayton cycle heat pump could be an interesting option. Here air of 1 bar is heated with waste heat to 50° C, is expanded in a turbine to 0,5 bar and 2° C and is then heated up to 73° C with waste heat of 90° C and compressed to 1 bar at 165° C. This heat will be used for drying in the production of milk powder. The turbo compressor will deliver 4 kg/s of hot air at 165° C and consumes 200 kW for a production of 500 kg milk powder per hour. The COP is calculated to be 2,75.

Also absorption heat pumps at least in principle allow the achievement of higher temperatures. The development of a high temperature absorption heat pump has received little attention thus far. There is considerable lack of knowledge regarding suitable working fluid pairs and their thermodynamic and thermal properties. To this end different working fluid pairs for heat pump operation above 120° C will be studied such as methyl amine-water, methanol-lithium bromide etc.

Finally a heat transformer will be constructed which is expected to produce 10 kW heat at 125° C and 10 kW at 30° C with a heat input of 20 kW at 85° C. The working fluid pair will be NH_3/H_2O.

5. CONCLUSIONS.

The objectives which we hope to achieve by the end of the second programme are :

- A clear insight in the possibilities of different types of heat sources

- To have sufficient data on absorption and ORC driven heat pumps to be able to judge their economic and technical feasibility

- To have a satisfactory control system for both compressor and absorption heat pumps

- The establishment of guidelines for heat pump COP determination

- Finally a start should have been made with the identification of non technical barriers for the introduction of heat pumps.

6. LIST OF AVAILABLE FINAL REPORTS.

Many projects resulted in final reports, a list of which is given below. They are indicated by a EUR number of which the final letters indicate the language of the report : EN = English, FR = French, D = German. The reports can be ordered from :

 Office for official publications of the European Communities
 P.O.Box 1003
 LUXEMBOURG.

Heat sources

. Investigation on using the earth as a heat storage medium
 and as a heat source for heat pumps   EUR 6835 EN

. Recovery of waste heat in sewers by heat pumps   EUR 7061 D

Conventional heat pumps and components

. Development of domestic heat pumps   EUR 7098 EN

. Air source heat pumps under Northern European climatic
  conditions   EUR 7388 EN

. Frosting and defrosting behaviour of outdoor coils of
  air source heat pumps   EUR 7281 EN

. Microprocessor based control system for heat pumps   EUR 7283 EN

. Heat pump models for microprocessor based control systems   EUR 7046 FR

. Analysis of the factors which determine the COP of a heat
  pump   EUR 7048 EN

. Heat pumps for individual rooms   EUR 7007 EN

Advanced heat pumps

. Development of an absorption heat pump fired by primary
  energy for domestic heating   EUR 6326 EN

. Design construction and testing of a prototype absorption
  heat pump fired with primary fuel   EUR 7064 D

. Directly fired heat pump for domestic and light commercial
  application   EUR 7063 EN

. The design and development of an absorption heat pump for
  domestic heating   EUR 7129 EN

. Heat pump operating with a fluid mixture   EUR 6848 FR

. Feasibility and design study of a gas engine driven high
  temperature industrial heat pump   EUR 6262 EN

Heat pump applications

. A gas engine driven heat pump using air as heat source
  for an apartment building   EUR 7133 EN

. Diesel heat pump for district heating plants and for the
  heating of large housing blocks   EUR 6740 EN

THERMAL TRANSFORMERS FOR UP-GRADING INDUSTRIAL WASTE HEAT

I.E. Smith, C.O.B. Carey

Cranfield Institute of Technology, U.K.

Summary

This paper discusses the principles of thermal transformers for the treatment of waste heat. A fraction of the waste heat is elevated in temperature, whilst the remainder is rejected at a lower temperature in an absorption cycle. Possible absorbent/working fluids are listed, and the one commercially available thermal transformer is described.

Held at the University of Warwick, U.K.
Symposium organised and sponsored by
BHRA Fluid Engineering

1. INTRODUCTION

The process of thermal transformation may be defined as one in which a
quantity of energy is supplied at a given temperature, T_m; a certain fraction of it is
rejected to a lower temperature, T_o, and the remainder is upgraded to a higher
temperature, T_h. A device or cycle which will achieve this is termed a thermal
transformer.

Before examining the mechanism of thermal transformation a brief review of
the thermodynamic principles may be helpful.

In conventional heat transfer processes the taking in of a quantity of heat
in at a temperature, T_m, and rejecting it at a lower temperature, T_o, is accompanied
by an increase in the entropy, that is to say the process is irreversible. If,
however, the energy is downgraded reversibly from T_m to T_o (i.e. with no increase in
entropy) then, by implication, the energy, or a further supply of energy, may be
upgraded over a similar temperature interval which need not be at the same temperature
level as that at which the down grading took place, provided there is no nett decrease
in entropy.

Evaporation and condensation are reversible processes and, apart from a
small temperature difference necessary to pass heat into an evaporator or out of a
condenser, the input and output of heat occurs isothermally. One example of the
application of this process is, of course, the heat pipe where evaporation and
condensation occur at the same temperature.

However, processes which are reversible at a single temperature level are of
little interest so far as transferring heat up and down temperature gradients is
concerned, and if heat is to be "pumped" more or less reversibly a difference in
temperature must be established between the evaporation and condensation phases of the
process.

This may be done by increasing the pressure of the vapour between
evaporation and condensation, as in the case of the vapour compression heat pump. To
compress a vapour requires the expenditure of mechanical energy, and this energy may
be derived from the expansion of a quantity of vapour in another part of the cycle.

Such a process is exemplified by the Rankine – Rankine cycle as is shown in
Fig. 1 which illustrates the principle components, viz, a boiler and condenser,
providing a mechanical output via an expander which in turn drives a compressor which
compresses the vapour produced by an evaporator, following which it is condensed at
the higher temperature.

In the role of a thermal transformer, waste heat would be fed into the
boiler and rejected to the environment via the condenser at a low temperature whilst
an additional supply of waste heat would enter the evaporator, from whence the vapour
would be compressed and condensed at the high temperature.

The poor thermal efficiency of the power cycle operating over a low
temperature range is, in principle, compensated for by the ability of the compression
cycle to elevate several times more heat than corresponds to its power input.

This development and application of low temperature power producing Rankine
cycles has recently been reviewed by O'Callaghan and Hussein (Ref. 1) and they are now
under active development in many places.

2. ABSORPTION THERMAL TRANSFORMERS

The type of thermal transformer that will be discussed here is the absorption
transformer, the principles of which were outlined by Altenkirch (Ref. 2) in 1918. It
depends for its operation on the absorption of the vapour evolved by an evaporating
pure liquid in a medium, which may be liquid or a solid, having a strong affinity for
the vapour. As a result of this affinity the pressure of the vapour above the
absorbing medium is lower than the vapour pressure of the pure liquid.

It follows that if the pure liquid and the absorbing medium are at the same
pressure then the temperature of the absorbent will be higher than that of the pure
liquid when the system is in equilibrium. If the temperature of the absorbent is

slightly less than this equilibrium value, then the pure liquid will evaporate and be adsorbed at a higher temperature, releasing heat at this temperature. This is, of course, the principle behind the familiar absorption refrigerator.

In the context of the thermal transformer this principle is depicted in Fig. 2. Waste heat at T_M is supplied, in the first instance, to the absorber where the working fluid is driven off from the absorbent and the pure fluid then condensed at a lower temperature, T_O, rejecting heat to the environment. Having accumulated a pure liquid condensate, the supply of waste heat is then transferred to this vessel which rises to a temperature T_M. Evaporation takes place, and the re-absorption causes the absorber temperature to increase to a temperature T_H, providing a useful thermal output.

Whilst such a cycle will accept waste heat at T_M continuously, the output of useful heat at T_H is clearly intermittent with the off periods being spent rejecting heat to the environment at T_O. This may or may not be desirable in an industrial process.

Were the heat required for de-sorption or absorption equal to the heat of evaporation/condensation then a coefficient of performance for such a device defined as the quantity of heat delivered at T_H to the quantity of heat provided at T_M would be 0.5, i.e. half of the waste heat would be elevated in temperature and half degraded.

However, fundamental to the process of vapour pressure lowering by physical-chemical absorption is the fact that it is accompanied by the evolution of heat (in addition to the latent heat released by the working fluid), and similarly desorption is associated with an additional input of heat. Thus the enthalpy change in the absorber/desorber is greater than that in the condenser/evaporator. This implies that the process must have a coefficient of performance of less than 0.5. Furthermore a number of irreversibilities are immediately apparant in the above cycle, the principle ones being the sensible heat required to elevate the absorbing material from T_M to T_H and the condensate from T_O to T_M. Another is the sensible heat carried in the vapour passing from the absorber to condenser during heat rejection and also a similar quantity when the vapour is transferred from the evaporation stage to the absorber.

3. CONTINUOUS OPERATION

The principal irreversibilities may be minimised and the cycle made to provide a continuous output by providing separate units for absorption, desorption, evaporation and condensation and circulating the absorbent and working fluid between them. The arrangement of the cycle is shown in Fig. 3 with one pump circulating the absorbing solution and a second pumping the liquid condensate at T_O into the higher pressure evaporator where the temperature is T_M. The provision of a heat exchanger which transfers heat from the hot liquid leaving the absorber to the cooler fluid entering from the desorber may reduce the downgrading of heat (an irreversibility) to small proportions. However, as has been recently pointed out (Ref. 3), the load on this heat exchanger can be several times that of the thermal output of the device, resulting in a high capital cost.

4. TEMPERATURE ELEVATION

The degree of thermal "lift" that can be obtained depends on the vapour pressure lowering that can be achieved by the absorbing medium. For many absorbents this varies in an inverse manner with the concentration of the working fluid absorbed in them. Such materials are termed bi-variant implying that the vapour pressure depends on both temperature and concentration. One well known example of a bi-variant absorbent is lithium bromide, and the vapour pressure/temperature/concentration diagram for water absorption is shown in Fig. 4. The vapour pressure curve for pure water is also shown.

The temperature elevation that is possible (in principle) can be judged by reading along a constant pressure line from the temperature of pure water to that temperature corresponding to a particular concentration in the absorbent, and it is clear that the lower the concentration of water in the absorbent, the greater will be the temperature elevation possible.

However, an upper limit to this temperature elevation is set by the solubility of the salt in water. Once the absorbing solution becomes saturated, crystallisation will occur and the absorbent concentration will no longer increase.

Thus two factors will influence the temperature elevation (a) the saturation point which will determine the maximum temperature elevation possible and (b) the absorbent circulation rate in relation to the vapour flow which will determine the minimum temperature to which heat may be pumped. The higher the circulation rate then the closer to one another will these temperatures lie.

A second class of absorbents are termed univariant and are generally solid crystalline materials which are capable of adsorbing liquids to form well defined crystal structures when "solvation" takes place.

An example of one such reaction that has been investigated by Offenhartz & Brown (Ref. 4) is calcium chloride forming a complex containing 2 moles of methanol per mole of the chloride according to the reaction :

$$CaCl_2{}_{(s)} + 2\ CH_3OH_{(g)} \rightleftharpoons CaCl_2.2\ CH_3OH_{(s)} - 103.8\ kJ.$$

The total enthalpy change in this reaction is roughly 1.35 times the latent heat of methanol. The vapour pressure of this complex is shown in Fig. 5, from which it can be seen that the complex and pure alcohol are in pressure equilibrium when there is a temperature difference of about 65°C between them.

With this class of absorbent the temperature "lift" that is available will not vary as the process of methanolation proceeds so long as unsolvated salt is present within the absorber, and this may be an advantage if a thermal output at a constant temperature is required.

A solid absorbent corresponds to the situation depicted in Fig. 2, in which it has already been seen that only an intermittent delivery of heat is possible, and a continuous delivery of heat is not possible unless two such units are employed in parallel. This will be disadvantageous for continuous industrial processes but in those which are batchwise in operation it may not be. Wentworth, Johnson & Raldow (Ref. 5) have considered the interesting idea of circulating an adsorbing solid as a suspension in an inert liquid, thus combining the advantages of a univariant absorbent with the mobility and steady state output associated with bivariant liquid absorbents.

5. SELECTION OF WORKING FLUIDS AND ABSORBENTS

The choice of working fluids and absorbents for thermal transformers is legion, and their selection is a bewildering one involving considerations that extend far beyond the simple parameters of temperature, performance and operating pressure. Chemical and mechanical long term stability, transport properties, corrosion and cost, toxicity and hazard aspects enter into their selection exerting a strong influence of the overall system design and cost.

Table 1 indicates some of the combinations that might bear consideration for thermal transformers with, where necessary, a reference cited.

The last two combinations represent absorbent-resorbent combinations in which the working fluid is transferred from one absorbent to another which has a strong, but lower, affinity for it. One advantage of absorbent-resorbent systems is that the full vapour pressure of the working fluid is never exerted, which, in the case of ammonia could be very high, and for hydrogen, impossibly so.

An interesting absorbent is Hexamethyl phosphorous triamide absorbing R22 which has recently been the subject of a patent application (Ref. 6) and appears capable of providing elevations of over 100°C. Little is known of this absorbent as yet, but it should be noted that the low latent heat of the halogenated hydrocarbons will require high circulation rates of both working fluid and absorbent per unit of thermal output, and this, in turn, will necessitate a sizable heat exchanger in the absorbent circuit.

It must be emphasised that the temperature elevations listed above are approximate and theoretical values. They do not allow for the fact that to drive the working fluid from the pure state into the absorbent there must be a driving force,

in terms of temperature, nor do they take into account that there must be a further
temperature difference to transfer heat into and out of the media. Finally, for
systems that derive or provide sensible heat to and from a fluid, the temperatures
within the cycle will be at the least favourable ends of the sensible heat change
within the fluid.

It will be noted that those absorbents capable of providing the greatest
elevations in temperature viz. sulphuric acid, sodium hydroxide are all severly
corrosive, whilst those in which ammonia is the working media entail high operating
pressures if evaporation takes place at the waste heat temperature.

6.   CURRENT STATE OF DEVELOPMENT

With such a longstanding knowledge of the principles of thermal transformers
it is pertinent to enquire why they have not yet adopted in those sections of industry
where they could be beneficial in terms of providing energy savings.

The answer lies in conservatism. Industry is necessarily conservative in
financing innovative ideas and unproven concepts since capital risk is involved.
Similarly, the consumer side of industry is cautious in investing in new concepts
until they have been tested and proven.

To the best of the author's knowledge there is but one company at present
marketing thermal transformers and this is the Industrial Section of the Institute
Francais du Petrole.

A recent paper by Cohen, Salvat & Rojey (Ref. 7) of the Institute outlined
the IFB "Thermosorb" process (Fig. 6) which employs ammonia as the working fluid and
water as the absorbent. The cycle is of the absorption - resorption type with water
as the common absorbent, but appears in differing concentrations at the different
temperature levels. As a result of this the pressure does not exceed 40 bar even
though the thermal output at a temperature as high as 220°C is possible.

Because an absorption - resorption process is employed rather than
condensation and evaporation of the pure working fluid, a rather modest temperature
lift is realised. Temperature elevations of 20-50°C only are claimed for the process
but comparatively high ratios of thermal output to input of 0.4 are realised. For a
process in which a certain critical temperature exists, e.g. steam raising or
sterilisation, to name only two examples, to realise a 40% recovery of heat is no
small achievement. Capital costs are stated to lie between £120-240 per equivalent
ton of heating oil saved per year, and with oil currently costing £100/tonne this
offers an attractively short pay-back period.

7.   THE FUTURE

It must be recognised that the operating characteristics of the materials
used in the IFP Thermosorb process are both well known and understood, and the process
may thus be regarded as the first generation of thermal transformers.

Undoubtedly the future will see the exploitation of other working fluids and
absorbents, once a knowledge of their characteristics has been obtained in order that
design exercises may be carried out.

One problem that remains in terms of producing a system of general
applicability to industrial processes is the diversity of these processes themselves.

Industrial processes embrace a wide variety of thermal requirements and
temperature levels. Indeed the waste heat produced by them may be available at a
constant temperature, e.g. from condensation, or over a temperature range if it is
available only as sensible heat, or a combination of both. What is now required is a
product that will have a wide applicability over a range of operating requirements
that will meet the needs of a large market.

8.   REFERENCES

1.   Hussein, M., O'Callaghan, P.W., and Probert, S.D.: "Solar activated Power Generator utilising a multi-vane expander as a prime mover in an organic Rankine cycle". Solar World Forum, ISES, Brighton (UK), Aug. 1981.

2.   Altenkirch, E.: Zeitschr.f.d. gesamte Kalteindustrie, 25, 49-53, 57-60, 1918.

3.   Smith, I.E., and El-Shamarka, S.E.: "Absorption Heat Pumps for Space Heating". 3rd International Conference on Future Energy Concepts, London, Jan. 1981.

4.   Offenhartz, P.O., and Brown F.C.: "Methanol based Heat Pumps for Storage of Solar Energy". Proceedings of 14th IECEC, Boston (USA), Aug. 1979.

5.   Wentworth, W.E., Johnston, D.W., and Raldow, W.M.: "Chemical Heat Pumps using a dispersion of a metal salt ammoniate in an inert solvent". Solar Energy, 26, 141-145, 1981.

6.   UK Patent, GB 2036782A, N.V. Philips, July, 1980.

7.   Cohen, G., Salvat, J., and Rojey, A.: "Valorisen de calories a bas niveau au moyen de cycles trithermes". Entropie, 84, 31-36, 1978.

8.   McBride, J.R.: "Chemical Heat Pump Cycles for Energy Storage & Conversion". Paper C2, International Conference on Energy Storage, Brighton (UK), May, 1981.

9.   Baaken, K.: "Systeim Tepidus, High Capacity Thermochemical Storage/Heat Pump". Paper C1, ibid.

10. Blytas, G.C., and Daniels, F.: "Concentrated Solutions of Ammonia in NaSCN". J. Amer. Chem. Soc., 84, 1075-1083, 1962.

11. Thieme, A., and Albright, L.F.: "Solubility of Refrigerants 11, 21 and 22 in organic solvents containing a nitrogen atom and in mixtures of liquids". ASHRAE. J., 71-75, 1961.

12. ASHRAE Handbook of Fundamentals, Ch.17.

13. Gruen, D.M., Mendelsohn, M.H. and Sheft, I.: "Metal Hydrides as Chemical Heat Pumps". Solar Energy, 21, 153-156, 1978.

14. Offenhartz, P.O.: "Chemical Methods for Storing Thermal Energy". Int. J. Solar Energy, 8, 48-70, 1976.

TABLE 1

Working Fluid/Absorbent	Temp. Elevation ^{o}C	Reference
Water/LiBr	70	–
Ammonia/Water	60–80	–
Methanol/$LiBr.ZnBr_2$	70	3
Water/H_2SO_4	120	8
Water/NaOH	110	–
Water/Na_2S	55	9
NH_3/NaSCN	80	10
R22/Di-methyl acetamide	70	11
Water/Silica Gel	40	12
R22/Hexamethyl phosphorous triamide	100	6
H_2/$LaNi_5$:$CaNi_5$	70	13
NH_3/$FeCl_2$:$CaCl_2$	55	14

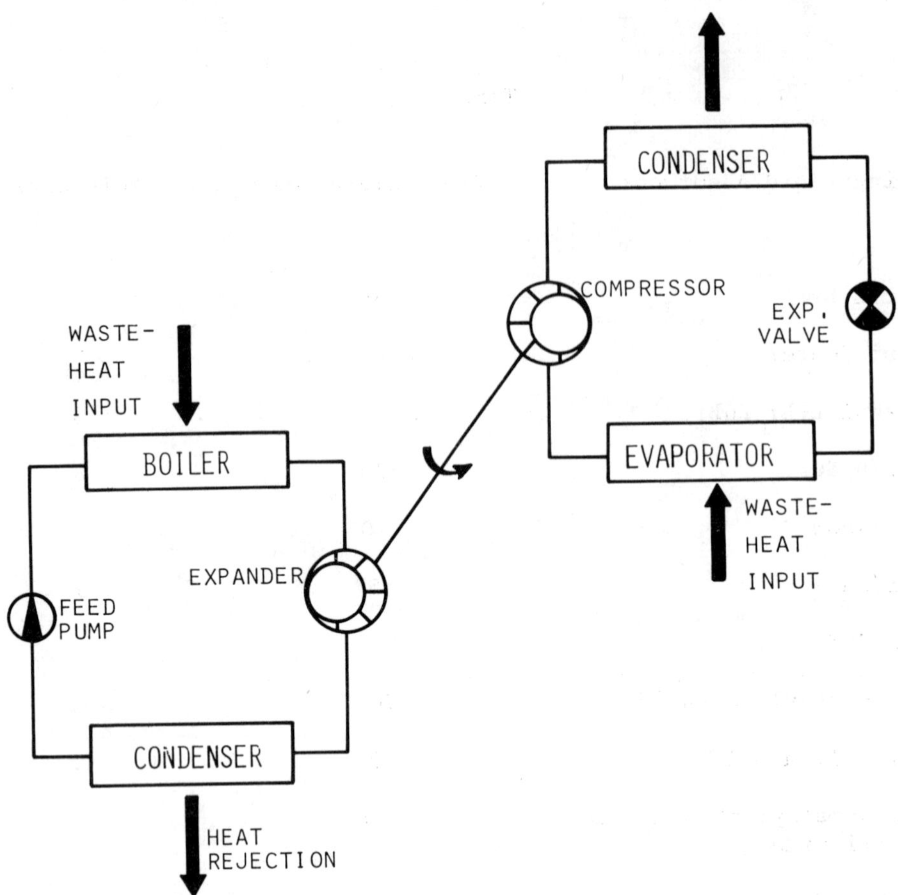

Fig. 1.   The Rankine – Rankine Cycle As A Thermal Transformer.

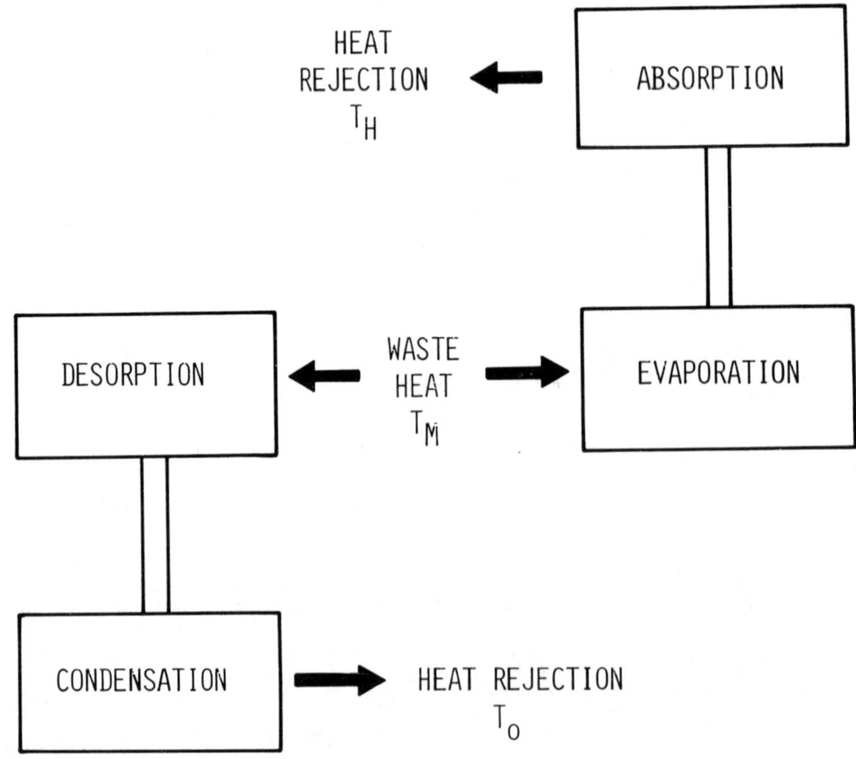

Fig. 2.   Principle of Thermal Transformation by Absorption/Desorption.

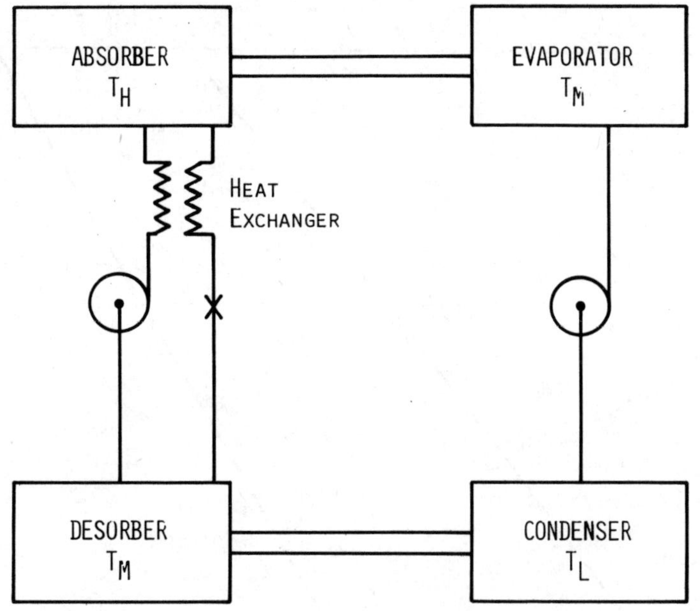

Fig. 3.   Circulating Thermal Transformer

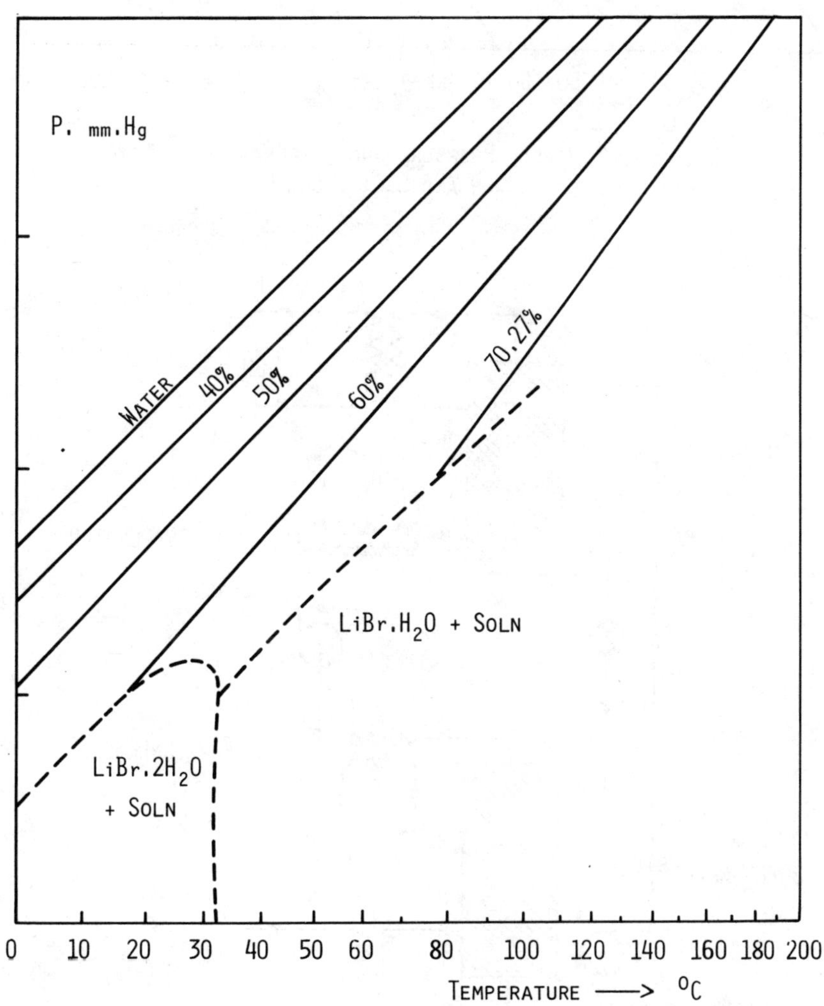

Fig. 4.   LiBr/H$_2$O System : Vapour Pressure Versus Temperature and Concentration (% By Weight)

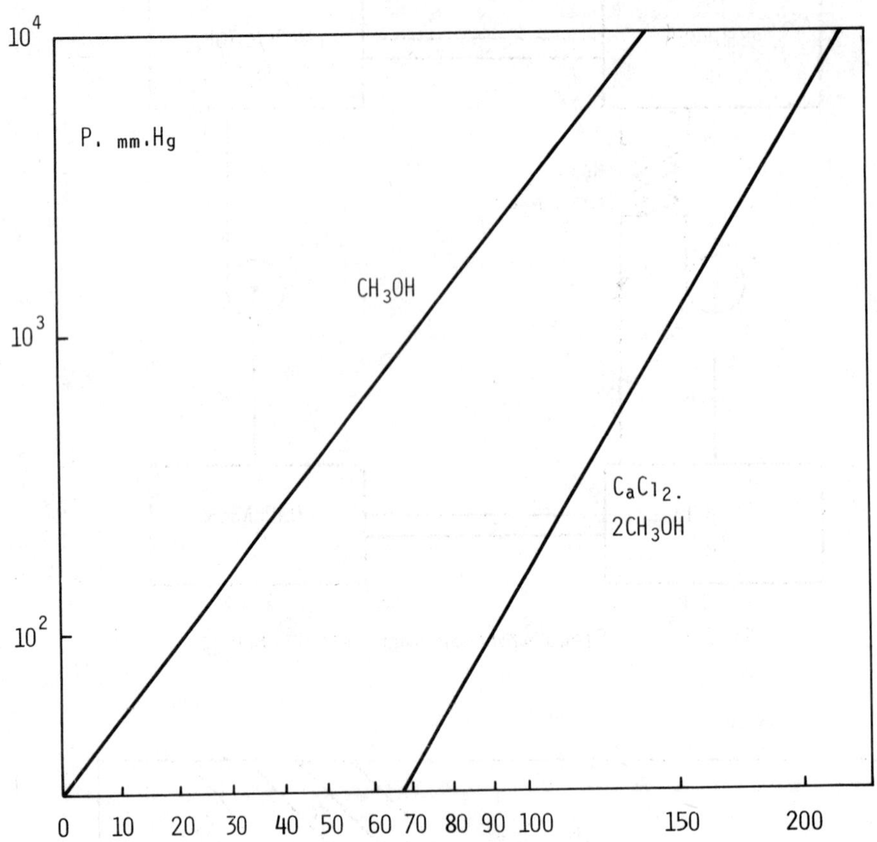

FIG. 5.   VAPOUR PRESSURE AND TEMPERATURE
CH$_3$OH AND CaCl$_2$.2CH$_3$OH

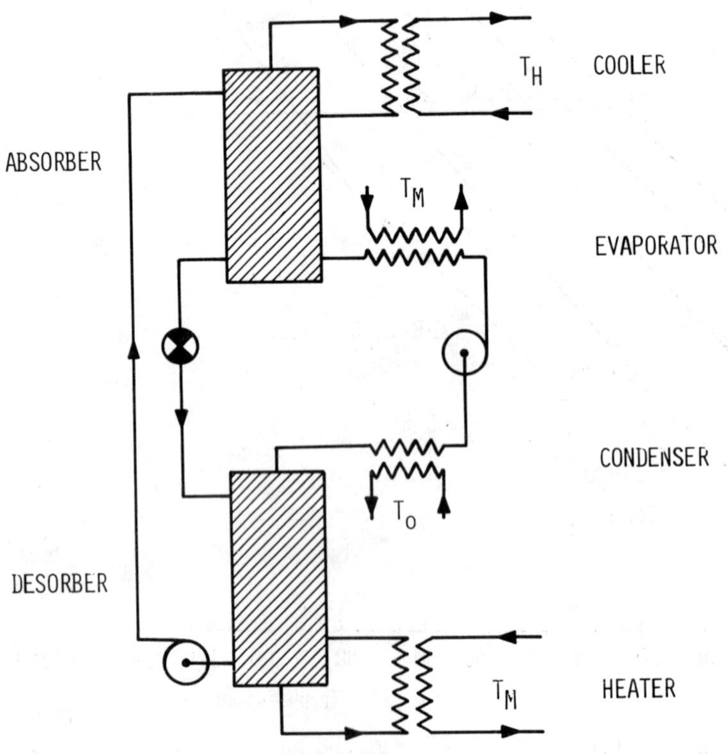

FIG. 6.   I.F.P. THERMOSORB PROCESS

THE SELECTION OF A WORK FLUID/LUBRICANT COMBINATION FOR A HIGH TEMPERATURE HEAT PUMP DEHUMIDIFIER

M.P. Bertinat, F.G. Drakesmith, B.J. Taylor

Electricity Council Research Centre, U.K.

Summary

A major problem in the design of a high temperature vapour compression heat pump is the choice of the work fluid/lubricant combination. The work fluid has to be thermodynamically and environmentally suitable, whilst the combination of it and the lubricant has to be thermally and chemically stable and provide adequate viscosity for the heat pump compressor.

The requirements of an $80^{o}C$ dehumidifier can be satisfied by the relatively low-cost combination of refrigerant R114 and a particular highly refined mineral oil. Such a dehumidifier was tested at ECRC by using it to dry 15 commerical loads of timber. Energy consumption was much less than with conventional drying.

For temperatures much higher than this, alternatives have to be found to both the standard refrigerant work fluids and mineral oils.

Held at the University of Warwick, U.K.
Symposium organised and sponsored by
BHRA Fluid Engineering

1. INTRODUCTION

Heat pump dehumidifiers have been used to reduce energy consumption in industrial drying processes for many years now, particularly in the timber-drying industry. In the past, these dehumidifiers have had a maximum working (i.e. drying) temperature of about 50°C which has rather limited their usefulness. A higher maximum temperature would obviously increase the range of applicability of this type of drying, so we set about developing a higher temperature machine, one that would be capable of drying at 80°C. This figure was chosen because it was high enough to encompass the majority of timber drying, which we saw as its biggest market, but not so high as to present overwhelming technical problems.

One of the major problems that did have to be overcome was that of finding a suitable work fluid/lubricant combination for use in a heat pump at this temperature. Work fluids suitable for low temperature use, such as R22, may be no use at higher temperatures for several reasons. For example, they may be thermally unstable, or they may have unacceptably high vapour pressures. Similarly, oils developed for low temperature use, where the problems are wax formation and work fluid flocculation, are likely to be quite unsuitable for high temperature use where the problems are low viscosity and thermal stability.

This paper describes how we went about selecting a satisfactory work fluid/lubricant combination for the 80°C dehumidifier and then goes on to consider briefly the requirements of still higher temperature machines.

2. BASIC REQUIREMENTS

The Key questions that need to be answered when choosing a work fluid/lubricant combination for a high temperature heat pump are:

   (i)   What are the properties required of an ideal work fluid?
  (ii)   What temperatures will the fluids be subjected to?
 (iii)   What substances might be good work fluids at these temperatures?
  (iv) What are the lubrication requirements of the system?
   (v)   What work fluid/lubricant combinations would meet these requirements?
  (vi) Are these work fluid/lubricant combinations chemically and thermally stable at the temperatures involved?

Let us now consider each of these questions in turn.

3. THE IDEAL WORK FLUID

In a vapour compression heat pump, the volatile work fluid is alternately evaporated at low pressure and temperature and condensed at high pressure and temperature. It absorbs heat at the evaporating temperature, T_E, absorbs work during compression from the evaporating pressure, p_E, to the condensing pressure, p_C, and rejects heat at the condensing temperature T_C, $(T_C > T_E)$. Ideally, the fluid should be:-

   (i)   thermodynamically favourable (see below);
  (ii)   non-toxic;
 (iii)   non-flammable;
  (iv) thermally stable;
   (v)   chemically inert;
  (vi) readily available;
 (vii) cheap.

For a fluid to be "thermodynamically favourable" it should have the following properties:

 (viii)   a critical temperature significantly higher than T_C in order that condensation does not occur too close to the critical point where the latent heat tends to zero;
  (ix)   a boiling point (at 1 bar) significantly less than T_E in order that the low pressure side of the system can always be at or above atmospheric pressure thus preventing any problems that might be caused if air or moisture were to leak into the system.
   (x)   a latent heat of condensation at T_C large compared with the work required to compress the gas from p_E to p_C so that a relatively high coefficient of performance (COP) can be obtained. (The COP determines the energy-saving potential of a heat pump, being defined as the ratio of the useful heat output to the high grade energy input.)

(xi)   a relatively large value of the product $\lambda\rho$, where λ is the latent heat at T_C and ρ is the vapour density at T_E and p_E, so that a relatively low specific compressor displacement (SCD) is required. (The SCD determines the size of compressor and hence the capital cost of a heat pump with a given heat output. It is defined as the volume of vapour entering the compressor per unit heat ouput from the condenser.)

(xii)   a saturated vapour pressure at T_C not so high as to impose design and safety limitations on the system.

(xiii)   a large value of dp/dh for saturated liquid, where h is enthalpy, so that only a small amount of subcooling (in energy terms) is required to prevent the formation of bubbles in the liquid line as a result of pipe pressure drop. (Such bubbles can result in partially empty evaporators and also damaged control valves.)

4.   WORK FLUID TEMPERATURES IN AN 80°C DEHUMIDIFIER

A dehumidifier takes air from the drying chamber or kiln, removes heat and water from it by passing it over the relatively cold evaporator, returns this heat (plus some extra from the compressor) to it by passing it over the relatively hot condenser, and then returns it to the kiln, hotter and dryer than when it left. Thus the required evaporating and condensing temperatures, T_E and T_C, for a dehumidifier working with a kiln at temperature T_K and humidity ϕ_K are governed by the fact that T_E must be significantly below the dewpoint of air at T_K and ϕ_K whilst T_C must be significantly above the temperature of the air returned to the kiln, which itself must be significantly higher than T_K.

For air at 80°C, the dewpoint varies from about 53°C to 70°C as the relative humidity varies from 30% to 70%. Hence for kiln conditions in this range, evaporating temperatures of about 45°C to 60°C are required, allowing for a temperature difference across the heat exchanger of about 10K. Condensing temperatures need to be about 95°C to 100°C. Thus the work fluid will be required to condense at upto 100°C and transfer heat through a temperature difference T_C-T_E of upto 50K.

5.   POSSIBLE WORK FLUIDS

The conventional halocarbon refrigerants are prime candidates for heat pump work fluids since they are generally non-toxic, non-flammable, readily available, and cheap. However, many of them are unsuitable for 100°C condensing because their critical temperatures are too low and/or their vapour pressures at this temperature are too high. Others are not commercially available.

The five that looked most promising, R11, R21[*], R113, R114 and R12B1, were compared thermodynamically by calculating how each would perform in an ideal vapour compression cycle (IVC cycle) between T_E=40°C and T_C=90°C. The IVC cycle, shown on a pressure-enthalpy diagram in figure 1, involves isentropic compression of vapour, isothermal condensation of this vapour, isenthalpic expansion of the resulting liquid, and isothermal evaporation of the expanded liquid. Just enough vapour superheat is introduced at either compressor inlet or compressor discharge to prevent liquid compression occurring, and all superheating and desuperheating occurs at constant pressure. There is no subcooling of the liquid.

Two IVC cycles are shown in figure 1 corresponding to the two types of fluid, which we shall designate types A and B. With type A fluids, the saturated vapour entropy decreases as the pressure rises, so saturated vapour entering the compressor emerges superheated after isentropic compression. With type B fluids, the saturated vapour entropy increases with pressure, so a certain amount of superheat is required in the vapour entering the compressor if it is to remain solely vapour during isentropic compression

Table 1 shows the results of these calculations for the five fluids. It is apparent that

(i)   the COPs do not vary much (5.4 to 6.2)

(ii)   the SCDs vary widely (0.34 to 1.39 m^3/MJ)

(iii)   the condensing pressures vary widely (3.5 to 13.9 bar) but all are practicable

* The toxicity of R21 is now in question.

R113 is the least attractive of the five, thermodynamically, having the highest SCD
and also the highest boiling point (48°C) which could lead to air ingress problems.
R21 and R12B1 look the most promising, with the lowest SCDs and good COPs. R114 and
R11 come somewhere in between, with R114 having the advantage of a 60% smaller SCD
but the disadvantage of a 10% smaller COP.

6. POSSIBLE LUBRICANTS

The two primary requirements of a lubricant for use in a high temperature heat pump
are that when mixed with the chosen work fluid at the appropriate pressures and
temperatures it should

   (i)   be chemically and thermally stable, and
   (ii) have the appropriate viscosity

There are many types of lubricant, each covering a wide range of viscosities. It is
logical therefore to assess the stabilities of the various types first, and then choose
the appropriate grade if possible to give the required viscosity.

Three basic types of lubricant were chosen for systematic study ranging from the
relatively cheap refined mineral oils, through the more expensive synthetic
hydrocarbons, to the very expensive synthetic fluorinated lubricants. It was
expected that cost and stability would go hand in hand, so the requirement was for the
cheapest type that would do the job.

7. THERMAL & CHEMICAL STABILITY

The chemistry of heat pump systems, particularly at high temperatures, is complex,
involving the work fluid, the lubricant, and traces of moisture and oxygen, together
with metals and metal oxides which may act as catalysts. It is important therefore
that not only must the work fluid be thermally stable, but also the work fluid/
lubricant combination must be chemically unreactive in the presence of the metals
and impurities found in the system at the temperatures involved. The next stage
was therefore to investigate the chemical stabilities of the thermodynamically
favourable work fluids in combination with the various types of lubricant.

The Chemical Simulator

Because of the large number of possible combinations of work fluids and lubricants,
it was obviously not feasible to test each combination in a working heat pump. We
thus set out to design a device which would simulate the chemical reactions that
occur in a heat pump at elevated temperature. The simulator needed to provide a cheap,
compact and practical method of obtaining a realistic measure of the chemical
performance and ranking order of the various fluid combinations in a heat pump over a
period of say five years, in as short a time as possible. (This technique represents
only one of several procedures developed at Capenhurst and, has been applied in
conjunction with others, such as sealed tube tests, degradation inhibition experiments,
chemical sampling from working system studies, etc., to obtain performance data on
these fluids.)

Figure 2 illustrates the system designed for the simulation experiments. The tests
were carried out in such a way that mixtures of potential work fluids and lubricants
were circulated from a cooled sump, via a metering pump, and injected into a copper
tube at 160° or 200° containing an aluminium and a steel spirally-grooved core. On
entering the heated region the volatile work fluid vaporised, passed up the copper
tube to a condenser and was then returned to the sump. The lubricant, meanwhile,
percolated downwards over the heated metal surfaces in an atmosphere of the work fluid
and returned to the sump, to begin a new cycle. During the course of the experiment
samples of the fluids were withdrawn and analysed for hydrogen ion, chloride ion,
fluoride ion and bromide ion, as well as for iron, copper and aluminium. At the end of
the test period the metal tube cores, and specimen metal flags in the sump were
dismantled and examined for corrosion, metal deposition and organic resin formation.
The lubricant was examined for changes in elemental analysis, viscosity, and colour
index as a measure of the degree of reaction and/or polymerisation which had occurred.

The simulator tests showed that R114 was the most stable of the five fluids whatever
the type of lubricant involved. They also showed that R11 and R113 were too unstable
for use even with the most stable lubricant and could be discounted. R12B1 and R21
were intermediate.

Both R114 (the most stable) and R21 (highest theoretical COP) were tried out in small-scale laboratory heat pumps and it was found that

(i)   the actual COPs of the two fluids were not much different;
(ii) the problems in using the more chemically active R21 would be considerable, partly because it attacks the sorts of materials used in seals and gaskets.

We therefore settled on R114 as the best work fluid.

This left us the problem of choosing a lubricant. The simulator tests indicated that with R114, a particular highly refined mineral oil might be sufficiently stable to do the job. If not, a more expensive synthetic hydrocarbon oil would probably be stable enough, whilst if all else failed, a synthetic fluorinated lubricant would provide the ultimate in stability.

8. VISCOSITY

The lubricant in a heat pump compressor is continuously exposed to work fluid vapour, and R114 is highly soluble in all three types of lubricant under consideration. As a result, the lubricant viscosity will be significantly reduced. The size of the reduction depends on the amount of R114 dissolved, which in turn depends upon the pressure and temperature obtaining. Thus in order to decide what grade of neat oil to use, we needed to know the viscosity-temperature-pressure relationships of R114/lubricant mixtures over the range of conditions likely to be encountered in the sump of the heat pump compressor.

We therefore developed an instrument that incorporated a modified suspended-level capillary viscometer and a pressure transducer, and which was capable of withstanding moderately high pressures and temperatures. Using this, we investigated several different samples of each lubricant under consideration, each sample containing a different proportion of R114 to lubricant. The pressure and viscosity of each sample was measured at various temperatures, enabling us to estimate the viscosity for any temperature and pressure over the range of measurements. The most useful presentation of the data obtained is as constant pressure curves on a viscosity-temperature plot, and figure 3 is such a plot for one of the lubricants considered. Note that it is dynamic or absolute viscosity, μ, that is plotted here (i.e. centipoise) although the capillary viscometer actually measures kinematic viscosity, ν, (i.e. centistokes), which is μ divided by density. This is an important point when dealing with synthetic lubricants because densities vary by a factor of 2 or more between the various types and it is μ that is the important parameter in lubrication, not ν. (For example, if a bearing requires a minimum viscosity of 5cP, the appropriate mineral oil (density $\approx$ 0.9kg/l) will need to be at least $5\frac{1}{2}$ cS but a synthetic polyperfluoro-ether lubricant (density $\approx$ 1.9kg/l) need only be $2\frac{1}{2}$ cS.)

Figure 3 is typical of all the R114/lubricant mixtures we investigated. The maxima exhibited by most of the curves result from the fact that both the viscosity of a liquid of constant composition and the solubility of a gas in a liquid generally fall with increasing temperature. Thus at low temperatures, where pressures are relatively high compared with the saturated vapour pressure of R114 and concentrations of R114 in the lubricant are relatively high, the changing solubility effect dominates and viscosity rises with temperature. At high temperatures, where solubilities are lower, the direct viscosity-temperature effect dominates and viscosity falls with temperature.

An analysis of the bearings in the compressor we intended to use suggested that a minimum viscosity of 2-3 cP was required for satisfactory lubrication. To be on the safe side we decided to double this figure and stipulate a minimum of 5cP at all times in the compressor sump. For the particular lubricant used to draw figure 3, this meant a maximum sump pressure of about $5\frac{1}{2}$ bar (in which case the sump temperature would have to be held at about 85°C). For a sump pressure of 5 bar, the sump temperature could be between 70°C and 95°C, corresponding to between 15 and 40K of superheat. Since evaporating temperatures of 60°C were envisaged, which with R114 imply sump pressures of about 5 bars, the use of this lubricant would require that sump temperatures be held within these limits.

Overall, the viscosity measurements showed that so long as sump temperature could be carefully controlled, any of the types of lubricants that we had found to be chemically stable could be used.

9. TESTING IN A DEHUMIDIFIER

With both simulator tests and viscosity measurements suggesting that the relatively cheap mineral oil/R114 combination might well be satisfactory, it had to be thoroughly tested in a real heat pump. A prototype dehumidifier was constructed and attached to a purpose-built humidity chamber. Using our chosen work fluid/lubricant combination, we found that overall COPs of 3 to $3\frac{1}{2}$ were obtainable with this machine at most of the kiln conditions that it had been designed to work at. Although significantly less than the theoretical (IVC) values, these COPs were considered satisfactory. Specific moisture extraction rates of upto 2.6kg/kWh were obtained, which compared well with the theoretical maximum value of 1.5kg/kWh that could be obtained in a perfect direct-heating system.

After a few weeks of running in and performance testing, the machine was run continuously for 500 hours over the whole range of working conditions. It was then subjected to 1000 stop/start sequences, each consisting of 15 minutes running and 5 minutes off, bringing the total running time upto 1000 hours. No problems arose during this time and a subsequent strip-down of the compressor revealed very little wear. Overall it appeared to be in good condition. We concluded that both our basic dehumidifier design and our chosen fluid combination were satisfactory.

Working in collaboration with Westair Ltd, we went on to test an improved version of the dehumidifier (which had a maximum specific moisture extraction rate of about 3.0kg/kWh) by using it to dry real timber. The timber-drying trials were designed, after agreement with representatives of the timber industry, to demonstrate the ability of the high temperature dehumidification system to dry timber in accordance with the standard drying schedules published in the Timber Drying Manual. Commercial loads of timber were used and as many varieties as possible were investigated. From the fifteen loads dried, we concluded that the system could produce dried timber of the same quality and in the same time as standard heat-and-vent kilns but at much lower energy costs.

10. HIGHER TEMPERATURES

The 80°C dehumidifier involved a maximum work-fluid condensing temperature of 100°C. For higher condensing temperatures, upto about 120°C, R114 is still a satisfactory work fluid, but for chemical stability and viscosity reasons a synthetic lubricant is called for. We have carried out wear tests on compressors in systems condensing at 120°C, and have found that with R114 and a particular synthetic hydrocarbon lubricant there are no problems at the temperatures and pressures involved.

However, at 120°C condensing, the theoretical COP of R114 is beginning to fall off since its critical temperature is only 146°C. Its vapour pressure is also getting rather high (20 bars). So for still higher condensing temperatures, say 150°C, a different work fluid must be used and few, if any, of the conventional refrigerants are suitable. Our exhaustive searches have revealed several possible alternatives, such as the perfluoro-alkanes and trifluoroethanol, but none which we could claim to be ideal. The best of them, combined with appropriate lubricants, are currently being evaluated in heat pump systems condensing at 150°C.

BIBLIOGRAPHY

For further details on various aspects of this work, see

1. Stirling, R & Drakesmith, F G   "Refrigerant R114: Viscosity and Solubility Measurements in Hydrocarbon and Synthetic Lubricants", paper B1-38, 15th International Congress of Refrigeration, Venice, 1979

    Drakesmith, F G & Stirling, R   "Chemical Testing of Fluids for High Temperature Heat Pumps", paper B1-39, ibid"

2. Drakesmith, F G   "Fluids for High Temperature Heat Pumps", paper no. 6 fo Symposium Papers 1981 No.3, Institution of Chemical Engineers (North Western Branch), Salford, April 1981

3. Driscoll, J L et al   "The Capenhurst Westair High Temperature Heat Pump Dryer", Electricity Council Research Centre report no. ECRC/R1440, June 1981

4.   "High Temperature Dryer Cuts Costs", Timber Trades Journal, pp24-25, 25 July 1981

5.   "High Temperature Kilning on Trial", Timber Trades Journal, pp28-29, 8 August 1981

TABLE 1   CALCULATED PERFORMANCE PARAMETERS FOR FIVE POSSIBLE WORK FLUIDS UNDERGOING IDEAL VAPOUR COMPRESSION (IVC) CYCLES CONDENSING AT 90°C AND EVAPORATING AT 40°C

FLUID	B.Pt ($^{\circ}$C)	Tcrit ($^{\circ}$C)	COP	SCD (m^3/MJ)	CONDENSING PRESSURE (bar)	SUCTION SUPERHEAT (K)	DISCHARGE SUPERHEAT (K)
R11	24	198	6.0	0.79	6.6	0	3
R21	9	178	6.2	0.40	10.7	0	15
R113	48	214	6.1	1.39	3.5	14	0
R114	4	146	5.4	0.46	11.5	14	0
R12B1	-4	154	5.8	0.34	13.9	0	3

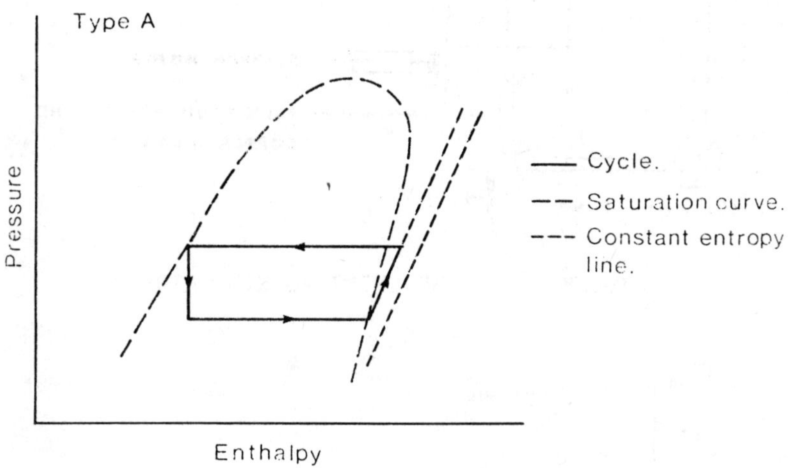

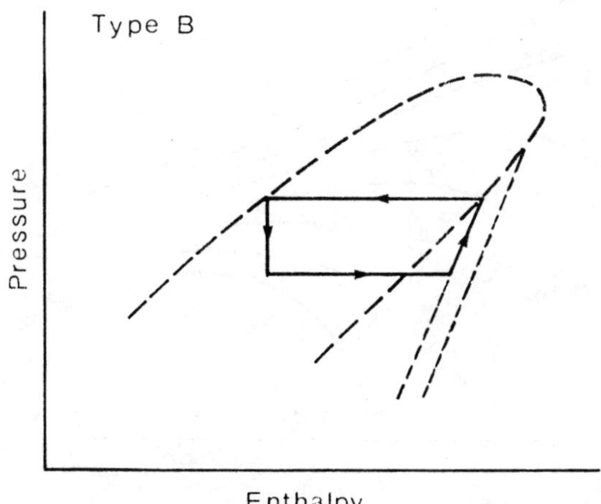

FIGURE 1 :   SCHEMATIC IDEAL VAPOUR COMPRESSION (IVC) CYCLES FOR TYPE A AND TYPE B FLUIDS

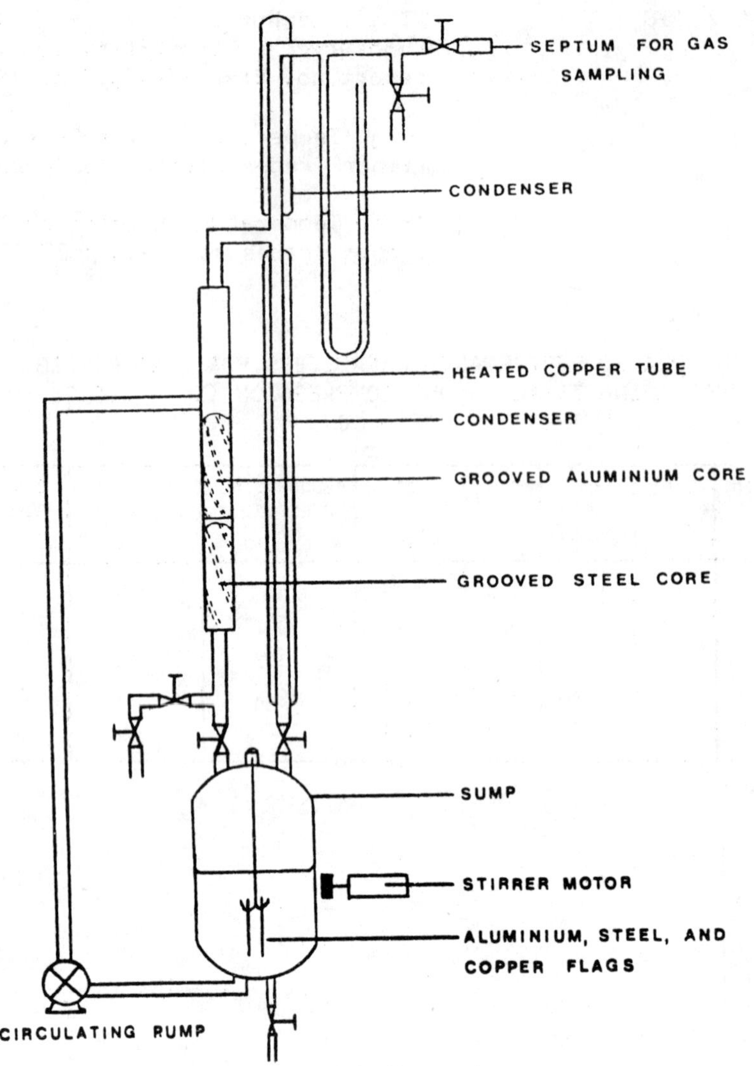

FIGURE 2 : THE CHEMICAL SIMULATOR

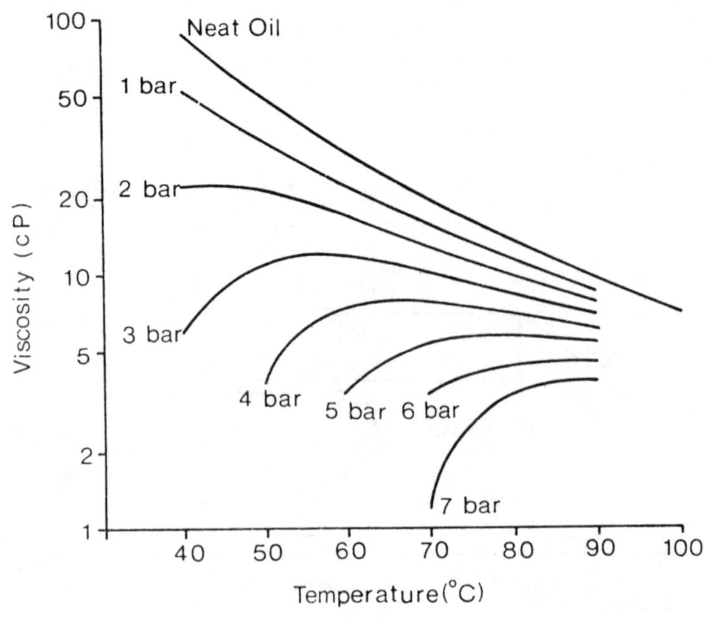

FIGURE 3 : VISCOSITY CURVES FOR A PARTICULAR MINERAL OIL/R114 COMBINATION

COMPARATIVE PERFORMANCE OF HEAT PUMPS VS COGENERATION DISTRICT HEATING SYSTEMS

P. Centrone

CNEN, Italy

E. Macchi and G. Angelino

Politecnico di Milano, Italy

Summary

Heat extraction in co-generating steam power stations is recognized to produce an important loss in plant power output. This waste of energy is compared with the power consumption of heat pump systems using surface water as heat source. In order to evaluate the energy effectiveness of water to water heat pumps the performance of some currently available basic components is analyzed. In particular the actual and potential overall efficiency of various types of compressors is found to range between 50 and 70% (including electric motor losses). Low temperature heaters performance is compared with that of ordinary radiators, showing the potential for a sizeable increase in specific output for some new configurations. Heat loads calculations show that de-rating ordinary heating systems to cover only a fraction (even if large) of the annual energy needs (for example 85%) allows an important reduction of heating surface temperature thus making heat pump systems viable also with comparatively poor performance heaters like ordinary radiators.

With reference to the Milan situation, water availability is analyzed and water to water heat pumps performance is computed assuming various types of final heaters and different engine efficiencies.

Even moderately efficient heat pump systems compare favourably with co-generation with respect to energy performance. Primary energy savings over conventional family size boilers range from about 30% in the case of low performance heat pumps, to over 50 in the case of good efficiency heat pump systems. The reference cogeneration system exhibits 41% primary energy saving.

Held at the University of Warwick, U.K.
Symposium organised and sponsored by
BHRA Fluid Engineering

1. INTRODUCTION

The combined production of power and heat is widely recognized as an effective means of energy conservation. When the heat consumer is a large city, the selection of the energy generation site outside the urban area removes an important polluting source from the most sensitive urban environment, possibly allowing the use of less expensive but more polluting fuels. Furthermore, outside the urban areas it is more likely that such conditions for power generation are found (cooling capacity power supply lines, etc.) that power can be produced at standard costs, also when the heat demand drops due to climate seasonal variations. In this way, plant capital costs are much more rapidly amortized. Recent plans for supplying the Milan area with an urban heating system were inspired by similar considerations.

A consequence of the increased distance between heat producing and consuming areas is the rise of the economic optimum top temperature for the heat carrier. At moderately high temperatures, the thermal power absorbed by the heat carrier from the steam power station is by no means a "waste heat", having a high thermodynamic cost which is reflected by the loss of power generating capacity which accompanies the heat extraction. Typically, one third of the plant power output is sacrified to heat generation.

For this reason, large central heating systems can be considered as important electric power consumers, and a form of COP (heat delivered to electric power loss ratio) can be calculated in any specified case.

It happens that this COP is not substantially different from the one that can be achieved through heat pumps using water as heat source. Provided adequate water resources are available within the town, a water based heat pump system has the potential of competing with a district heating system in terms of primary energy consumption. However if, in this way the problems connected with the laying of a hot water distributing network in an existing urban tissue are removed, other problems are risen, related to the lower temperature potential of heat pumps and to their capital and water connection costs.

The aim of this paper is to contribute to the evaluation of the merits of in-town water to water heat pump systems.

2. WATER AVAILABILITY

Rainfall and snowfall are abundant in the northern part of Italy and give rise to ample inland water resources, both on surface (rivers and lakes) and underground. Many important Italian towns have direct access to river waters in quantities much beyond any foreseeable energy requirement. The southern part of Italy is comparatively less rich of inland water. Moreover, both the North and the South of Italy are in a very favourable position for exploiting the comparatively warm waters of the Mediterranean Sea for energy unes (a particularly attractive situation is, for example, that of the city of Venice (Ref. 1).

Within such a general picture, the city of Milan, toward which is expecially focused this work, seams not to have at its disposal abundant surface water resources, at least in relation to its 1.7 million people population. No large rivers or lakes are found in the urban area. Most of the available surface water comes in through two irrigation canals, formerly used also for barge navigation. Two minor rivers cross the town. An important stream of ground water is continuously pumped from underground for the population needs. All these limited water resources are characterized by flowing directly within the city aside or under the city roads, so that an ample and complicated canal system lies under the populated area (see Fig.1). Overall winter water flow is around 50 m^3/s with minimum temperatures of 4.5 to 10°C. Most of this water has no further uses after crossing the town. Assuming an average heat extraction corresponding to a 3°C temperature drop, a heating capacity of about 800 MW base load (1500 MW peak, with conventional integration) could be obtained through an electric heat pump system, covering the needs of about 1/3rd of the entire population.

Winter irrigation water supply to the city could possibly be increased, the limiting factor being not the water availability (agricultural winter uses of water are modest(*)) but the carrying capacity of the canals which could be improved by appropria-

---

(*) winter use of irrigation water in Lombardy is, in fact, an energy use. Comparatively warm water, distributed continuously on the fields, prevents the freezing of the soil and causes an increase in grass crop. The cost of this form of "heat" to the farmer is extremely low. A heat pump system could efford to pay the unit energy such more, having the economic potential of promoting the exploitation of surplus winter carrying capacity of irrigation water ways.

te actions.

3. <u>LOW TEMPERATURE HEATERS PERFORMANCE AND DEVELOPMENT</u>

It is well known that is not economically justified to meet peak heat demand by means of the heat pump plant. A peaking system relying on electric power or fossil fuel must be provided. In most cases, such peaking (and, possibly, back-up) plant must be inclusive of the related final heaters. Consequently the heat-pump - connected heat distributing units must handle only a fraction of overall heat requirement.

In the case of the Milan area, a conventional minimum ambient temperature of -5°C is usually adopted for heat needs calculations, while the average temperature for the heating season (mid October to mid April) is +5.9°C. This indirectly implies that a moderate fraction of the peak power,continuously supplied, is sufficient to deliver a large fraction of the overall heating load. As shown in Fig. 2, for instance, a system with a maximum capacity rated at 53% of the conventional peak heating load supplies about 85% of the seasonal heat need. Such a system covers the base load in the period mid November-end of February and is employed at part load outside that period (see Fig. 3). The selection of a moderate capacity for the heat pump system makes it easier to over-size the final heaters, which is beneficial for heat pump performance. Using floor heating panels, for instance, a peak load average water temperature of 35°C is requested to supply a required 100 W/m^2 peak heating load (Fig. 4), while committing to the same panels a heating duty reduced to 53% implies a lowering of the mean water temperature to 29°C. However, undersizing the heat pump system reduces its effectiveness as an energy conservation device.

Considering old buildings heated by standard radiators (water distributing systems are traditionally used in Italy) and assuming that a heat pump system replaces the original boiler, the heat capacity level to be assigned to the heat pump system is to be optimized. The water temperature requested to transfer that capacity to the surroundings through the existing radiators can be computed, as done in Fig. 5 (radiator outputs at reduced loads were taken from Ref. 2). In order to supply 85% of the annual energy consumption , a water inlet temperature of 53.5°C is requested (against 75°C for peak load) which is still marginally acceptable for an energy-effective heat pump. At 70% annual load supply, only 47°C water inlet temperature is required.

Considering the adoption of heat pumps in new buildings, or assuming that heating units replacement is economically justified, any effort should be made to employ low temperature heaters. Recent regulations about energy conservation,reducing the admissible heating loads, contribute directly to a better acceptability of the loss of capacity caused by a lower temperature heat carrier.

Floor heating panels are perfectly adequate to heat pump intrinsic requirements of base-load, low temperature operation. Natural convection heaters of various kinds are under development, which promise a reduction of mean water temperature by 15 to 20°C for a given output. The performance of one of such heaters, commercially avai<u>lable</u>,is reported in Fig. 4 (curve A)(Ref. 3).

In the research work about low temperature convectors at the Milan Polytechnic, among others,the basic configuration illustrated in Fig. 6 was considered. Fresh air taken from the ambient at floor level is forced through a continuous fin heat transfer matrix, owing to the draft induced by a chimney superimposed to said matrix. Overall heater height was limited to about 1 m both for reasons of space encumbrance and for containing room vertical temperature gradients. With this constraint, the basic variables of the heater configuration are fin height, fin spacing and fin thickness.

The influence of these variables, as obtained from a computer program which solves the relevant heat and mass transfer problem is illustrated in Fig. 7 (Ref.4). Laboratory tests basically confirm the predicted theoretical behaviour.

The following observations are suggested from the inspection of Fig.7.

a) close fin spacing is detrimental for unit's output, no matter the amount of heating surface which is enclosed in the matrix (for a 1 mm spacing, the optimum output is approximately half that achievable with 3 mm spacing);

b) for each fin spacing, an optimum value of the fin length in the flow direction (dimension C of Fig. 6) exists, beyond which the performance of the heater rapidly deteriorates, due to the loss of mass flow produced by friction;

c) optimum matrices are characterized by wide fin spacings, connected with long fin lengths (up to about 200 mm for 5 mm spacing). These proportions are not usually found in commercial heaters of similar basic configuration.

The output of this type of heaters, as obtained from tests performed in a thermostatic room, is reported in Fig. 4 (curve B). At a mean water temperature of 50°C they offer a performance similar to that given by an ordinary radiator of similar dimensions at 70°C. An ordinary radiator supplying 85% of the annual load at 50°C mean water temperature could be replaced by a convector of this type of similar dimensions and equal output working with a mean water temperature of 38°C.

Considering forced convection convectors, ordinary equipment is already adequate to the low temperature operation which is consistent with the adoption of heat pumps.

The work performed at Milan Polytechnic, however, suggested the following observations:

a) ordinary equipment is usually designed for a very poor fan and electric efficiency (the ratio of the hydraulic power delivered by the fan to the electric power input is often well below 10%);

b) fan power levels much below usual values are sufficient to improve dramatrically the performance of natural convectors, also reducing, at the same time, the acceptance problems connected with the generation of comparatively cool, high speed air streams within living spaces.
As shown in Fig. 8, which reports the computed performance of a heat transfer matrix of the type of that illustrated in Fig. 6, excellent thermal power densities are achievable for a heat output/ideal fan power ratio of 1200. The development of very low power fans and electric motors of good efficiency could greatly improve the acceptability of forced convection heaters;

c) as in the case of natural convection heaters, optimum heat transfer matrices are characterized by moderate fin spacings and long fin heights in the flow direction.

4. COMPRESSORS PERFORMANCE AND AVAILABILITY

High efficiency in the compression phase is of primary importance for the actual attainment of good heat pump thermodynamic performance. Although the heat pump is a well-known machine, and there is an apparent abundance of technical data about its performance, very confusing results are in general obtained in attempting to derive a correlation of compressor efficiency as a function of most important parameters, say pressure ratio and power. In fact, widely scattered and unreliable data were obtained from a survey of performance given by various manufacturers; moreover, only overall system, rather than single component performance are commonly quoted. A large number of small commercial heat pumps exhibit a rather poor (40 ÷ 50%) overall compressor efficiency (defined as ideal power for isentropic compression/actual electric power consumption). However, some evidence that high efficiency can be obtained even in small reciprocating compressors can be found in the technical literature: for instance, Ref.5 quotes a compression efficiency as high as 86.6% (dropping to 69% including the electric losses of a 80% efficient motor) for a "good performance" reciprocating compressor of relatively small size (2274 W_{el}), operating with a condensation-evaporation temperature difference of 47.2°C. High compression efficiency are inferred also from COP measurements given in Ref. 6 for a "state-of-the-art" heat pump, having a heating capacity of about 10.3 kW and a condensation-evaporation temperature difference of about 50°C.

The heating capacity of the above mentioned heat pumps are appropriate for a single family house or apartment. In Italian towns, most people live in relatively large apartment buildings, heated by centralized systems: typical heating loads to be covered by heat pump for these applications would range between 25 ÷ 500 kW_{th} which lead to compressor power of 8÷150 kW_{el}.

A 90 kW helical rotary compressor, directly coupled to a 2 pole hermetic motor, using a "positive pressure" refrigerant is quoted by the manufacturer (Ref.7) to have an overall efficiency[*] of about 73%. The same compressor could be derated to lower heating loads and power consumptions, if used with a more expanded refrigerant (for example R-11), with a little penalty on efficiency. For hot water generation at temperatures above 40°C, two stages compressors can be necessary, a costlier solution which would allow the adoption of vapour extraction during expansion, with beneficial effects on COP.

---

[*] computed by COP data given in Ref. 7, for assumed minimum temperature differences in the heat exchangers of 5°C.

Centrifugal compressors are also attractive for the considered application: besides their inherent advantages over a volumetric machine (small encumbrance, reliability, simple mounting, low vibration level, etc.), they are potentially low-cost efficient machines.

The operating characteristics which affect the design of a centrifugal compressor are represented in Fig. 9 for some common refrigerants: isentropic specific works in the range of 20÷30 kJ/kg are required at the evaporating and condensing temperatures considered for this application, to be compared to about 200 kJ/kg necessary to compress air from ambient temperature with a pressure ratio of σ. This means that low peripheral speeds, and related safe mechanical stresses can be adopted for the rotors working with refrigerants. The required pressure ratios are below 8, for condensing temperatures not exceeding 50°C.

To demonstrate that efficiency of about 80% can be achieved by single stage supersonic centrifugal compressors of small size, the Fig. 10 was prepared: it can be seen that high efficiency is obtained for high pressure ratio, small size compresors (100÷150 mm), with modern aerodynamic design. According to similarity rules, these high efficiencies can be obtained with any working fluid: in fact, the compressors described in Ref. 9 and Ref. 10 were designed and tested for refrigerants. In other words, a single stage centrifugal compressor working with a refrigerant would have the same aerodynamic problems already solved for advanced air compressors, but would operate at much lower peripheral speeds, or, if the comparison is carried out for the same wheel size, at a much lower rotating speed.

It is of interest to calculate the actual size and rotating speed of single stage centrifugal compressors at temperatures useful for the considered application: if the compressor is designed at optimum specific speed (N_s = 120 with Balje' unit, Ref. 15), and wheels with radial blades are assumed, the results shown in Fig. 11 are obtained: with R-113, a wheel of reasonable size (R = 53 mm) running at 35500 rpm is found for low shaft power (10 kW). For 100 kW power, R-113, R-11 and R114 could be chosen as working fluids, yielding outer rotor radii of 167, 103 and 73 mm and speed of revolution of 11200, 20000 and 22500 respectively. These compressors, if well designed, could have an adiabatic efficiency of about 80%.

Including mechanical and electric losses, their performance would be similar to the one quoted above for helical compressors.

The experience gained with air compressors in other fields (turbochargers, small gas turbines) has shown, that, if produced in large quantities, their cost could very attractive.

Eventually, for larger units, industrial two stage-centrifugal compressors could be adopted, with overall efficiency above 70% and possibility of vapor extraction. It can be concluded from the above analysis that overall compression efficiencies of about 70% are feasible and within the state of the art for the applications considered in this paper. A lower efficiency level (about 50%) would be however more representative of existing commercial heat pumps of limited capacity.

5. ENERGY PERFORMANCE OF HEAT PUMPS-COMPARISON WITH CONVENTIONAL SYSTEMS

In this section, the results of a series of calculations concerning the yearly energy performance of various heating systems, including the proposed water to water heat pumps, are presented and discussed. Since most systems use both electrical energy and heat generated by fuel burning as energy input, an equivalence ratio between these two forms of energy must be selected for a meaningful comparison. Since in Italy the average efficiency of electric power generation is 37% (Ref. 16), this value will be assumed to convert heat into electric power. This assumption penalizes "electric" solutions, when compared to systems using more valuable fuels than the ones used in power stations. The comparison is carried out for heating systems using different procedures of heat generation, namely: cogeneration, conventional boilers and heat pumps. While calculations were carried out for single representative values of various parameters affecting energy balances for the first of two cases, two extreme options on heat pump performance, named "high" and "low" efficiency respectively, were selected. Moreover, two options (radiators and floor panels) were considered for the final heaters, owing to the great influence of condensing temperature on the system performance.

The considered options and related hypotheses made in calculations are summarised in the following :

a) <u>Large urban heating system, fed by out-of-town cogenerating power stations.</u> Reference is made to data given in the feasibility study (Ref. 16) of a urban heating project for the Milan area. In this study, it is proposed to supply a large fraction (about 1300 MW_{th}) of the thermal needs of Milan by converting two existing large power stations into cogenerating plants. In each 320 MW station, an electric power loss of about 107.4 MW_{el} is predicted, for generating 424 MW_{th} at the selected hot water temperatures (160/65°C). The generated heat is then conveyied by a 20 km feeder to the city distribution line. Peaking stations, using conventional boilers,are foreseen on the city boundaries.

The following assumptions are made in Ref. 16[*]:

- Electric power loss/heat generation ratio of the power plant = 0.253
- Hot water temperatures in the feeder = 160/65°C
- Hot water temperatures in the distribution system = 120/60°C
- Overall pressure drop in the feeder = 40 bar
- Overall pressure drop in the distribution system = 20 bar
- Design overall efficiency of the pumping stations = 0.75

- Yearly average overall thermal losses = 7%
- Yearly average overall specific pumping consumption = 3.5% (kW_{el}/kW_{th})
  (2.36 at design point)
- Yearly average heat produced by cogeneration/overall heat requirement = 88%
- Yearly average efficiency of peaking boilers = 90%

With the above assumptions, the "equivalent" COP of the urban heating system is given by:

$$COP = \frac{\text{Yearly averaged useful thermal energy}}{\text{yearly averaged loss of electric energy}} = \frac{1}{0.253} \; (1-0.07)(1-\frac{0.035}{0.37}) \; 0.88 +$$

$$+ \; 0.12 \; \frac{0.90}{0.37} = 3.22$$

Besides the high temperature solution for the heat carrier, an energy-conserving, non conventional option was also evaluated, featuring 105-65°C water temperatures for the main loop. Under this assumption, however, the ducts must be duplicated and pumping losses, which are already important in the high temperature solution, become much more significant. Furthermore, the economic optimisation of similar systems for which there are detailed data, favours, systematically, high water temperatures, which hence appear the most likely to be eventually selected.

b) <u>High efficiency (η_c = 0.70) water to water heat pumps</u>. With reference to data on heaters given in Section 3, two options are considered, both consistant with the assumption of replacing boilers with heat pumps in existing buildings, without changing the final heaters system:

2a) floor heating panels, working at a mean water temperature of 29°C.
   (an energy balance similar to that relating to floor heating panels
   is obtained also using high-performance forced-convection heaters,
   which, in some cases, could replace in old buildings, at reasonable
   costs, existing high temperature radiators).

2b) conventional radiators, working with water inlet temperature of 47°C.

For both options, it is supposed to commit 53% of the nominal (-5°C outdoor temperature) heating duty to the heat pump system.
Heat pump performance is computed according to curves of Fig.12, making allowance for auxiliary consumptions (5%). Minimum temperature differences and water temperature variations were assumed both equal to 5°C in the condenser and to 3°C in the evaporator, at nominal conditions. The peaking heat capacity (15% of yearly heat requirement) is committed either to electric resistance heaters (options 2a1 and 2b1) or to fuel burning systems having an assumed average efficiency of 65% (options 2a2 and 2b2).

c) <u>Low efficiency (η_c = 0.50) water to water heat pumps.</u> The same hypotheses and options as above were considered.

d) <u>Conventional boiler</u>. An yearly average boiler efficiency of 70% was assumed.
(lower values typically 60 to 65%, are often quoted as
representative of actual average boiler performance).

The **results obtained** are summarized in Tab. 1. It can be seen that, when **compared to large district** heating systems fed by cogenerating plants, "high efficiency"
heat pumps enjoy better (if connected to floor heating systems) or similar (if
connected to conventional radiators) performance. "Low efficiency" heat pumps are
superior to cogenerating systems, only if used with floor heating panels. Part of
the positive results obtained through the use of heat pumps is due to the assumed
regulation mode which commits to the heat pump the base load and rely on conventional
systems for peaking. The heating mode for the peaking capacity is not critical:
even direct electric heating does not heavily penalize the system performance.

6. CONCLUSIONS

Water to water heat pump systems seem to have an energy saving potential similar
to that of out-of-town cogeneration, both exhibiting the basic advantage of low-cost
fuel burning capability. Furthermore, both systems imply comparatively large capital
investments, which are followed in short time by a return money flow, only in the case of
distributed heat pumps. The disturbance impact on the city life, provided water is available directly within the urban area, should favour the heat pump system.
In many historical towns water ways had, in the past, an important economic
function, which dwindled with time, and was followed, in recent years, by their physical degradation. Recognizing their energy potential could represent the basis for an
economical re-evaluation and physical reclamation.

(*) It is believed that these assumptions are representative of similar urban cogenerating
systems.

Tab.1 Yearly delivered heat/yearly equivalent electric energy consumption
ratios obtained for various heating   options for the Milan climate; primary energy savings with respect to conventional boilers
(with an assumed yearly average efficiency of 70%) are given in the
second column.

Option	Description	Yearly delivered heat/ Yearly equivalent electric energy consumption	Primary energy saving with respect to option 4
1	Large urban heating system, fed by cogenerating power stations	3.22	41.2%
2a1	- High efficiency heat pumps - Floor heating panels - Peaking capacity by electricity	4.88	61.2%
2a2	- High efficiency heat pumps - Floor heating panels - Peaking capacity by fuel burning	5.01	62.2%
2b1	- High efficiency heat pumps - Conventional radiators - Peaking capacity by electrical	3.18	40.5%
2b2	- High efficiency heat pumps - Conventional radiators - Peaking capacity by fuel burning	3.31	42.9%
3a1	- Low efficiency heat pumps - Floor heating panels - Peaking capacity by electricity	3.68	48.6%
3a2	- Low efficiency heat pumps - Floor heating panels - Peaking capacity by fuel burning	3.81	50.3%
3b1	- Low efficiency heat pumps - Conventional radiators - Peaking capacity by electricity	2.57	26.3%
3b2	- Low efficiency heat pumps - Conventional radiators - Peaking capacity by fuel burning	2.70	30.0%
4	Conventional boilers	1.89*	-

(*) fuel energy consumption is conventionally converted to electric energy consumption
at an efficiency of 0.37.

7.   <u>REFERENCES</u>

1   <u>Angelino, G. and al.</u>: "A proposal for a low pollution central heating system for the city of Venice". In: Proc. 6th International Congress of Climatistics (Milan, Italy: Mar. 2-6, 1975), REHVA, 1975, Paper IV - 04, 31 pp.

2   ASHRAE Guide and Data Book, New York, 1966, p. 860.

3   Technical literature by Deléage S.A., 35045 Saint Malo, France.

4   <u>Brioschi, S. and Digonzelli, L.</u>: "Investigation on thermal performance of low temperature convectors", graduation thesis, Milan 1980 (In Italian).

5   ASHRAE, Guide and Data Book, New York, 1966, p. 644

6   <u>Baxter, Van D.</u>: "ACES: Final performance report - December 1, 1978 through September 15, 1980", Technical Report ORNL/CON-64. Oak Ridge, April 1981, 134 pp.

7   Anon.: "50 Hertz 'PEX-H' hermetic package chillers", Dunham-Bush Technical Bulletin, FORM NO.6040B, March 1972.

8   <u>Dean, R.C.</u>: "Advanced radial compressors" LS50, Von Karman Institute for Fluid dynamics, Brussels, May 1972, 61 pp.

9   <u>Poulain, J. and Janssens, G.</u>: "Centrifugal compressor wheel without bending stresses on blades" ..(Roue de compresseur centrifuge sans flexion dans les ailes). In AGARD-CP-282 on "Centrifugal compressors, flow phenomena and performance", Brussels, May 1980, 9 pp. (In French).

10 <u>Sovrano, R. and Avram, P.</u>: "Experimental investigation of the near- surge flow in a high performance centrifugal compressor", O.N.E.R.A. Technical Paper T.P. n° 1976-2, 9 pp.

11 <u>Chapman, D.C.</u>: "Model 250-C30/C28B Compressor Development", In AGARD-CP-282 on "Centrifugal compressors, flow phenomena and performance", Brussels, May 1980, 6 pp.

12 <u>Schorr, P.G. and al.</u>: "Design and development of small high pressure ratio, single-stage centrifugal compressors", in "Advanced centrifugal compressors", ASME, New York, 1971.

13 <u>Jansen, W. and al.</u>: "Improvements in surge margin for centrifugal compressors". In AGARD-CP-282 on "Centrifugal Compressors, flow phenomena", Brussels, May 1980, 16 pp.

14 <u>Chevis, R.W. and Varley, R.J.</u>: "Centrifugal compressors for small aero-automotive gas turbine engines", in AGARD-CP-282 on "Centrifugal compressors, flow phenomena and performance", Brussels, May 1980, 17 pp.

15 <u>Baljé, O.E.</u>: "A study of design criteria and matching of turbomachines. Part B - Compressor and pump performance and matching of turbo components", Journal of Engineering for Power. Trans. ASME Series A, <u>84</u>, 1, January 1961, pp. 83-114.

16 Anon.: "Proposal for District Heating in Lombardy - Feasibility Study of Phase 1" (Proposta di piano di teleriscaldamento della Regione Lombardia - Progetto di fattibilità della 1^a fase)", Lombardia Risorse, January 1981, 157 pp. (In Italian).

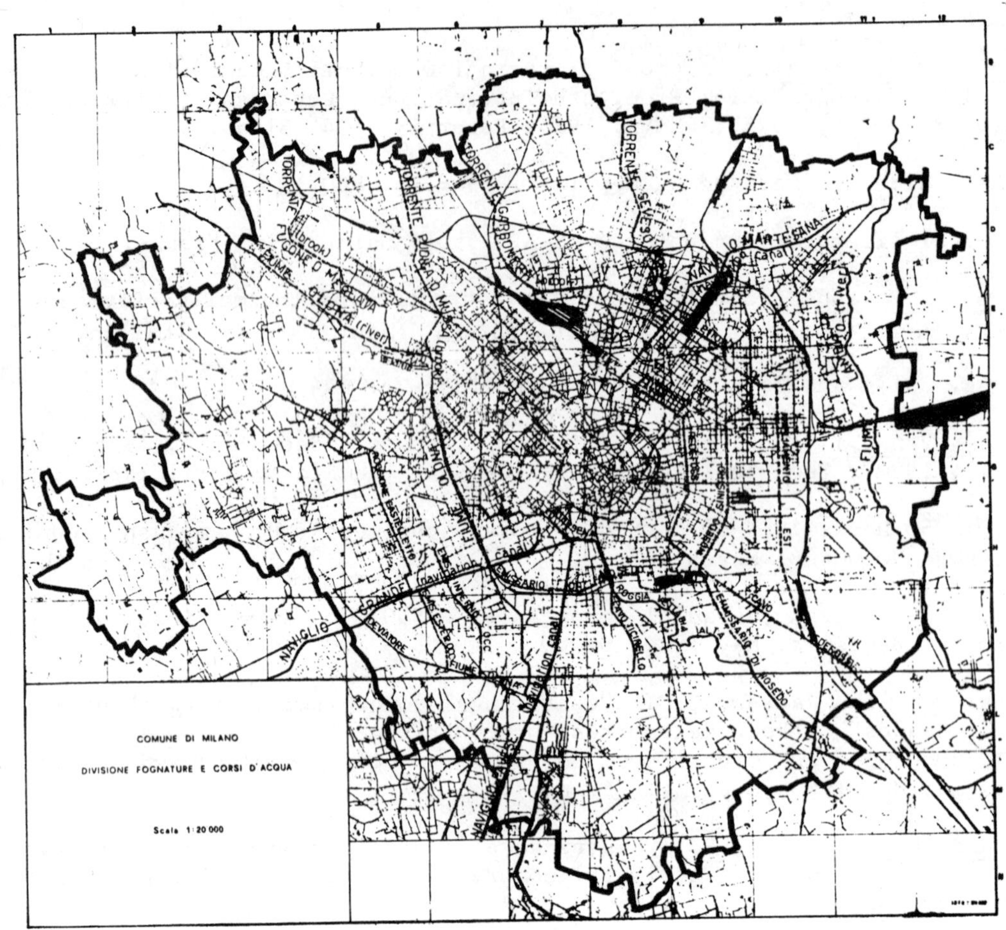

Fig. 1   Water ways network around and under the city of Milan

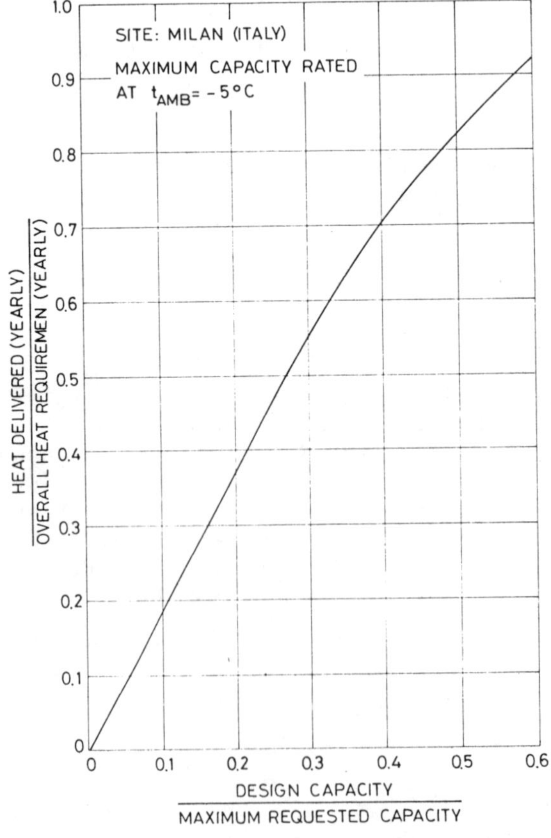

Fig. 2   Annual heat delivery as a function of the design capacity

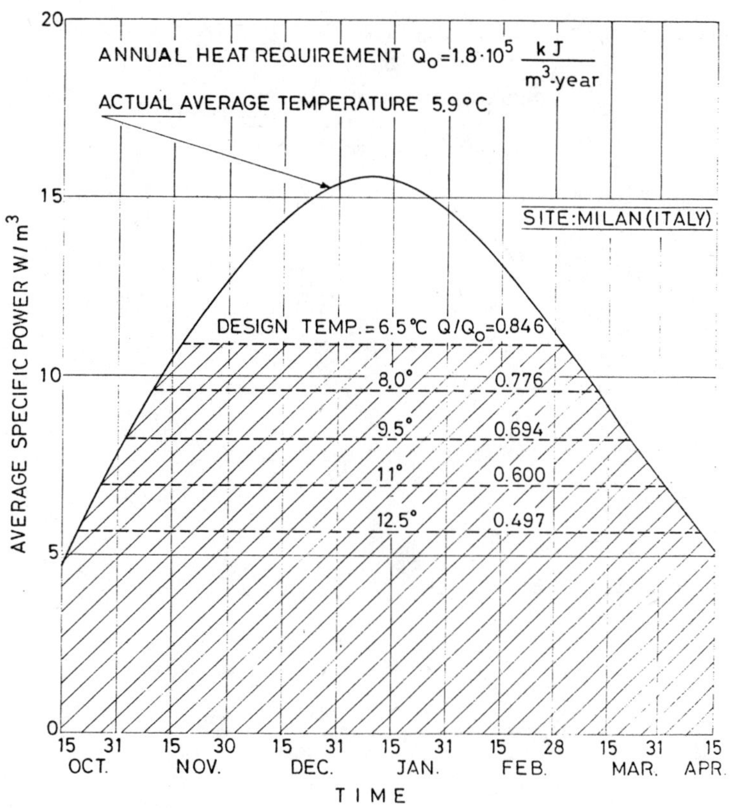

Fig. 3   Average specific power required by domestic heating during the year in Milan

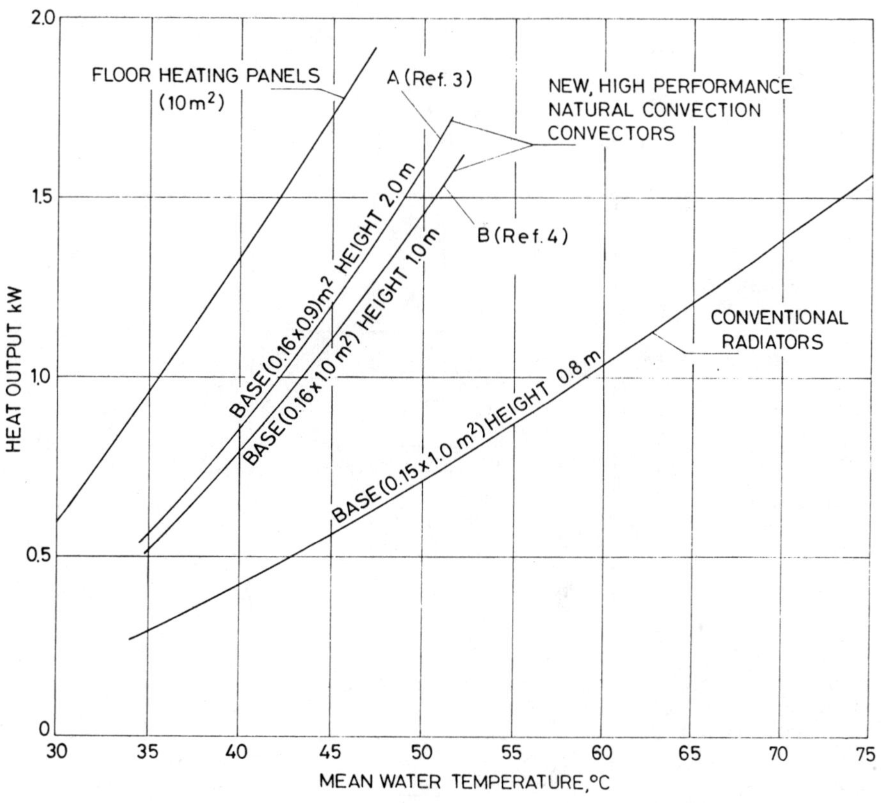

Fig. 4   Heat output of various heaters as a function of mean water temperature

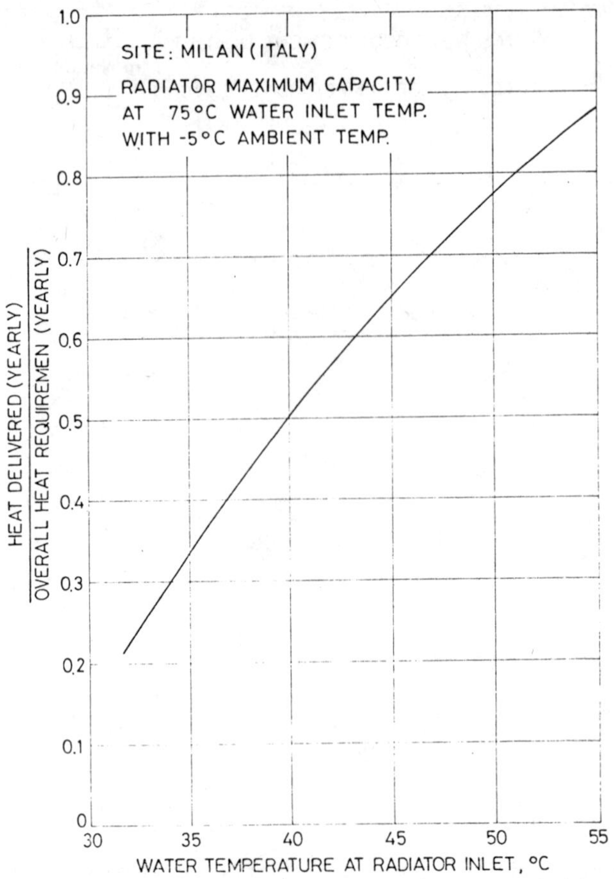

Fig. 5   Fraction of the yearly heat requirement delivered by a conventional radiator as a function of the water temperature
at radiator inlet

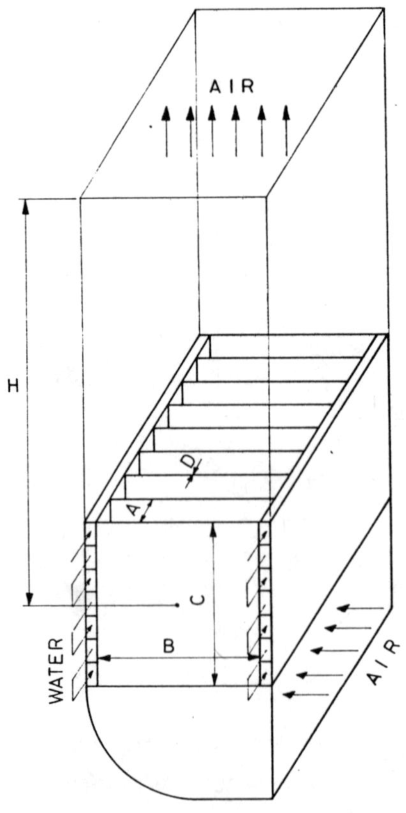

Fig. 6   Basic configuration of a low temperature convector

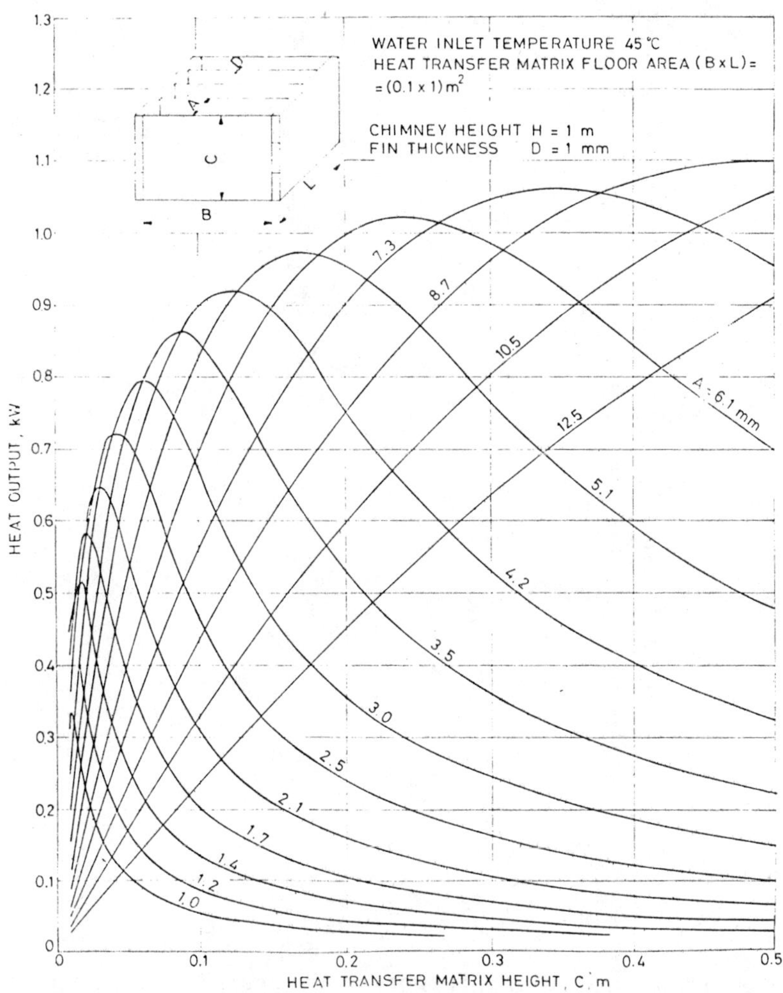

Fig. 7   Computed performance of natural draft low temperature convectors for various geometric configurations

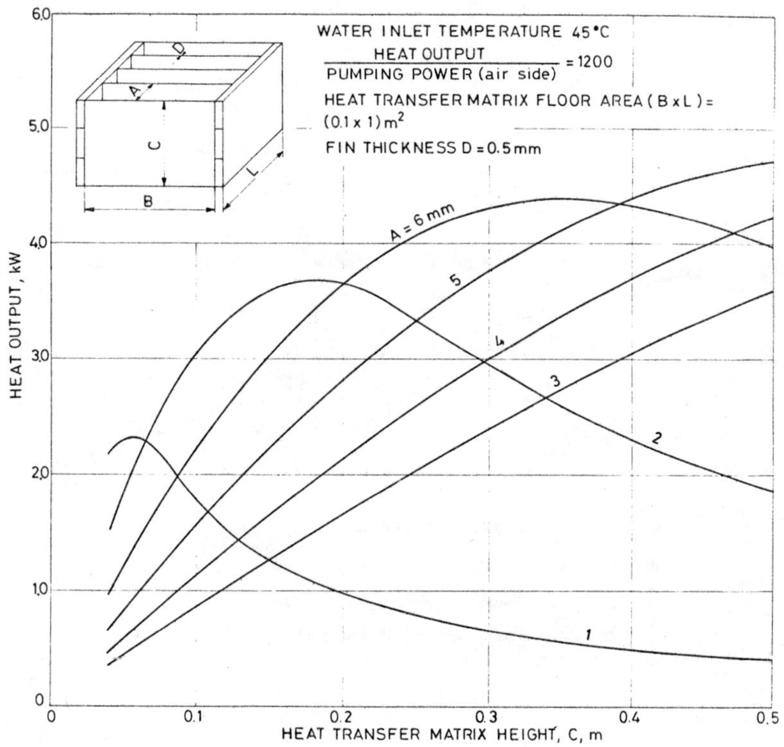

Fig. 8   Computed performance of forced convection convectors for various geometric configurations

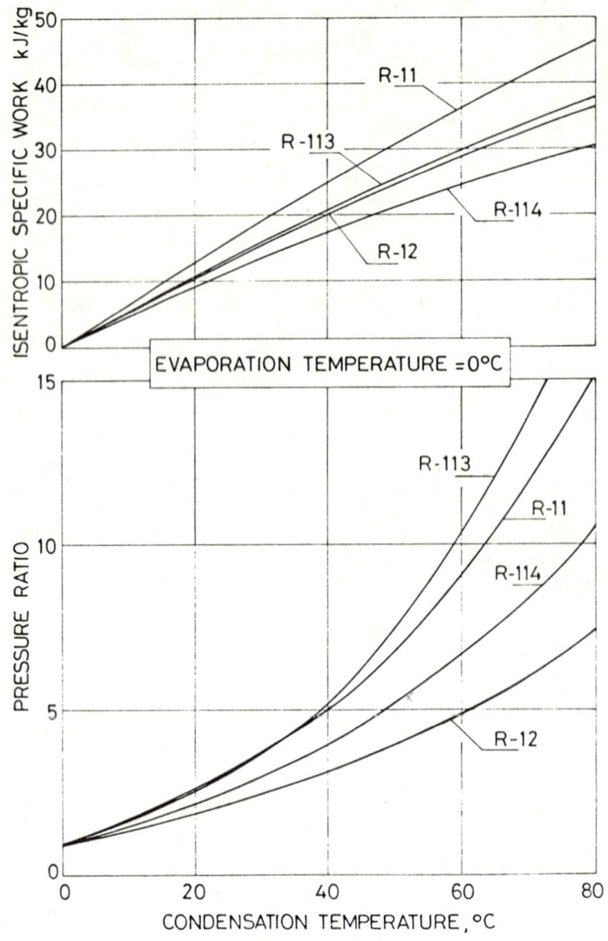

Fig. 9   Isentropic specific work and pressure ratio for compressors
using common refrigerants

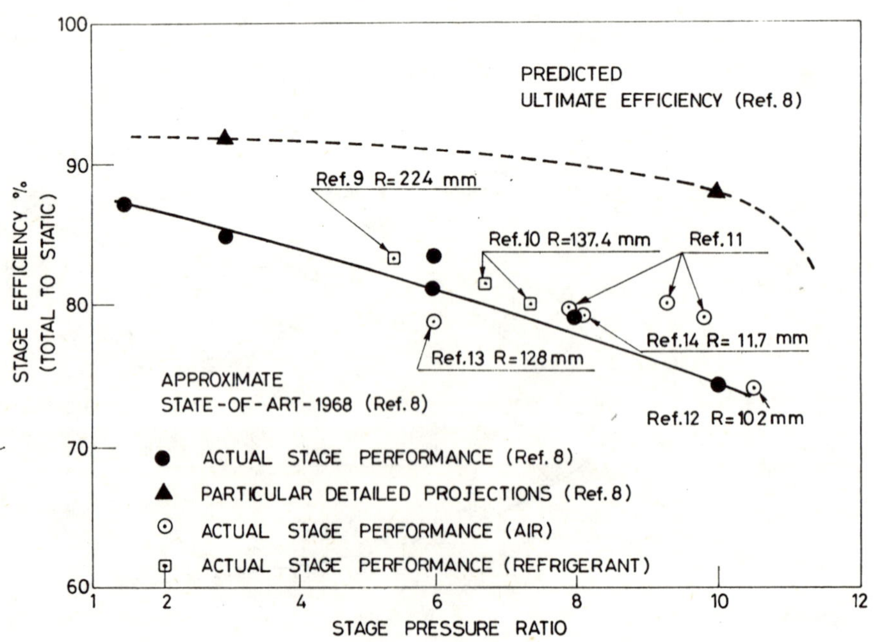

Fig.10   Efficiency of small advanced single stage centrifugal compressors. R = outer rotor radius.

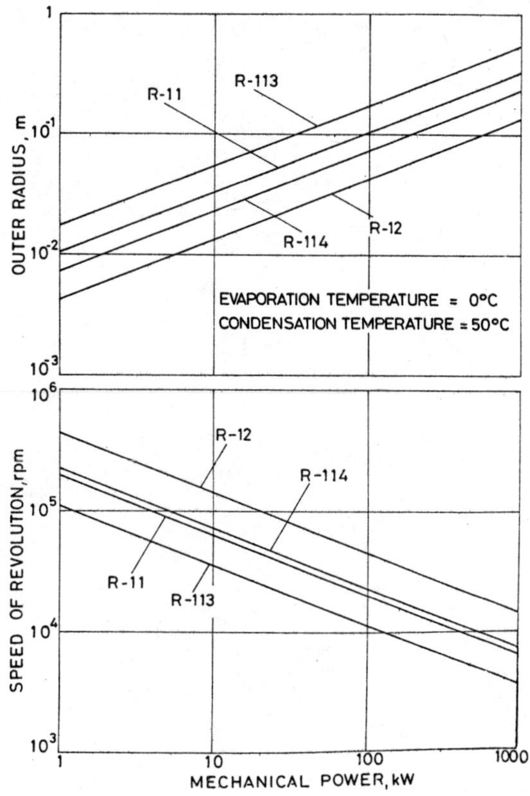

Fig.11   Speed of revolution and wheel outer radius of optimized single stage centrifugal compressors working with various refrigerants.

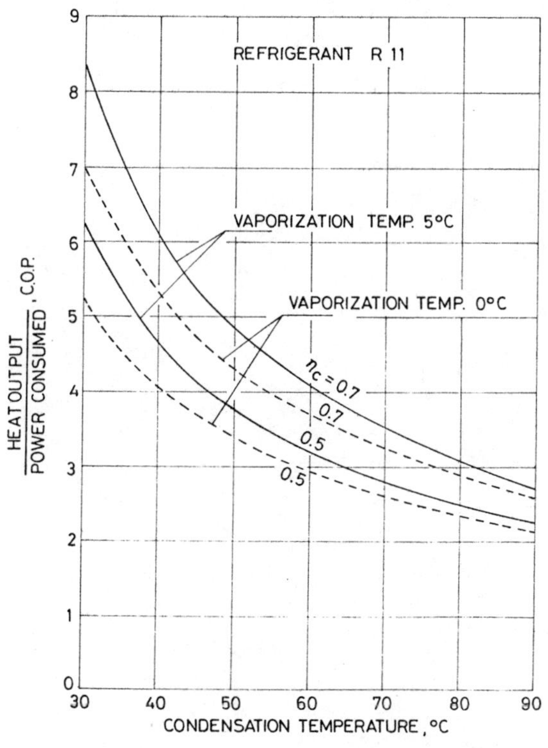

Fig.12   Heat pump COP as a function of vaporization and condensation temperatures and of compressor efficiency.

TECHNICAL AND ECONOMIC EXPERIENCE OF LARGE HEAT PUMP PLANTS

C. Jacobsen

Stal Refrigeration AB, Sweden

Summary

This paper outlines the design parameters, the technical and economic situation at the design stage and the experience gained throughout the plant operating time.

The plants have been selected from a large number of plants in operation both in Sweden and elsewhere, with a view to providing up-to-date, factual information on interesting plants using various heat sources and for different heat sink applications.

The heat sources selected for this paper are:

- Cooling water from refrigerating plant condensers

- Discharge lines in ammonia refrigerating plants

- Process water from aluminium extruders

- Treated sewage water

- Untreated sewage water

The selected plants provide indisputable evidence that industrial heat pumps can be made both reliable and highly profitable. As energy prices and regulations vary from one country to another, recalculation will be necessary.

Held at the University of Warwick, U.K.
Symposium organised and sponsored by
BHRA Fluid Engineering
©BHRA Fluid Engineering, Cranfield, Bedford MK43 0AJ, England.

INTRODUCTION

Theory and practical experience do not necessarily walk hand in hand. If their paths diverge, serious economic consequences may result, since the investment cost for a heat pump system frequently exceeds that of other energy conversion systems.

Occasionally, unfavourable heat pump operating results are reported, the plant having been found unprofitable, or unreliable, or to work with a lower COP than anticipated or promised. In such cases the heat pump technology is never the culprit. In all probability, the heat pump has been used under conditions for which it was not designed, in systems which can never be profitable or were designed by people lacking the requisite know-how. Since such information will seldom pave the way for future energy savings, appraisal of a few well designed plants may be highly valuable.

Although heat pumps have been designed and supplied by the company I represent for several decades prior to the 1973-74 energy crisis, experience gained after that period will be emphasised in the present context.

The heat sink, or the consumer side of the energy generated at the heat pump condenser, has been one or a combination of the following categories: district, block or central heating systems, tap water or industrial cleaning water production, industrial process water production and industrial processes.

The art of handling the various types of heat source and of controlling variations in quality is very difficult to master. In this context, quality is understood to mean: flow rate, continuity, temperature and temperature stability, corrosivity and type and quantity of foreign matter. Various types of evaporators and heat exchangers are being and have been utilized, depending on the heat source available. The most common ones are illustrated in Fig. 1.

On many occasions, the availability of one particular heat source may not satisfy the requirements and a combination with other sources may be beneficial. Over the years, most of the available heat sources have been used for heat pumps and several plants are now in operation.

Factors outlined in this introduction are examined in detail, including background, the energy requirements, the energy savings and the economical results for seven selected plants.

TECHNICAL AND ECONOMIC EXPERIENCE OF LARGE HEAT PUMP PLANTS

All costs given below are in Swedish kronor (SEK).
Rate of exchange pr. october 1, 1981: SEK 10 = £1.

PLANT A:

Reference name:	Dafgård, Sweden
Design heating capacity:	765 kW
Compressor motor power:	250 kW
Design leaving hot water temperature:	70°C
Heat source: Cooling water from refrigerating plant condensers	
Heat source temperature:	25°C

Back in 1975, Dafgård, one of Sweden's major producers of frozen food for the retail trade, catering establishments and industry, decided to expand their production facilities. This also necessitated an extension to the freezing, cold-storage and heating facilities.

Additional heating capacity was needed to produce hot water for cleaning the new process area and for heating new offices and auxiliary areas. The solution for this additional heat requirement appeared quite simple: why not supplement the two oil-fired boilers with an additional boiler?

It was asked, however, if this could be considered a sensible solution in view of the already high heating costs.

Let us consider the requirements and solution a little more closely, bearing in mind the fact that the existing boiler room was already heavily occupied and the chimney fully utilized:

The total annual heating requirements, including the new extension, would be:

Days requiring heating of premises only: 1,080 MWh/year
(see Fig. 2)

Days requiring heating of premises and production
of hot water for cleaning purposes (see Fig. 3):   3,320 MWh/year

Total annual heating requirement: 4,400 MWh/year

The refrigerating plant, however, discharges substantial quantities of heat energy from the condenser(s), previously transferred into the cooling water system and to the outside air. The problem was solved by installing a heat pump utilizing the condenser heat of the refrigerating plant as a heat source and generating heat at the required temperature level.

The heat pump, type VSV 51E, was connected to the existing system (see Fig. 4) according to Fig. 5. The water-cooled condensers of the refrigerating plant route the heat to an indoor basin (frostfree) and to cooling towers where the surplus not utilized by the heat pump as a heat source is discharged. The heat pump utilizes the same condenser water at a temperature level of approx. 25°C as a heat source thus reducing the load on the cooling towers. This gives less contamination and less maintenance on the water side of the system.

The condenser side of the heat pump is connected to the heating system in order to produce hot water at temperature levels according to Fig. 6, as a function of production and ambient temperatures. In this particular plant the following energy balance was attained:

—   Heat energy available from heat pump: 3,860 MWh/year

—   Heat energy required from oil-fired boilers: 540   ”

—   Total annual requirement: 4,400   ”

Provided the total annual heating demand would have been satisfied by using three oil-fired boilers (each of 830 kW) at 70% boiler efficiency, an annual consumption of some 630 m^3/year of oil would be required. Use of the heat pump enabled the oil consumption to be drastically reduced to approx. 77 m^3/year, with a saving of some 550 m^3/year. The energy required to operate the entire plant totals:

Electrical energy for heat pump compressor and for pumps: 1,177 MWh/year

Oil energy (77 m^3/year):   778   ”

Total annual energy requirement: 1,955   ”

From this, the annual COP-value can be calculated:

—   Heat pump system only (3860/1177):   3.28

—   Total system (4400/1955):   2.25

Among other significant factors, however, is the economic situation. Let us therefore examine the investment calculation providing the basis for the decision made back in 1975:

INVESTMENT CALCULATION

HEAT PUMP	OIL-FIRED BOILER
INVESTMENT COSTS	INVESTMENT COSTS

HEAT PUMP		OIL-FIRED BOILER	
Heat pump + plumbing and connection to boiler hot water system	500,000	Not necessary to establish	
Deduct Government energy subsidy	- 165,000		
	335,000		
A write-off period of 7 years and a 15% interest rate gives an annual cost of	80,000		
OPERATING COSTS		**OPERATING COSTS**	
Electricity (for running the heat pump 1,177,000 kWh) Fixed costs, annually	51,000	629 m^3 (138,500 imp. gals) of oil at SEK 500 per m^3	315,000
1,177,000 kWh at 0.07 SEK per kWh	82,500		
Additional heat 77 m^3 (17,000 imp gals) of oil at SEK 500 per m^3	38,500		
Maintenance	8,000	Maintenance	5,000
Annual operating costs	180,000	Annual operating costs	320,000
Total annual costs	260,000		

From the above values we can calculate the pay-back time as applicable in 1975:

— With government subsidy $\dfrac{335,000}{320,000 - 180,000} = 2.4$ years

— Without government subsidy $\dfrac{550,000}{320,000 - 180,000} = 3.9$ years

and with 15% interest included for the pay-back time 3.2 and 5.5 years respectively

There are, however, other interesting issues to be considered.

— What has happened since the decision was made in 1975 ?

— What about the economic situation ?

— What about reliability ?

— What about maintenance and service ?

To begin with, oil prices have escalated since then from SEK 500 to approx SEK 1,850 today (1,850/500= 3.7) compared with a roughly 75% increase in electricity costs. Consequently, the gain in operating cost has significantly increased and the real pay back-time has become shorter than originally calculated. In 1980, five years after project commencement, Dafgård anticipates an annual saving in the order of SEK 750,000, bearing in mind that their total investment was only SEK 335,000! Energy prices are hardly likely to decrease and consequently continued and improved profitability may be anticipated in years to come.

Reliability is of the utmost importance in a food processing plant, for hygienic reasons and obviously also for economic reasons. Since two oil-fired boilers are operated in parallel with the heat pump the company will have a good basis for comparison.

The heat pump has been highly reliable, partly because a screw compressor is used, and Dafgård reports lower maintenance costs for the heat pump system than for the oil-fired boiler system. Apart from scheduled maintenance, the only part of the system demanding attention is the cooling tower installation, a factor reduced in connection with the introduction of the heat pump.

Back in 1975 it was easy to opt for the heat pump solution for economic reasons, and it was easy to select a heat pump with a screw compressor in view of reliability and maintenance.

PLANT B:

Reference name:	Arla/Götene, Sweden
Design heating capacity:	805 kW
Compressor motor power:	250 kW
Design leaving hot water temperature:	70°C
Heat source: Ammonia gas in refrigerating plant discharge line	
Heat source temperature: Ammonia condensing temperature:	30°C

At one of Sweden's largest and most modern dairies, Arla/Götene, emphasis was placed on reducing costs, maintenance and energy consumption. In 1976-77, when energy prices again escalated, investigations to determine if energy was recoverable from the ice accumulation plant were initiated. The refrigerating plant connected to the ice bank system is equipped with evaporative condensers and discharges approx. 570 kW to the outside air at a condensing temperature of approx. 30°C.

For the production of dried milk powder, heating is required continuously throughout the year, concurrently with the ice bank duty. If the heat initially extracted from products by the refrigerating plant is recovered, it may be reused in the process by preheating the air for the drying towers, utilized in the dried milk powder production.

Since evaporative condensers were used in the ammonia refrigerating plant, the evaporator for the heat pump VSVH 51E was chosen for ammonia condensing on the outside of the tubes in a shell-and-tube heat exchanger and with refrigerant R12 evaporating inside the tubes. The plant hook-up is illustrated in Fig. 7.

The investment calculation on which the decision made in 1977 was based is as follows:

BEFORE		AFTER	
OPERATING COSTS		**OPERATING COSTS**	
		Heat pump	
Oil		Electricity consumption 970,200 kWh	
541 m^3 at			
800 SEK/m^3 =	432,800	Fixed charge 215 SEK/kW	53,000
		Unit costs	
		970,200 kWh at 0.125 SEK/kWh	121,275
Service and maintenance ?		Service & Maintenance	
		(3% of total investment)	33,000
		Other operating costs	4,975
			213,000
Total annual			
operating costs	432,800		
		INVESTMENT COSTS	
		Heat pump	459,000
		Pre-heating coils &	
		installation of same.	641,000
		Reduction for government	
		subsidy (55%)	- 605,000
		Net investment	495,000
		Amortisation over 6 yrs. based on 15% annual interest	
		Annual cost	130,000
		Total annual cost	343,000

From the values above the following calculations can be made:

Total energy saving: $\dfrac{(541 \times 10 - 970.2)\, 100}{541 \times 10} = 82\%$

COP of heat pump at 70% boiler efficiency: $\dfrac{541 \times 10 \times 0.7}{970.2} = 3.9$

Pay-back time with government subsidy: $\dfrac{495,000}{432,800 - 213,000} = 2.25$ years

Pay-back time without government subsidy: $\dfrac{1,000,000}{432,800 - 213,000} = 5.0$ years

and with 15% interest included for the pay-back time 2.9 and 8.1 years respectively.

Since 1977, when it was decided to install the heat pump system, energy prices have escalated rapidly and shorter pay-back times than those calculated above have been experienced.
In 1980, the average oil price was some 70% higher than in 1977 and so far this year yet another 30% above the 1980 level. Consequently, a net gain of some SEK 525,000 was expected already for 1980, bearing in mind that the net investment was SEK 495,000!

With the double duty between the ice bank and the drying towers, reliability is essential and a heat pump equipped with a screw compressor has given outstanding results.

PLANT C:

Reference name:	Gränges, Sweden	
Design heating capacity:	1206 kW	1905 kW
Compressor motor power:	315 kW	650 kW
Design leaving hot water temperature:	52°C	95°C
Heat source: Process water from aluminium extruders		
Heat source temperature	21°C	52°C

Gränges, the large Swedish company whose products include semi-manufactured aluminium sections, uses cooling water at various process stages, especially at the extruders.

Formerly, water taken from a nearby lake was used for cooling purposes, causing maintenance problems at various stages throughout the year. This situation is shown in Fig. 8.
Several buildings of varying age are spread over the site and a heating load must be supplied for heating of the premises, for ventilation supply air and for tap water production. An internal centralized heating system existed and functioned as a small district heating system.

In 1980, two heat pumps were incorporated in the system with a hook-up according to Fig. 9. The energy rejected from the extruders is stored temporarily in two accumulating tanks and serves as heat source for heat pump No. 1 (HP 1). This unit, VSVH 57, works with refrigerant R12 and is designed for a maximum heating capacity of 1400 kW. Depending on requirements, the leaving hot water temperature can be regulated between +40°C and +70°C.

The heat carrying system from HP1 is connected to two parallel circuits, one for the ventilation air duty and one serving as a heat source for the second heat pump (HP 2). This latter unit, VSVH 73, works with refrigerant R114 and is designed for a maximum capacity of 2000 kW, and a maximum leaving hot water temperature of +105°C.

Two different modes of operation may be chosen.

In the first case, only HP 1 is in operation and all the heat energy generated is transferred to the heat exchangers of the ventilation system. The temperature level of the leaving hot water may be chosen between + 40°C and + 70°C, either by manual setting or automatically governed by an outdoor sited thermostat.
In the second case, both HP1 and HP2 are in operation. Depending on the load on the heating system (a function of the outdoor temperature), the leaving hot water temperature of HP2 may be governed between + 70°C and + 105°C. To assure optional conditions for the entire system, the leaving hot water temperature of HP1 is adjusted automatically, controlled by the actual temperature of HP2.

Each heat pump is equipped with a screw compressor, an electronically governed, motorized expansion valve system and an electronic control system, constituting the complex governing programme.

Including the energy required for compressor motors and pumps, and depending on the mode of operation, the COP of heat will be:

Case no 1, HP 1 only:	Between 4.5 and 4.8
Case no 2, HP 1 and HP 2 together:	Between 2.0 and 2.3

It is estimated that the heat pump plant will save roughly 100 m^3/month consuming 375 Mwh/month of electric energy in 1980 and 415 Mwh/month in 1981. The plant was commissioned in April, 1980.

The following data are of interest:

Total investment cost	SEK 4,350,000
Government subsidy	SEK 2,000,000
Net total investment	SEK 2,350,000

Estimated oil saving in 1980 (8 months):   800 m^3
Estimated electricity consumption in 1980:   3,000 MWh
Estimated oil saving in 1981 (12 months)   1,200 m^3
Estimated electricity consumption in 1981:   5,000 MWh
Oil price: approx.   SEK 1,950/m^3
Electricity price 1980:   SEK 0.16/kWh
Electricity price 1981:   SEK 0.18/kWh

From the above, the pay-back time can be calculated at:

With government subsidy: 1.55 years
Without government subsidy: 3.3   ”
Energy saving: 62.5%

The maintenance required besides routine service have been the heat source side due to fouling from the extruders.

The plant operation have given outstanding results and a repeat order have been commissioned.

PLANT D:

Reference name: Sala Heby, Sweden
Design heating capacity: 3,062 kW
Compressor motor power: 1400 kW
Design leaving hot water temperature: 70°C
Heat source: Treated sewage water
Heat source temperature: 8°C

In the municipality of Sala, a district heating system was already established in which use was made of a 40 MW centralized oil-fired boiler plant in the vicinity of the municipal sewage treatment plant.

In 1980, efforts were made to save energy and improve profitability. Work then commenced on building a heat pump plant, utilizing treated sewage water as the heat source, and if possible with sufficient capacity to satisfy hot tap water requirements during the summer months only. For part of the production period, however, the sewage water flow rate is inadequate and during such times water is taken from a small nearby river and mixed with the sewage water.

Consequently, the heat source temperature tends to be very low for part of the year and for this reason the choice had to be made from among ”freezeproof” types of evaporator. The water quality itself excluded materials or combinations including copper, steel and even stainless steel. For these reasons, and in view of costs, Sala Heby opted for an evaporator with flat, horizontal, hot-dip galvanized tube bundles, with the water sprinkling on the outside of the tubes and refrigerant R12 evaporating on the inside. The plant also includes a screw compressor, the largest in the world, driven electrically by a 1,400 kW motor. The site lay-out with the central heating plant, the sewage treatment plant, the river and the heat pump plant is illustrated in Fig. 10 and the hook-up in Fig. 11. Fig. 12 shows the chosen evaporator. The following data may be found from Fig. 13:

a)   Heat pump capacity and total capacity throughout the year.
b)   Hot water temperature throughout the year.
c)   Heat source temperature throughout the year.

The annual COP of heat, including the compressor and pump motors, has been calculated at 2.7 and it is estimated that some 20% of oil consumption can be saved, being some 2500 m^3/year. The total annual electric energy requirement amounts to 8890 MWh/year.

The total investment cost, including pile-driving, ground work, construction, water wells, pumps and piping, heat pump and engineering amounts to SEK 5.0 million, of which the heat pump accounts for SEK 3 million.

When the calculation includes 17% interest rate, a write-off period of 15 years and 2% maintenance costs calculated on the total investment an energy price of 0.10 SEK/kWh will be obtained, and a pay-back time of 2.35 years when the interest is included for the pay-back time.

PLANT E:

Reference name:	Skurup, Sweden
Design heating capacity:	325 kW
Compressor motor power:	100 kW
Design leaving hot water temperature:	38°C
Heat source:	Untreated sewage water
Heat source temperature:	8°C

Back in 1978, work commenced on construction of 132 houses in a small residential area including a centralized heating system with an oil-fired boiler plant. A pump station with an accumulator tank was already situated next to the planned boiler plant, collecting all the untreated sewage water from the neighbourhood before transferring it to the central treatment plant. Utilizing this untreated water as a heat source a heat pump was designed — probably the first of its kind in the world — and was used partly for testing purposes. Some of the design data are as follows:

Number of houses:	132
Heated area:	10.500 m^2
Heating load at -16°C ambient temperature:	560 kW
Heating load at 0°C ambient temperature:	300 kW
Hot tap water load, 45°C:	100 kW
Temperature to/from radiators at 0°C:	45/35°C
Sewage water flow, maximum:	900 litres/min
Sewage water temperature, minimum:	8°C
Heat pump capacity: Condenser load	320 kW
Gas cooler for hot tap water	30 kW
Electric motor power	110 kW
Calculated annual COP of heat:	3.5

The plant hook-up, shown in Fig. 14, uses a screw compressor working with R12 as refrigerant and an evaporator working with water inside the tubes in a closed shell-and-tube heat exchanger.

Untreated sewage water collected in accumulator 1 is pumped through a strainer from whence clean water is conducted to accumulator 2. From here, the water is pumped through the heat pump evaporator and back to the outlet side of the strainer, used to flush the separated impurities into accumulator 3 and finally back into the untreated sewage system.

Since the plant was originally intended to be used partly for testing, some special provisions and arrangements were made. Measuring data on temperatures, flow rates, pressures, electricity consumption and generated heat energy, both from the heat pump and from the oil-fired boilers, are collected and transferred to a comprehensive computer system. Consequently, the investment cost of this particular plant is higher than normal and the calculation of pay-back time is omitted.

Unfortunately, the construction of the 132 houses was delayed and consequently the heat load for the heat pump was reduced during the initial stage. The heat pump supplier had guaranteed not only the maximum capacity but also the COP throughout the year. In order to test the plant at an early stage under alternative and controlled conditions, a false load was introduced into the system. This enabled the plant to be tested simulating the load at various temperatures and under conditions that only 25%, 50% and 100% of the originally planned houses had been completed. The test results are shown in the following diagrams.

Fig. 15:   Capacity versus ambient temperature

Fig. 16:   Leaving hot water temperature versus ambient temperature

Fig. 17:   COP versus ambient temperature

From Fig. 17 it is evident that, provided all houses are connected, the COP will exceed 4 for about 4,000 hours a year and that the guaranteed values were reached under all conditions. When only some of the planned houses were connected, a substantial reduction was obtained both in the COP-value and in the slope of the curves. This demonstrates the importance of correct designing and the relationship between actual load and heat pump capacity. Despite the lower load during the initial stage, the plant has worked very successfully. Hitherto, a couple of unanticipated situations have arisen, adding experience to the already comprehensive know-how in the field of heat pump applications.

On one occasion a road tanker dumped a full load of fat and fatty matter into accumulator tank No. 1. The system was inspected and the evaporator opened on the water side but no trace of fat could be detected. This accident verified the theory that fatty substances do not clog to the evaporator surfaces at the low temperatures concerned.

Mainly due to the delay in the building program, the heavy testing program and the fact that the plant was commissioned as a pilot plant this particular plant have not been profitable. New plants, however, will be cost effective.

PLANT F:

Reference name:	Kodammarna/Gothenburg, Sweden
Design heating capacity:	65 kW
Compressor motor power:	28 kW
Design leaving hot water temperature:	45°C
Heat source:	Untreated sewage water
Heat source temperature:	7°C

Plans are under way for designing a large heat pump utilizing untreated sewage water as the heat source, since this is available closer to the heat energy consumers than treated water. As part of the preparations, a small heat pump was installed in 1980 in a pump station for untreated water. The primary intention was to test the plant to determine the minimum requirements on the evaporator and sewage-water sides of the system in order not to invest more than necessary on the larger project.

The plant hook-up is illustrated in Fig. 18.

In this case a vertical shell-and-tube evaporator together with a semi-hermetic reciprocating compressor working with R22 was chosen.

SIGNIFICANT FACTORS:

Heat energy available from heat pump: 85%:	191 MWh/year
Heat energy required from oil fired burners: 15%	34 MWh/year
Total annual requirement:	225 MWh/year
Total annual COP:	3.2
Energy prices:	
Electricity:	0,271 SEK/kWh
Oil: 1400 SEK/m^3: (at 75% efficiency)	0,187 SEK/kWh

Operating costs:
Before:
Oil: $225 \times 10^3 \times 0.187 =$ 42,075 SEK/year

After:
Oil: $225 \times 10^3 \times 0.15 \times 0.187 =$ 6400 SEK/year

Electricity:
$225 \times 10^3 \times 0.85 \times 0.271/3.2 =$ 16,300 SEK/year

Total 22,700 SEK/year

Total investment: 285,000 SEK

Pay-back time: $\dfrac{285,000}{42,075 - 22,700} = 15$ years

The extraordinary feature of this plant is the absence of filters before the evaporator on the untreated sewage water side. The already existing strainer with 30 mm between the bars is all that reduces the content of undesirable impurities before the water enters the top of the evaporator, where the mixture of water and impurities is distributed to the inside of a number of vertically fitted pipes and leaves the evaporator at the bottom, returning to the outlet of the pump station.

The plant was commissioned in the summer of 1980 and has been working entirely satisfactorily. Obviously, impurities of various kinds will separate and clog at some specific points in the circuit. So far, the necessary cleaning work has been done manually and has not proved any inconvenience to the employees. The station is normally unmanned. For testing purposes, however, thermometer pockets were installed in the sewage water line in front of the evaporator and clogged the passage when impurities accumulated on them. If considered necessary, automatic cleaning arrangements can be introduced.

Based on the larger plants conditions the figures proves to be cost effective and the test plant has proved to fullfil the requirements.

PLANT G:

Reference name:	NNP/Östersund, Sweden
Design heating capacity:	1,363 kW
Compressor motor power:	2 x 250 kW
Design leaving hot water temperature:	65°C
Heat source: Cooling water from refrigerating plants and condensers	
Heat source temperature:	18°C

NNP, a large dairy company, extended their production facilities back in 1978-79 to include a new refrigerating plant. The management realized that considerable amounts of heat energy would be generated at the condenser side of this plant and wondered if this could be recovered. On the other hand, the available energy exceeded the dairy's own internal requirements and consequently an agreement was reached with the local district heating system authorities. Two heat pumps were installed and connected to the system as shown in Fig. 19. The refrigerating plant features water-cooled condensers which discharge the heated water into a 180 m^3 accumulator basin at a temperature of approx. 18°C. This water serves as heat source for the two heat pumps which, in their evaporators, chill the water down to approx. 10°C before dumping it into another 180 m^3 basin from which the refrigerating plant condensers are fed.

The two heat pumps are connected in series on the water side in order to improve the COP-value. The hot water leaving the heat pump condensers is distributed in a parallel-connected circuit for internal consumption and for district heating purposes.

Use of the heat pump enabled the oil consumption to be drastically reduced and even other significant factors were observed:

Heat energy available from heat pumps:	8500 MWh/year
Electric enegy for heat pump compressor:	2500 MWh/year
Electric energy for pumps etc.	150 MWh/year
Electric energy reduction on refrigerating plant	100 MWh/year
Total electric energy requirement:	2550 MWh/year
COP = 8500/2550=	3.35

For the dairy the oil consumption will be reduced by 880 m^3/year and for the district heating system by 205 m^3/year with a total of 1085 m^3/year.
Calculated on the present energy price level nearly 1.5 million Swedish Kronor/year will be the profit.

Due to low and constant cooling water temperature to the refrigerating plant a lower condensing temperature will be obtained than with cooling towers only.
Higher efficiency, reduced maintenance will be gained as well.

DISCUSSION

A variety of new heat sources may be used for large heat punp installations.

The technology and the capability of handling these sources as well as the experience gained from plants using these new sources will be of utmost importance for the plants of tomorrow. Thus failures at this end may be disastrous, not only with regard to profitability, but for the trade as such.

For some of the plants described here the elapsed time of operation has been too short for a complete survey and consequently a continuous follow-up will be required.

The energy prices of the compared system will naturally be of importance for the profitability and, as they differ from country to country, other pay-back times than those stated here may be experienced.

CONCLUSIONS

Although only a few heat pump installations have been described and analysed in this paper, the results from numerous installations verify several important facts:

— Heat pump plants are reliable if properly designed.

— Maintenance costs will normally not exceed the cost of an oil-fired boiler installation of equal capacity.

— The heat pump system can be cost-effective when properly designed.

— Higher energy prices will increase profitability.

— Heat pumps are already available in the required capacity range and adapted to the various heat sources available.

— Know-how is already available for comprehensive plant analyses and design.

REFERENCES:

The plants covered in this paper and entitled to government subsidy are thoroughly described in reports being the base for the grants by:

Swedish Counsil for Building Research
S.t Göransgatan 66
S-112 33 STOCKHOLM
Sweden

Distribution through:

Svensk Byggtjänst
Box 7853
S-103 99 STOCKHOLM
Sweden

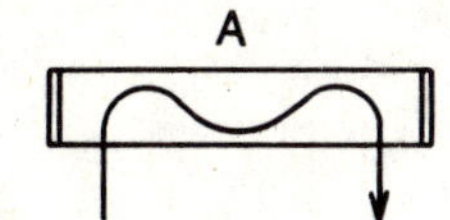
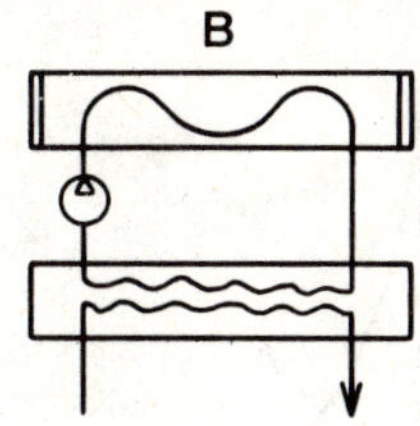
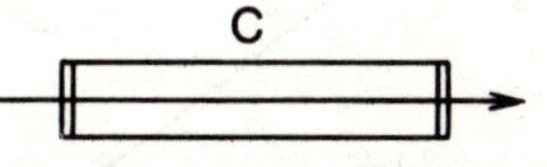
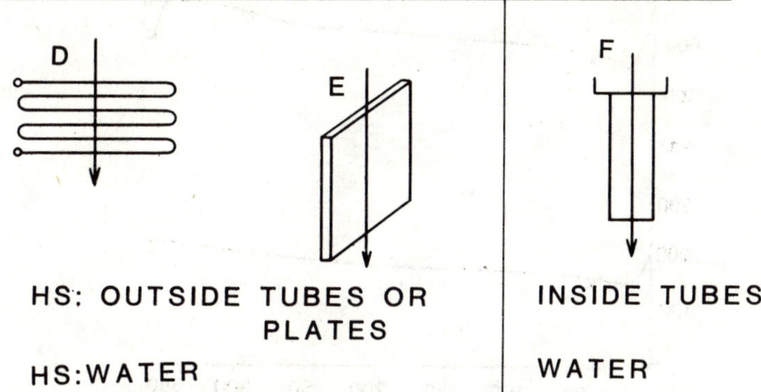

Fig. 1   EVAPORATOR TYPES VERSUS HEAT SOURCES

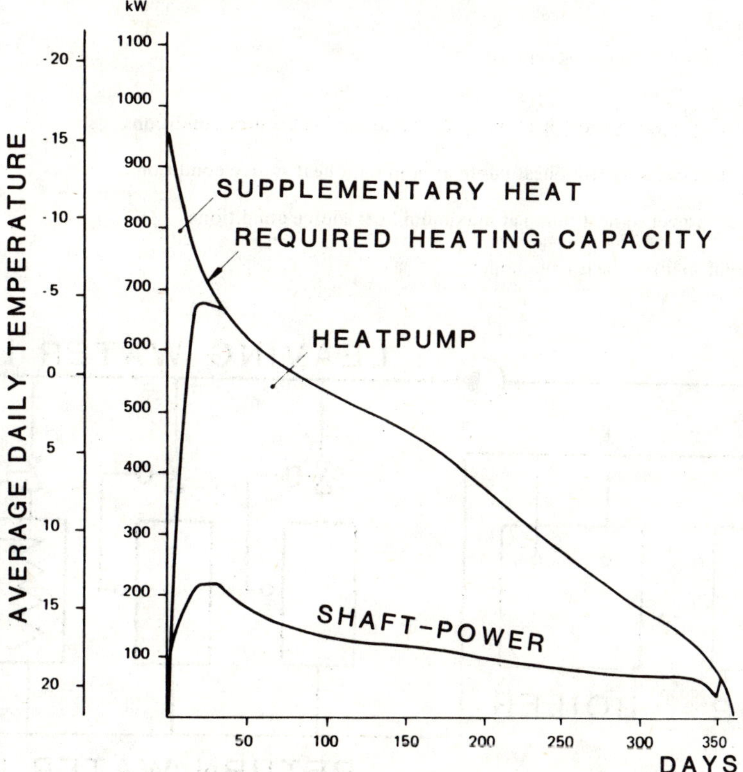

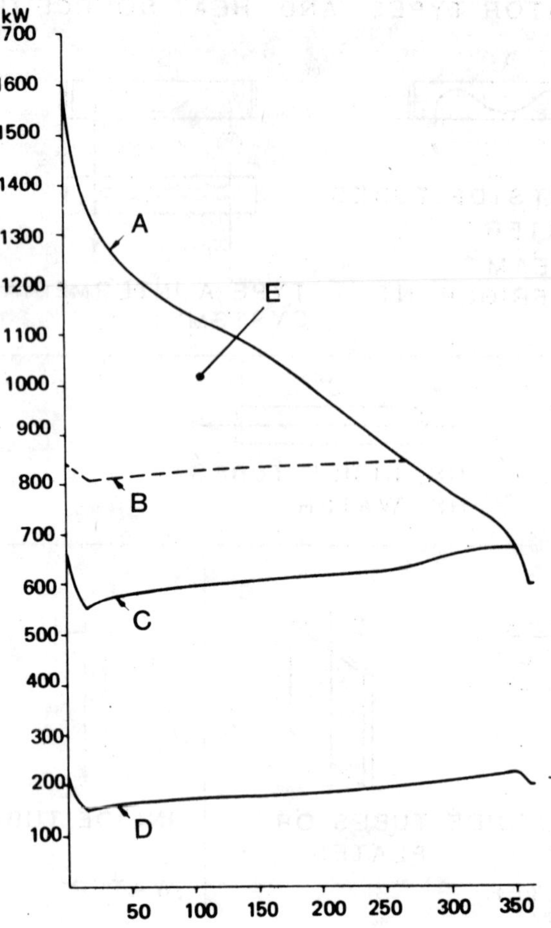

Fig. 3   CAPACITY REQUIREMENTS FOR DAYS REQUIRING HEATING OF PREMISES
AND PRODUCTION OF TAPWATER FOR CLEANING PURPOSES.

A.   Required total heating capacity

B.   Heating capacity from heat pump at maximum heat source conditions

C.   Heating capacity from heat pump at minimum heat source conditions

D.   Shaft power of heat pump at maximum heat source conditions

E.   Supplementary heat from boilers

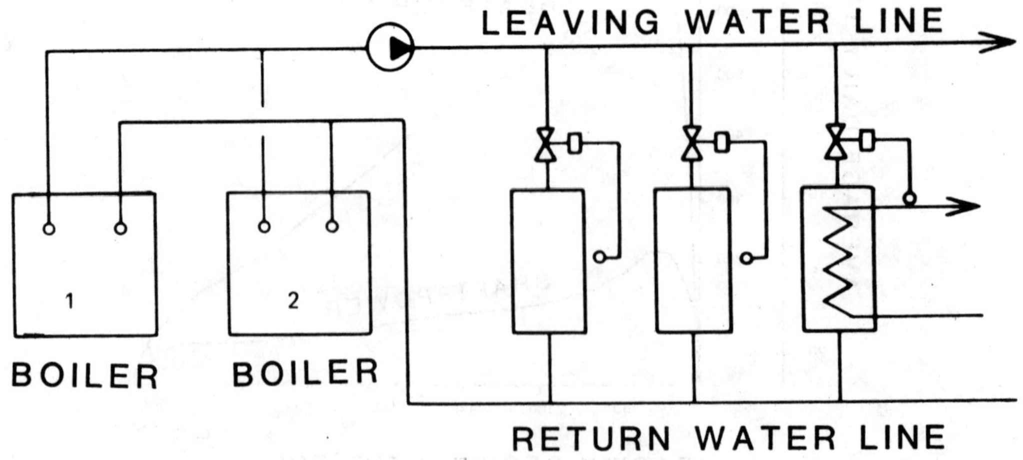

Fig 4   EXISTING HEATING SYSTEM

298

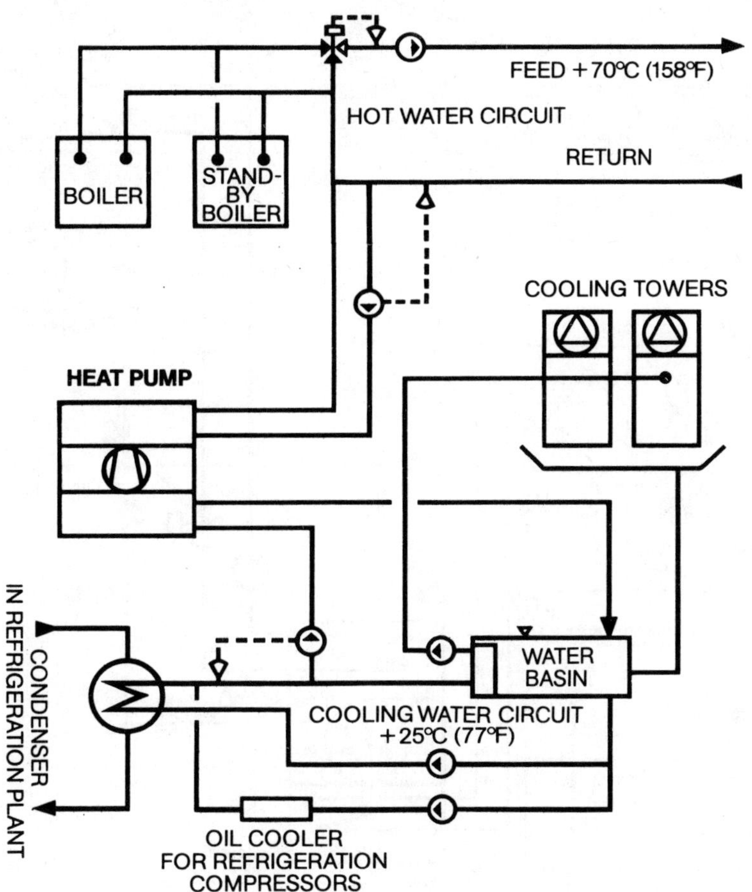

Fig. 5   FUNDAMENTAL HOOK-UP FOR PLANT A

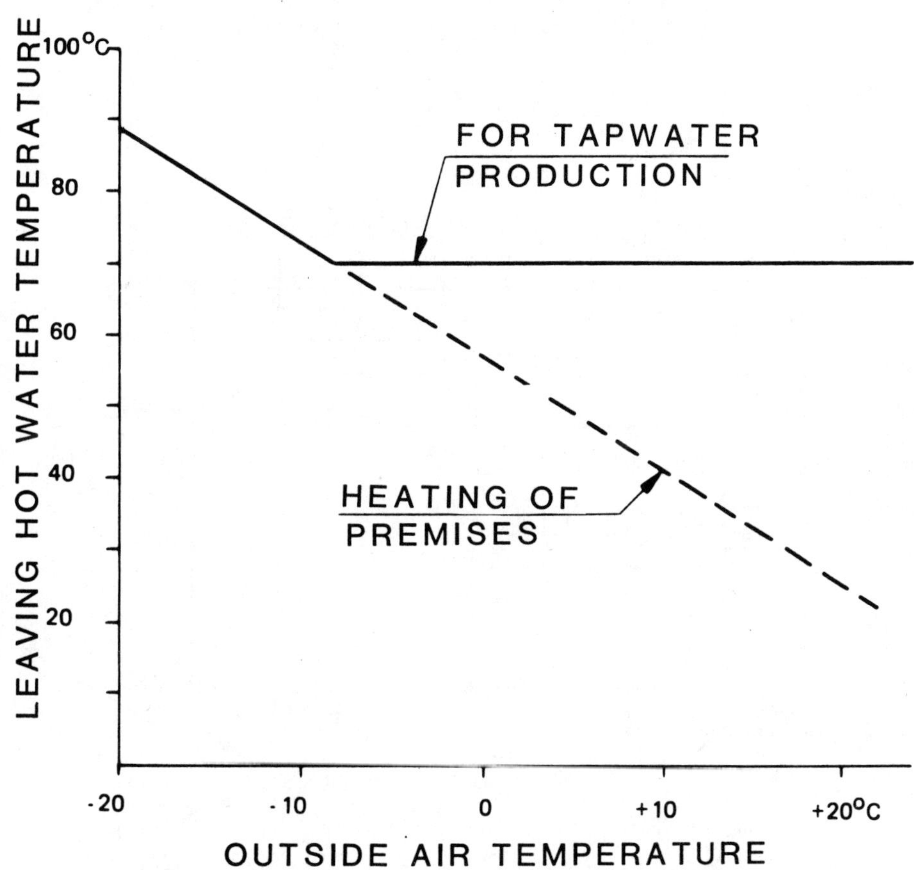

Fig. 6   LEAVING HOT WATER TEMPERATURE VERSUS OUTSIDE AIR TEMPERATURE

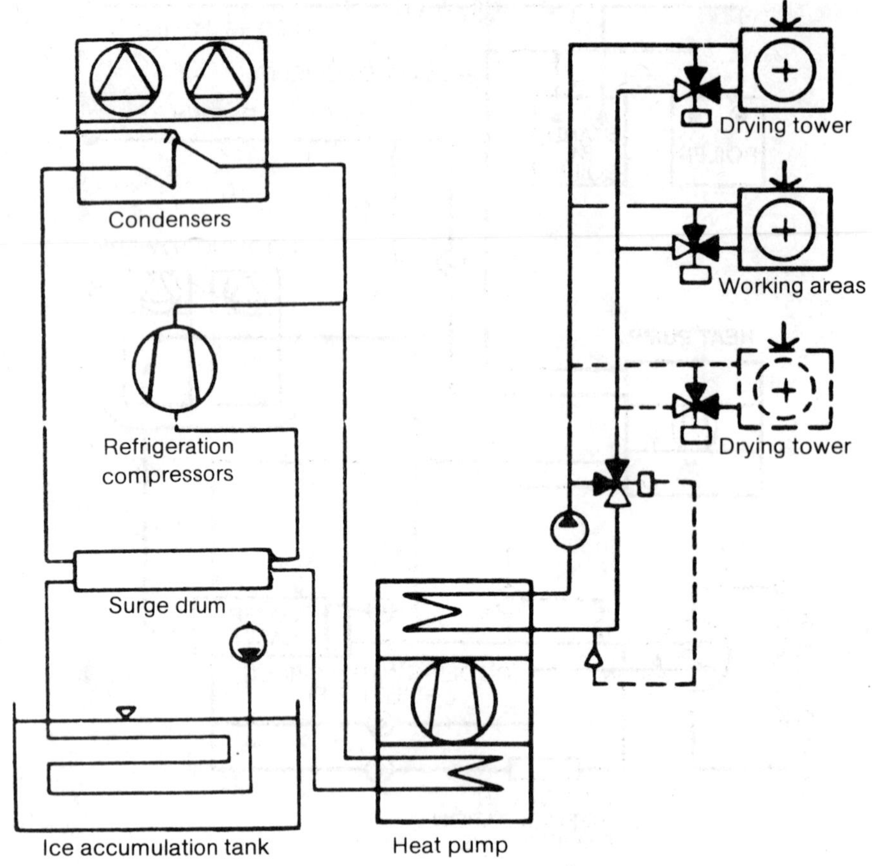

Fig. 7   FUNDAMENTAL HOOK-UP FOR PLANT B

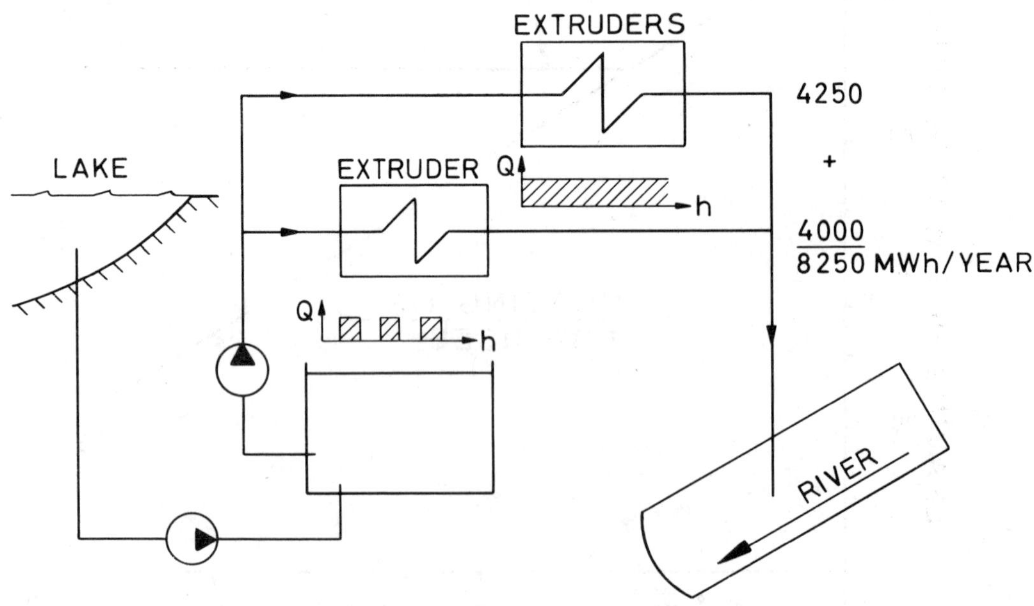

Fig. 8   PREVIOUS COOLING SYSTEM FOR PLANT C

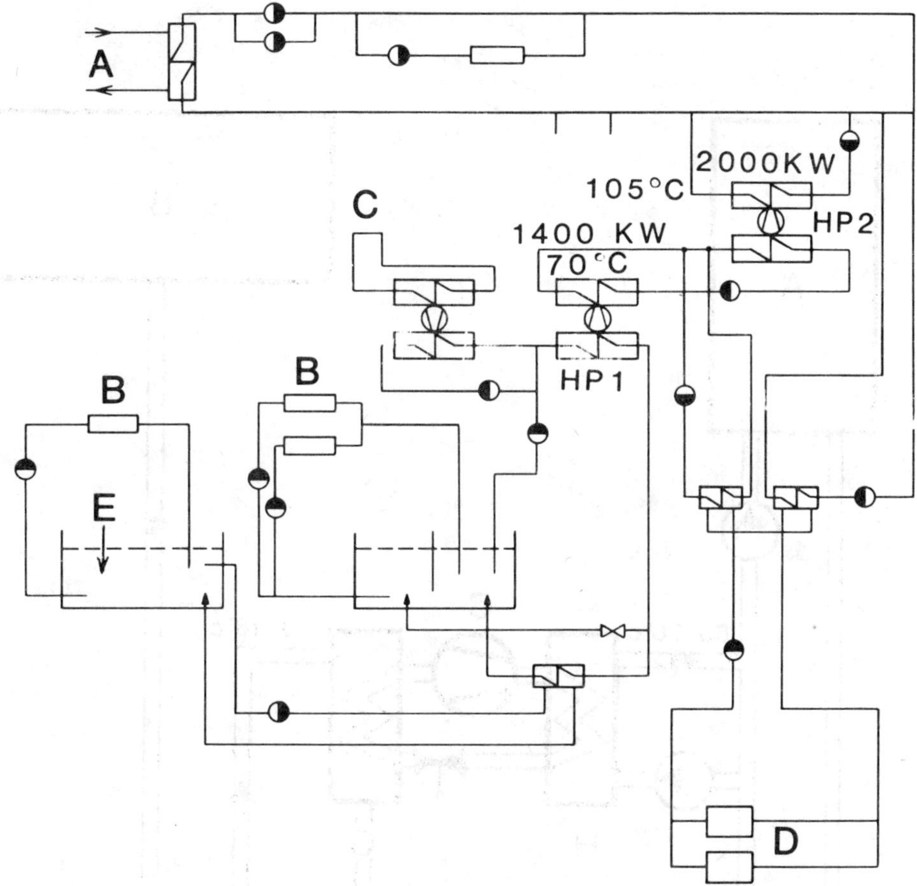

Fig. 9   NEW COOLING AND HEATING SYSTEM HOOK-UP FOR PLANT C

    A.   To central heating plant

    B.   Extruders

    C.   Air conditioning

    D.   Ventilation

    E.   Lake-water

HP1:Heat pump (R12)

HP2:Heat pump (R114)

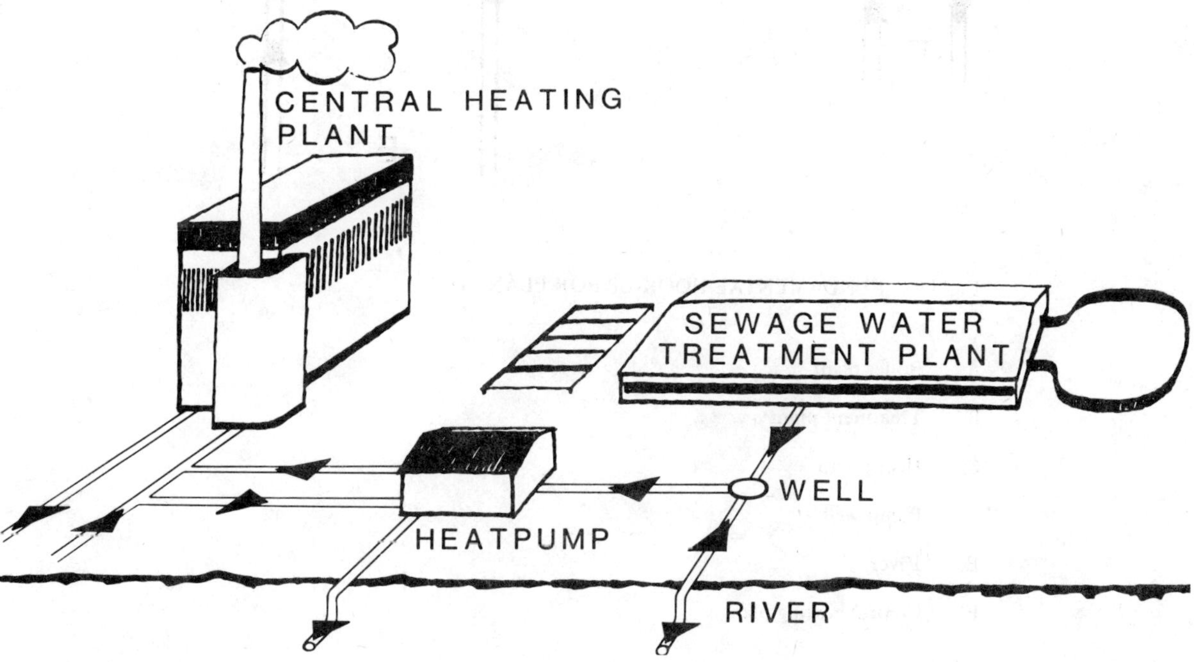

Fig. 10   SITE LAY-OUT FOR PLANT D

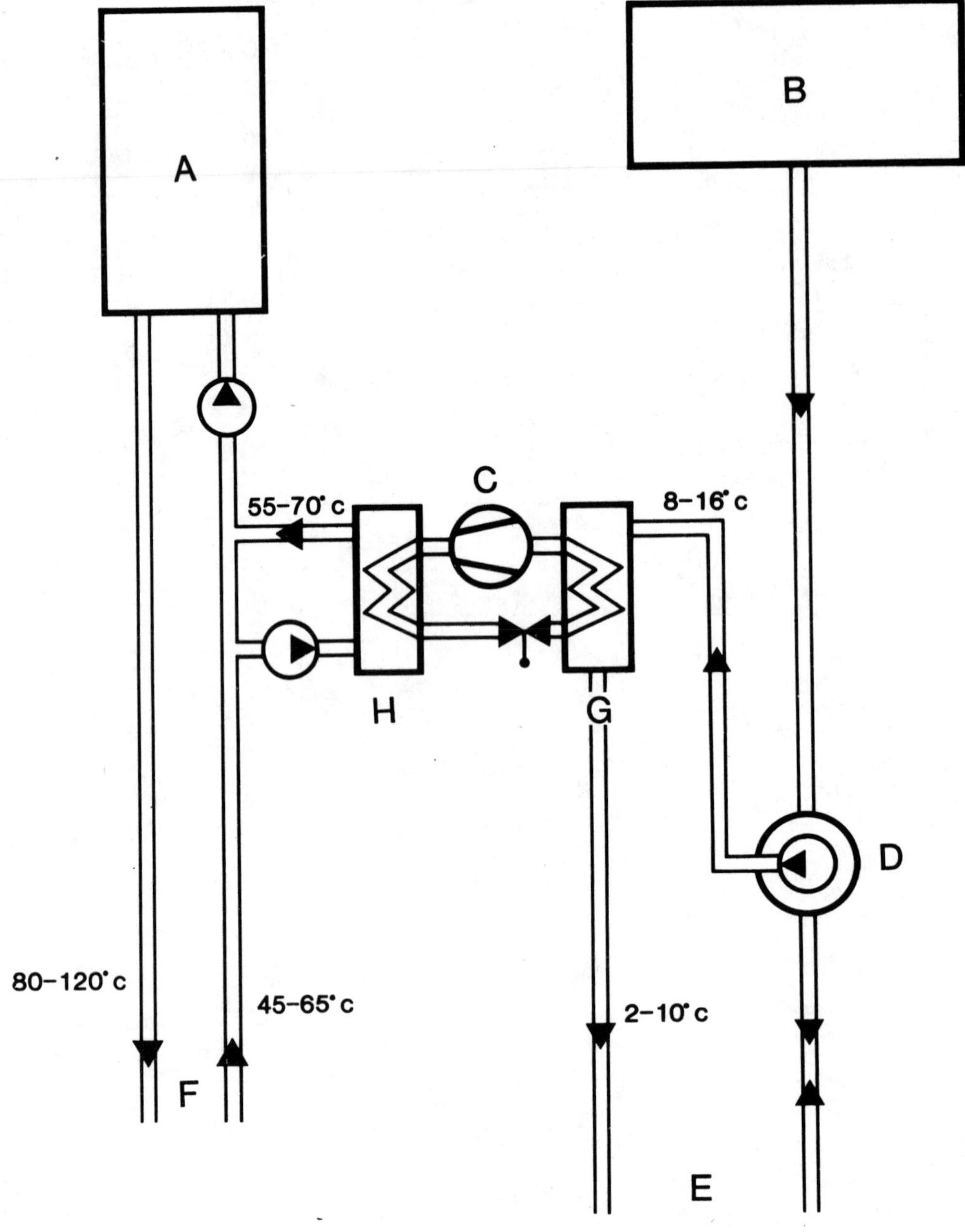

Fig. 11   FUNDAMENTAL HOOK-UP FOR PLANT D

A.   Boiler plant

B.   Treatment plant

C.   Heat pump

D.   Pump well

E.   River

F.   Central heating plant

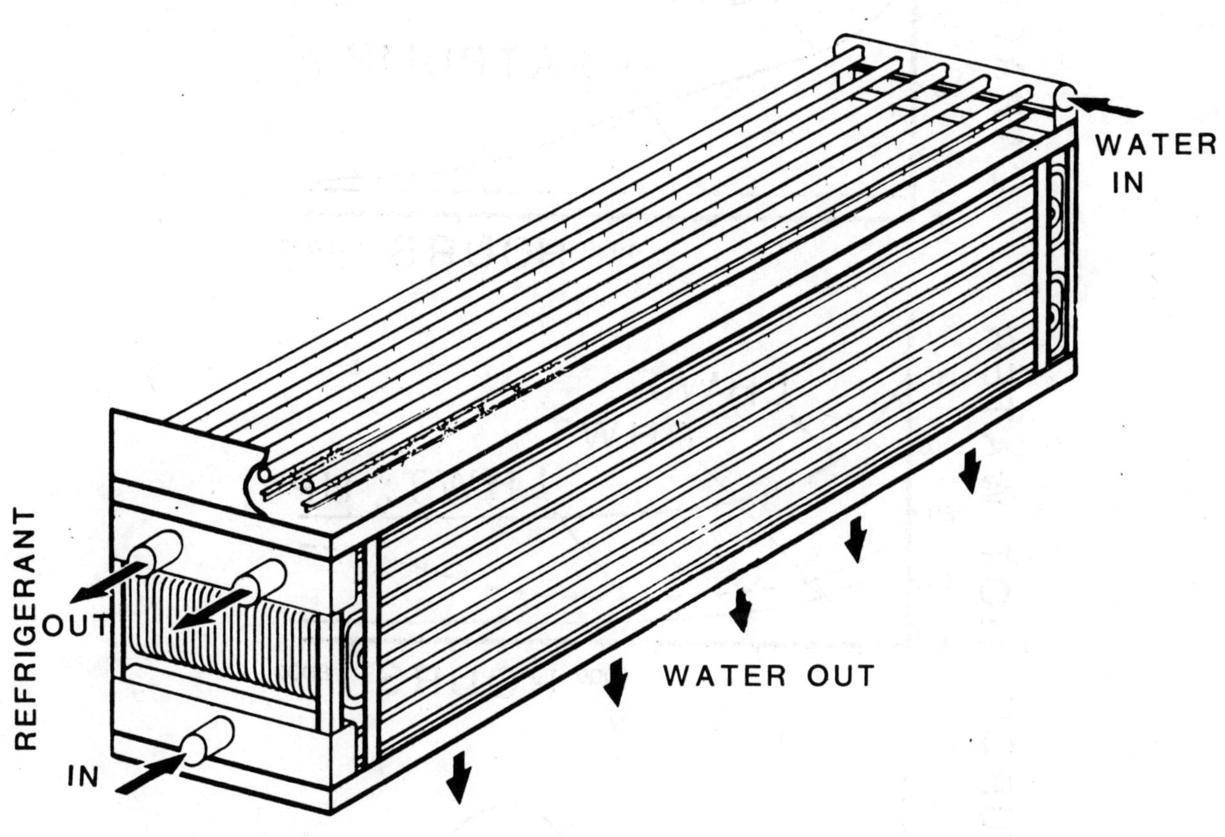

Fig. 12   EVAPORATOR
LOW TEMPERATURE TYPE

CAPACITY: 3000 KW

POWER INPUT: 1000 KW

ANNUAL COP: 2,7

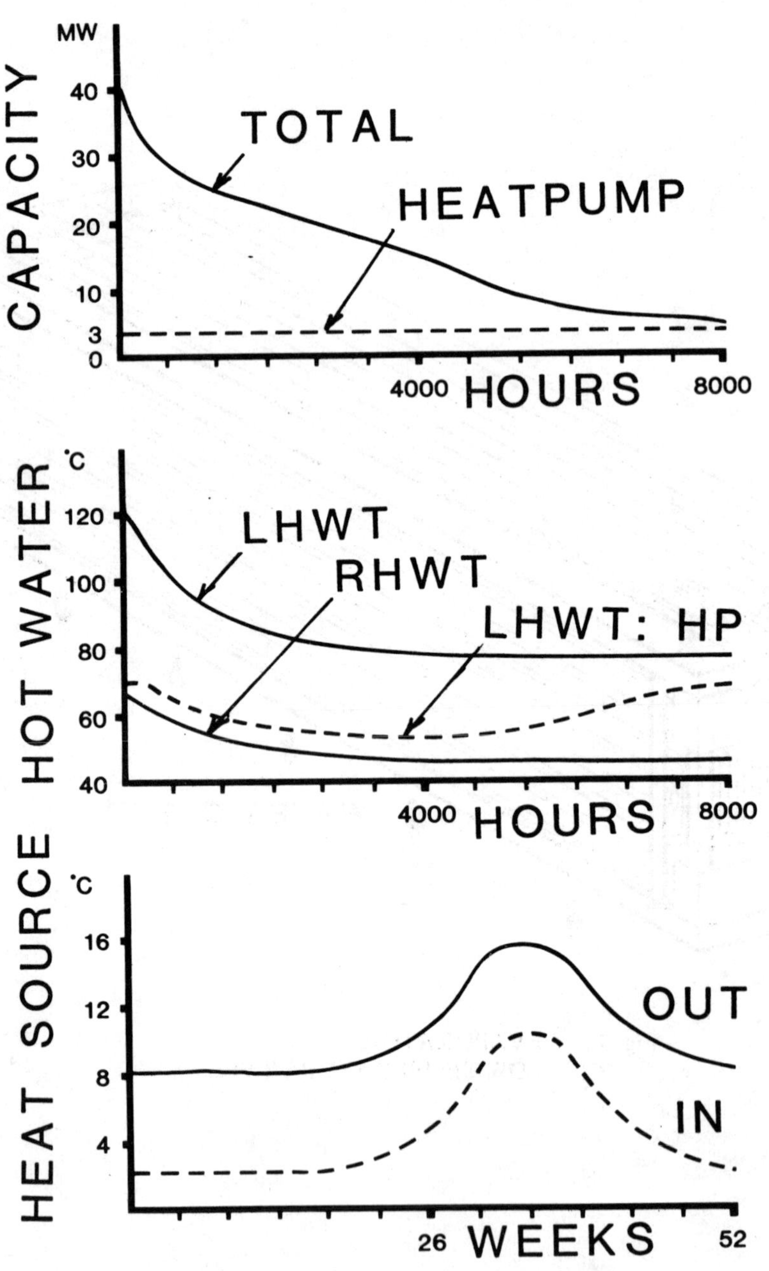

Fig. 13   TECHNICAL DATA FOR PLANT D

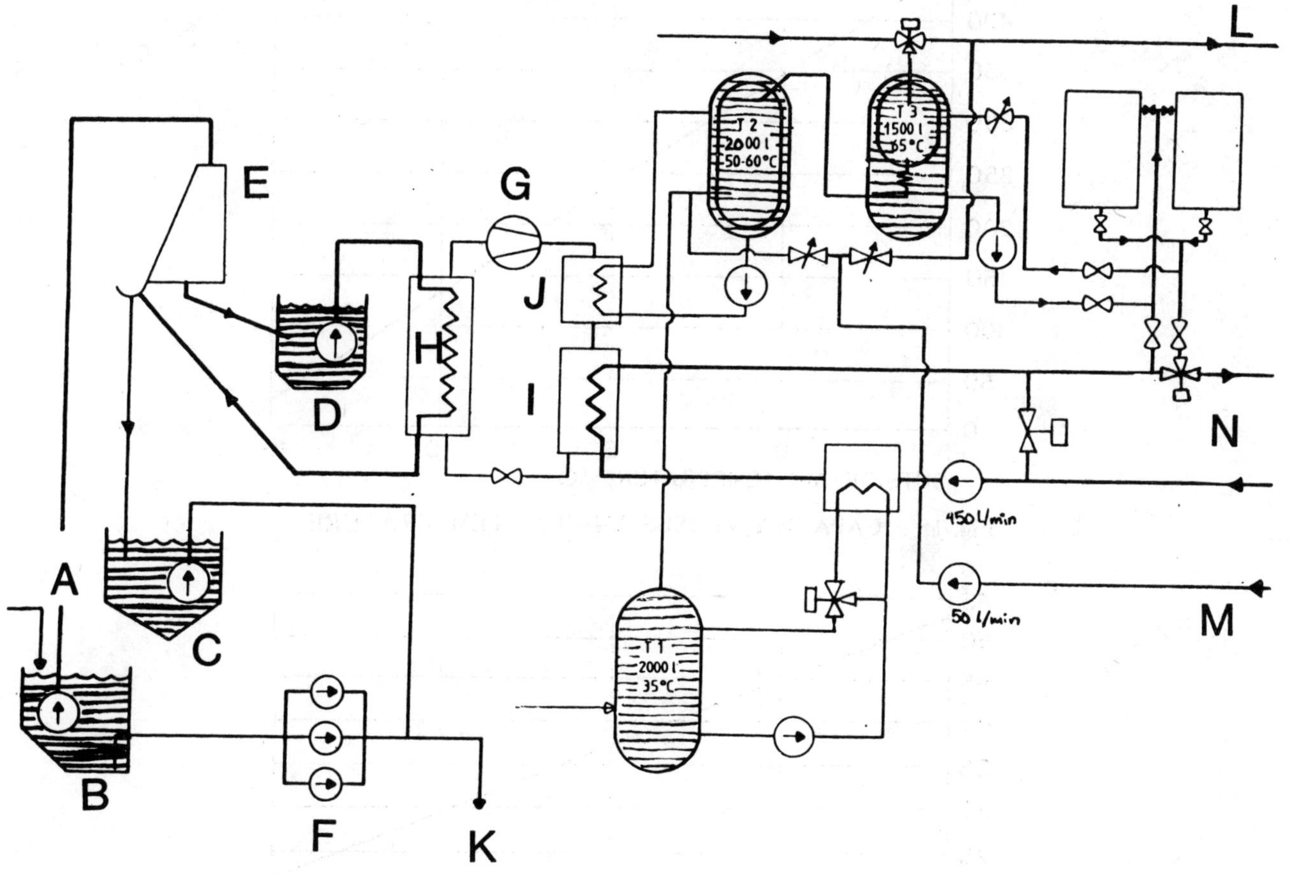

Fig. 14   FUNDAMENTAL HOOK-UP FOR PLANT E

A.   Untreated sewage water supply

B.   Basin No. 1 untreated water

C.   Basin No. 2 untreated water

D.   Basin No. 3 filtered water

E.   Filter

F.   Sewage water pumps

G.   Compressor

H.   Evaporator

I.   Condenser

I.   Gas cooler

K.   To treatment plant

L.   Tap water circuit: leaving

M.   Tap water circuit: return

N.   Radiator circuit

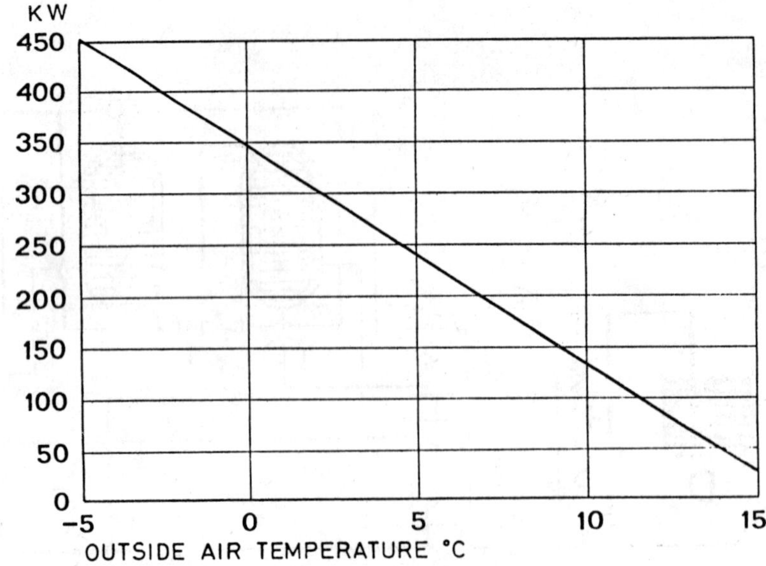

Fig. 15   CAPACITY VERSUS AMBIENT TEMPERATURE

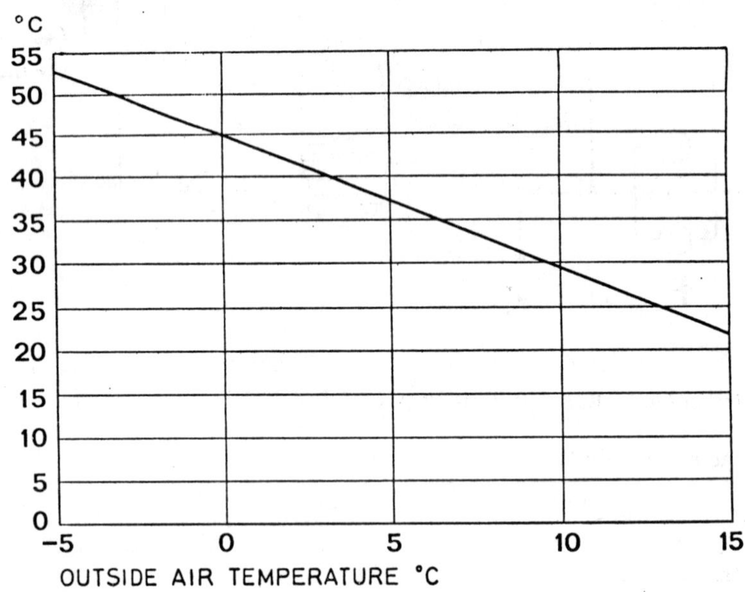

Fig. 16   LEAVING HOT WATER TEMPERATURE VERSUS AMBIENT TEMPERATURE

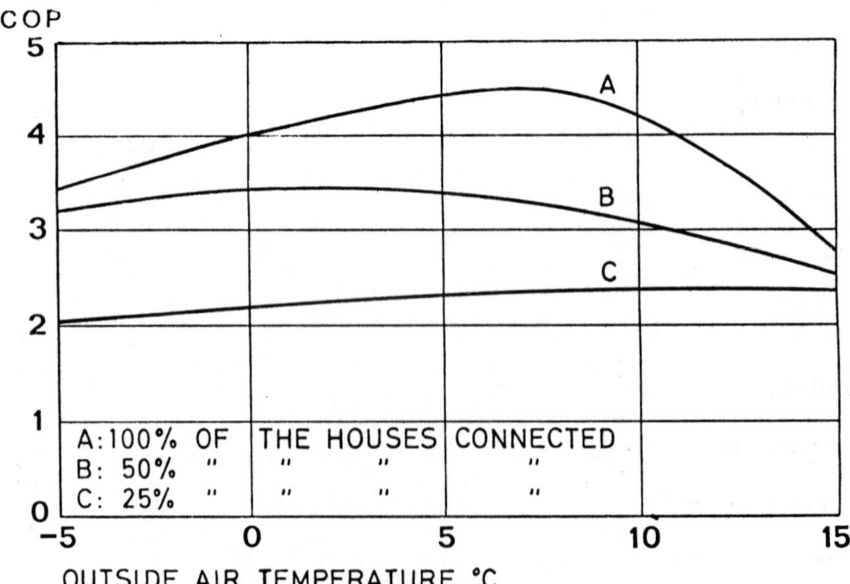

Fig. 17   COP VERSUS AMBIENT TEMPERATURE

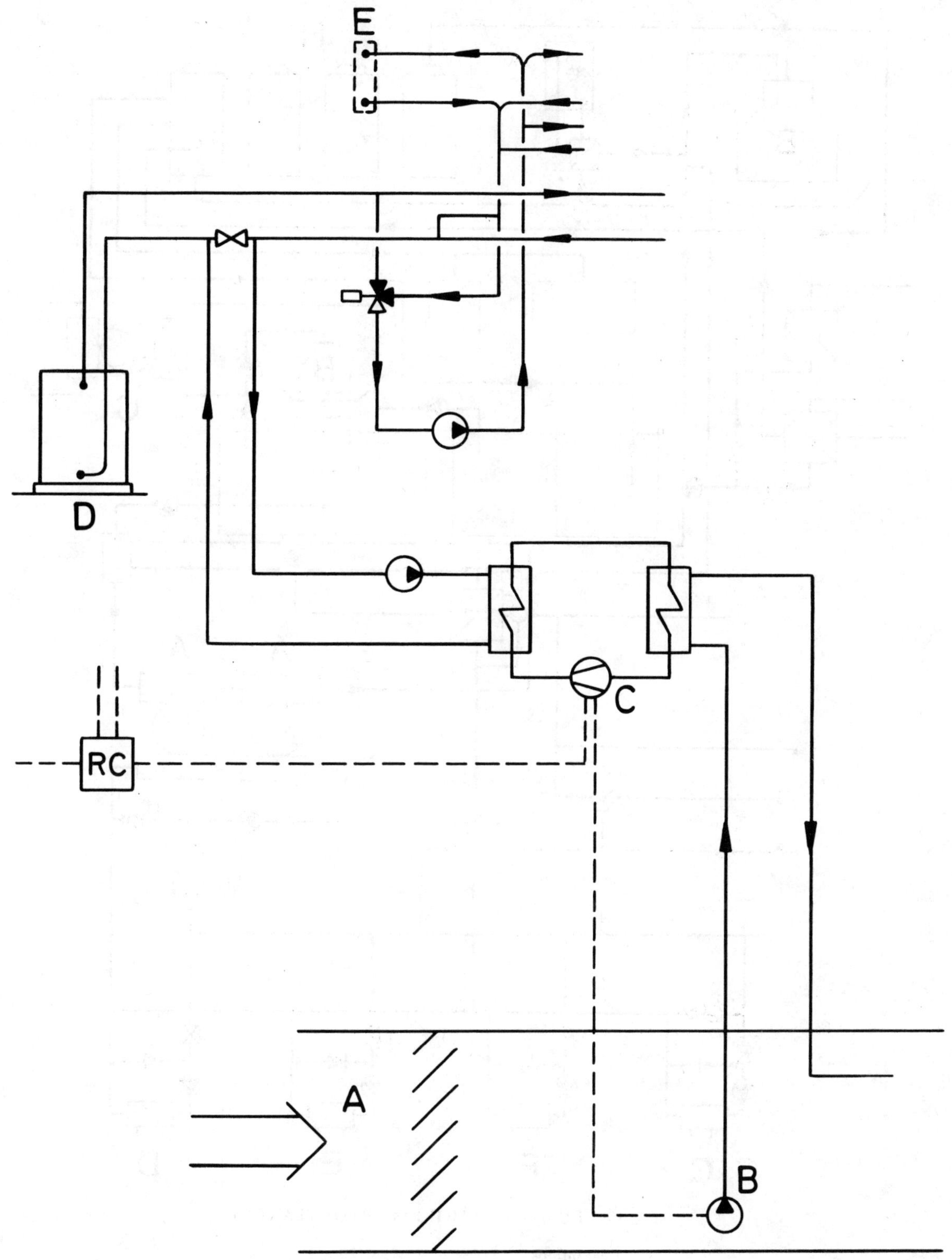

Fig. 18   FUNDAMENTAL HOOK-UP FOR PLANT F

A.   Strainer

B.   Sewage water pump

C.   Heat pump

D.   Boiler

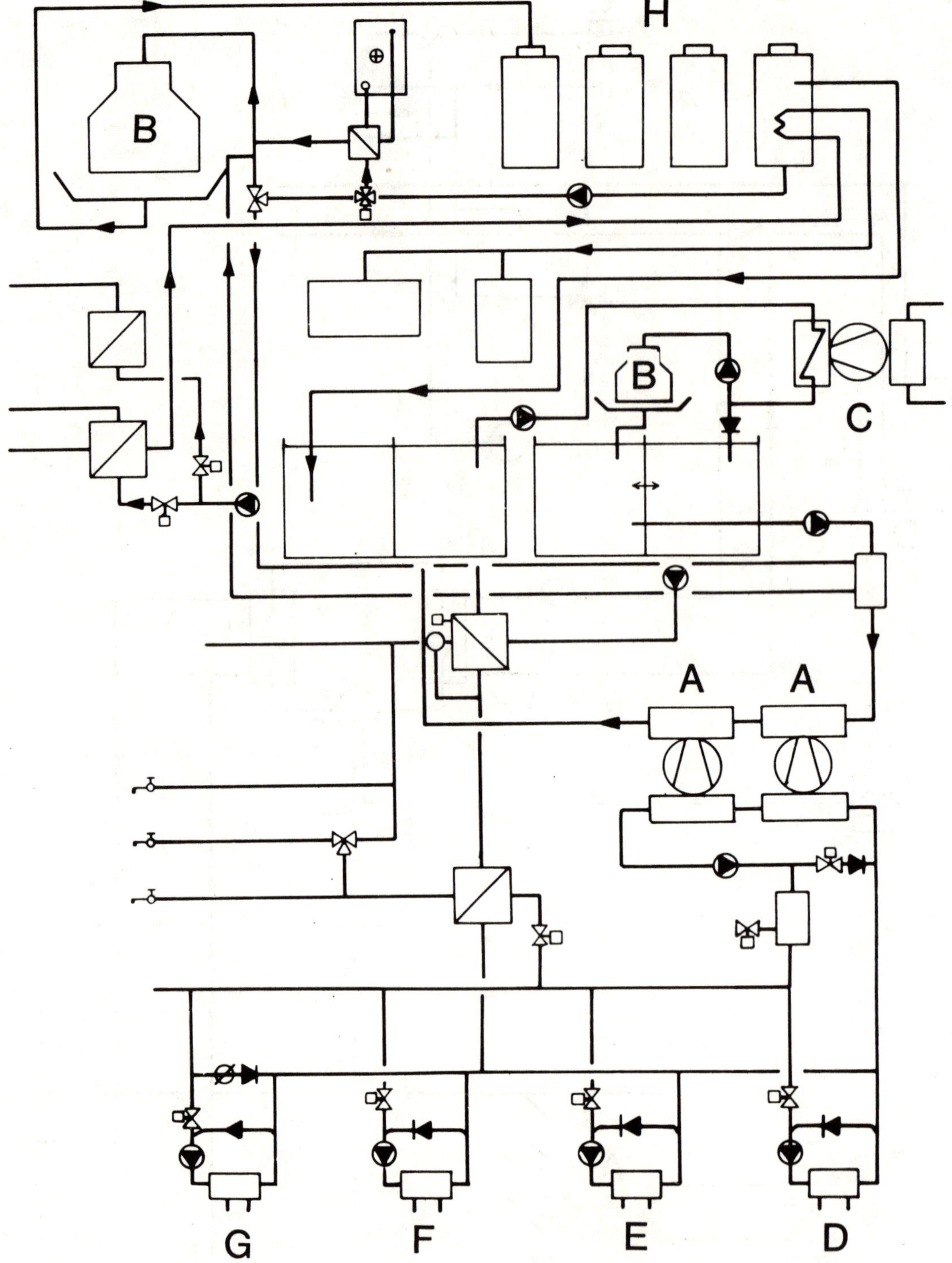

Fig. 19   FUNDAMENTAL HOOK-UP FOR PLANT G

A.   Heat pumps

B.   Cooling towers

C.   Refrigerating plants

D.   Dairy

E.   Cheese factory

F.   Other buildings

G.   District heating system

H.   Evaporation plant (milk powder)

SOME INVESTIGATIONS OF THE USE OF SHEET METAL FACTORY ROOFING IN A SIMPLE SOLAR HEAT PUMP SYSTEM

J. R. Waters

Coventry (Lanchester) Polytechnic, U.K.

Summary

This paper is concerned with an industrial application in which the roof of a factory building is used as an uncovered solar collector in a series connected solar heat pump system.

Held at the University of Warwick, U.K.
Symposium organised and sponsored by
BHRA Fluid Engineering
©BHRA Fluid Engineering, Cranfield, Bedford MK43 0AJ, England.

NOMENCLATURE

E_A = energy consumed by auxiliary heater

E_F = free energy entering system

E_{HP} = energy consumed by heat pump compressor

E_{PM} = equivalent primary energy

I = solar intensity

T_a = external ambient temperature

T_c = average collector temperature

U = thermal loss coefficient

α = solar absorption coefficient of collector surface

τ = solar transmission coefficient of collector glazing

η = efficiency

F_S = collector thermal performance factor

1. INTRODUCTION

Heat pumps are frequently used to upgrade the heat that is collected in the store of a solar energy system, and there have been many studies of solar heat pump systems, for example by Bessler and Hwang (ref.1), Terrell (ref.2), Hume (ref.3) and Lawaetz (ref.4). However, most previous work has been concerned with covered solar collectors and with systems suitable for domestic dwellings. This paper is concerned with an industrial application in which the roof of a factory building is used as an uncovered solar collector in a series connected solar heat pump system.

The possible advantages are:-

    (i)   uncovered solar collectors are cheap;

    (ii)   factory roofs provide a large collector area;

   (iii)   the system does not suffer from the plumbing, fixing, freezing and over-heating problems associated with covered collectors;

   (iv)   it is suitable for retrofit to existing buildings.

2. BACKGROUND CONSIDERATIONS

2.1 Solar Collectors

The usual expression for the efficiency of a flat plate collector is

$$\eta = F_s \left(\tau\alpha - U. \frac{(T_c - T_a)}{I} \right)$$

(Note: F_S is the symbol used by the U.K. branch of the International Solar Energy Society)

and figure 1 is a typical plot of η versus $\frac{T_c - T_a}{I}$ for uncovered, single glazed and double glazed collectors. It can be seen that a solar collector is operational (i.e. has a positive efficiency) only when $\frac{T_c - T_a}{I} < x$, where x is the intercept on the horizontal axis, which implies that each collector requires a minimum solar intensity, I_{min}, given by

$$I_{min} = \frac{T_c - T_a}{x}$$

In the U.K., approximately 60% of the incoming radiation is low intensity diffuse radiation, and therefore I_{min} must be kept low for a collector to be effective. This can be achieved by using single or double glazed collectors for which x is large. However covered collectors are expensive, and, as has been shown by Wozniak (ref.5), may also generate additional costs due to their fixing requirements to the building. Alternatively, I_{min} can be kept low by designing a system in which $T_c - T_a$ is very small, in which case an uncovered collector will be satisfactory. For example, taking F_S as 1, $T_c - T_a$ as 3 ^{o}K, and the intercept, x, as 0.06 for an uncovered collector, the value of I_{min} is 50 Wm^{-2}, which is exceeded even on most dull days in winter. Furthermore, when the solar intensity, I, rises above 200 W/m^2, the uncovered collector becomes more efficient than either of the covered collectors.

The utilisation of the roof surface itself as a solar collector has been attempted in many ways. In the case of industrial buildings, low first cost and simplicity are important, and so an approach which uses the normal roof construction with a minimum of modification is most likely to be acceptable. The sheet materials which are often used are ideal as uncovered collectors, and it is only necessary to add a delivery pipe at ridge level to trickle water over the roof surface.

2.2 Heat Pumps

Figure 2 illustrates the theoretical coefficient of performance of a single stage vapour compression machine using R12. The calculation used equations given by Duminil (ref.6), and included the following assumptions:

Superheating of vapour at compressor inlet   10 $^{\circ}$K

Sub-cooling of liquid at condenser outlet   5 $^{\circ}$K

Mechanical efficiency of compressor, η_m 0.9

Indicated efficiency of compressor, η_i $1 - 0.05r$ where r is the compression ratio.

To be economical, a heat pump must achieve a coefficient of performance which is above a minimum value. This minimum value depends on the power source to the compressor, and on the alternatives to the heat pump which are available for a given application. Taking 2.5 as a typical minimum coefficient of performance, and assuming that delivery temperatures of up to 70 $^{\circ}$C may be required, it can be seen from figure 2 that the minimum evaporator temperature is about 6 $^{\circ}$C. This implies a heat source with a minimum temperature of approximately 10 $^{\circ}$C. Such sources are rare in temperate climates.

2.3 Solar Heat Pumps

There are several methods of configuring a heat pump with a solar collector, the most common being:

(i)   The parallel solar heat pump.

The heat pump draws its heat from an independent source (e.g. external air) and operates in parallel to the solar collector.

(ii)   The series solar heat pump

The heat pump evaporator draws energy from the solar storage tank, and the condenser supplies the load.

(iii)   Dual source system

The heat pump has two evaporators, one placed in the solar store, the other placed in an independent source. The heat pump uses whichever source will provide the highest coefficient of performance.

In all three cases, controls may be added to allow the collector to supply the load directly whenever its store is hot enough, and to switch in an auxiliary heater when neither collector nor heat pump can satisfy the load. Freeman, Mitchell and Audit (ref.7) have carried out a theoretical study of the application of three systems to the annual heating load of houses in the U.S.A. Table 1 has been computed from their results, and shows the percentage of the annual load which is met by free energy E_F, auxiliary energy E_A, and heat pump compressor energy E_{HP}. The predicted annual coefficient of performance of the heat pump is also shown. The final column in the table is an estimate of the primary energy consumption, expressed as a percentage of the delivered energy, and calculated by (see Appendix):

$$\text{Primary energy, } E_{PM} \simeq 3\,E_{HP} + \frac{1}{0.7}\,E_A$$

This equation assumes an electrically driven heat pump, and a fossil fuel auxiliary heater of 70% efficiency. The table shows that a pure solar system uses least primary energy, and that the series solar heat pump is next best. However, these results refer to a covered collector. The use of an uncovered collector would reduce the performance of all the solar systems, but the reduction would probably be most severe for the pure solar system, the parallel solar heat pump and the dual source solar heat pump, because these three systems all draw energy directly from the solar collector. The series solar heat pump may be expected to be less affected by the change to an uncovered collector, because the continual chilling of the collector tank will help to maintain collector efficiency, whilst simultaneously the supply of heat to the evaporator will help maintain the coefficient of performance of the heat pump. For this reason, the series solar collector was selected for the application described here.

3. THE INDUSTRIAL SOLAR HEAT PUMP SYSTEM

3.1 Description of the System

Figure 3 is a schematic diagram of the system. The roof is a conventional industrial roof of sheet material, and the rainwater gutter and downpipe are standard components. The ridge pipe may be any suitable size pipe, with spray holes drilled directly into it, or with attached spray nozzles. This pipe may be mounted externally (as on the prototype) or concealed by the ridge flashing. The insulation beneath the roof surface must, in the U.K., satisfy the Building Regulations, which currently stipulate a maximum thermal transmittance of 0.6 $Wm^{-2}\,^{o}K^{-1}$;this is adequate for the purposes of solar collection. The return from the rainwater gutter is taken into a small catch tank fitted with a standard siphonic overflow device, and an on-off valve on the return to the solar storage tank. When it is raining, the valve is normally off, and the rainwater fills the catch tank until it siphons out, the resulting turbulence being sufficient to remove most of the debris (leaves, etc.) brought down from the roof. The heat pump evaporator and condenser coils are mounted directly in the low temperature and high temperature storage tanks respectively. In the prototype the heat pump compressor will be electrically driven, and the compressor and its motor will be mounted outside the storage tanks (to simplify access and monitoring). A modulating expansion valve and a variable speed compressor will provide considerable flexibility in operation.

The load circuit contains an auxiliary heater, mounted after the high temperature storage tank, to maintain the correct delivery temperature into the load when the output from the heat pump is insufficient.

The microprocessor monitors temperatures in the solar collector, the heat pump and the load circuit, and it also monitors the compressor speed. From these inputs it generates an optimum control strategy.

3.2   Expected System Performance

A crude estimate of the system performance can be obtained by combining the efficiency curve for an uncovered solar collector from figure 1 with the COP curve for a condenser temperature of 73 ^{o}C from figure 2. The result is shown in figure 4, which suggests that the performance characteristic of the system is a relatively weak function of solar storage tank temperature with a broad peak. This further suggests that the system will tend to stabilise at a maximum operating efficiency, and that, for a given set of external parameters, the maximum efficiency will be characterised by a particular temperature of the solar collector storage tank. Figure 4 was computed for continuous operation in average conditions, and so the equilibrium temperature of about 14 ^{o}C may be expected to be typical.

3.3   The Prototype System

A prototype system is being constructed with the following general specification:-

Solar collector (roof) area   25 m^2

Solar collector storage tank   1 m^3

High temperature storage tank   0.1 m^3

Rated maximum output into load (nominal)   10 kW

Heat pump compressor power input (maximum)   2 kW

Microprocessor-controller Rockwell Aim   65

The collector, collector storage tank, and the whole of the collector circuit is complete and is operational. The rest of the system has been designed and is under construction. A large proportion of the software for the Aim 65 has been written and tested.

4.   RESULTS OF MEASUREMENTS ON THE SOLAR COLLECTOR

Measurements have been carried out on the solar collector circuit with the objective of evaluating:

(i)   the efficiency of solar collection

(ii)   the principal factors affecting efficiency

(iii)   the practical problems of using normal roof materials for solar collection.

The panel was mounted on the roof of a laboratory building in Coventry, U.K. (latitude 51 ^{o}N) at a pitch of 30 ^{o}, and facing due South. For these initial

measurements, the circulating pump was switched by an on-off controller fed by the
difference in temperature between the solar panel and the storage tank.   To improve
stability, the controller incorporated an adjustable offset of +5 $^\circ$C (i.e. switch on
occurred when the panel temperature rose more than 5 $^\circ$C above the tank temperature),
but additional measurements were conducted with an offset of 1.6 $^\circ$C.   The controller
dead band was approximately $\pm$ 0.2 $^\circ$C.   Solar radiation was measured with a dome-type
thermopile solarimeter, and copper – constantan thermocouples were used to measure
temperatures around the system.   Wind velocities were obtained from an official
meteorological station approximately 10 miles away.

Results were analysed in hourly intervals, and two efficiencies were calculated.
The first was a "crude" efficiency, and was the energy delivered to the storage tank
as a percentage of the incident solar radiation, and the second was a "corrected"
efficiency in which allowance was made for heat losses from the panel and all
connecting pipes.   The second of these is applicable to the case when the storage tank
temperature is close to external ambient temperature, and is therefore the efficiency
that may be expected with the heat pump in operation.   The figures obtained were:

> "crude" efficiency, mean 11%, R.M.S. deviation 8%

> "corrected" efficiency, mean 44%, R.M.S. deviation 14%

The low value for the "crude" efficiency shows that most of the solar energy
collected was lost to the surroundings.   The individual results for each hour showed
considerable scatter about the mean, and attempts were made to correlate the results
with a number of variables, including external air temperature, incident radiation
intensity, the temperature difference between the panel and the storage tank, and wind
speed.   Only the last of them produced a significant correlation, suggesting that
forced convective heat loss from the panel surface is the dominant factor.   The only
other major factor affecting efficiency appeared to be the area of roof surface which
is wetted by the circulating water.

The area wetted by the water depends on:

(i)   the pitch and depth of the corrigations of the sheet material which is used,

(ii)   the design of pipe and nozzle system which delivers the water onto the roof,

(iii)   surface tension effects, which tend to cause the water to collect into
narrow streams as it runs over the roof surface.

The "corrected" efficiency, because it allows for all heat losses, is a measure
of the product $F_s \pi \alpha$.   Since $\pi = 1$ for uncovered collectors, and because $\alpha \simeq 0.8$ for the
surface of this particular panel, the figure of 44% suggests that $F_s \simeq 0.55$.

The experimental work did not reveal any serious practical problems.   Evaporative
loss of circulating water from the panel surface was small, and the arrangements for
dealing with rainwater and debris proved satisfactory.   However, contamination of the
circulating water by atmospheric pollutants was not measured.

5.   PREDICTION OF SYSTEM PERFORMANCE

A computer model of the system has been devised in order to predict the possible
energy savings.   Full details of the model are given in reference 8, but its principal
characteristics are:-

1.   It predicts system performance as a function of time in discrete time steps
(in essence, it is an implicit finite difference model).

2.   The main inputs are incident solar radiation, external air temperature, and
the load pattern, all expressed as time series to suit a range of typical operating
conditions.

3.   The main outputs are cumulative totals of the energy delivered by the system
to the external load, and the 'purchased' energy entering the system from the auxiliary
heater, heat pump compressor and circulating pumps.

The principal assumptions are:-

1.   The solar panel switches on whenever the panel surface temperature is higher
than the solar storage tank temperature (the thermostat is assumed to have an offset
the size of which may be chosen as part of the input data).

2.   The heat pump COP may be expressed as a function of the temperatures of the two storage tanks.   Functions giving the theoretical COP of R11 and R12 have been constructed assuming that the heat pump operates on a simple vapour-compression cycle with saturated liquid at the condenser outlet and saturated vapour at the compressor inlet.   The calculation assumes η_m = 0.9, η_i = 1-0.05r, and that at condenser and evaporator there is temperature difference of 5 ^{o}C between the refrigerant temperature and the storage tank temperature.

3.   The heat pump switches on only when all of the following conditions are satisfied:

(i)   there is a load demand on the high temperature circuit,

(ii)   the solar storage tank temperature is greater than 5 ^{o}C,

(iii)   the high temperature storage tank is at or below a certain maximum temperature (normally chosen to be just above the temperature at which the auxiliary heater switches on),

(iv)   the value of the heat pump COP, estimated from the instantaneous values of the temperatures of the two storage tanks, is above a minimum value (which is selected as part of the input data).

4.   The auxiliary heater switches on only when the temperature of the water leaving the high temperature storage tank is less than its design value.

Several methods of presenting weather data to the model have been tried.   The method which has proved most useful consists of tables of 24 hourly values of temperature and radiation, each table representing one day of a particular type of weather, at a particular month of the year.   For each month three types of day are defined:

(i)   Hot day.   Radiation values are the average for days with the highest 2% radiation intensities in that month.

Temperatures are mean monthly values plus two standard deviations.

(ii)   Average day.   Radiation values are the average for all days.

Temperatures are mean monthly values.

(iii)   Cold day.   Radiation values are the average for days with the lowest 2% radiation intensities in that month.

Temperatures are mean monthly values minus two standard deviations.

The activities which take place within the building determine the load pattern which the solar heat pump system must satisfy.   Three load profiles have been prepared representing heavy, medium or light loads;   these are equivalent, respectively, to an average daily demand, per unit area of panel, of 252 $W.m^{-2}$, 110 $W.m^{-2}$ and 46 $W.m^{-2}$.

For each simulation the model is allowed to run until the output values are constant for successive days (i.e. equilibrium has been reached).   The results, therefore, refer to a succession of days of identical weather and unchanged load pattern.

Tables 2, 3 and 4 show the results for a set of simulations for an industrial solar heat pump, with the specification described in paragraph 3.3, fitted to a building with a low pitch (i.e. less than 10 ^{o}) roof in London.   The input data common to all these simulations was:

(i)   "corrected efficiency" of the solar collector ($F_S\,\alpha$)   40%

(ii)   minimum COP of the heat pump   0

(iii)   heat pump refrigerant   R12

(iv)   design output temperature of auxiliary heater   65 ^{o}C

(v)   maximum temperature of high temperature storage tank   70 ^{o}C

The energy flows in the tables are expressed as a percentage of the final delivered energy into the load, and the primary energy has been computed in the same manner as in table 1.   A primary energy consumption of 143% is the same as could be achieved by the auxiliary heater acting   alone.   The principal conclusions to be drawn from the results are:-

(i)   only in the worst weather conditions is it impossible to collect any free energy,

(ii)   the free energy proportion is comparatively insensitive to changes in the magnitude of the load - increasing the load by a factor of 5.5 approximately halves the free energy contribution,

(iii)   the diurnal and annual variation in the solar storage tank temperature is considerable - this could be reduced by increasing the size of the storage tank, but it is not certain that this will give an overall improvement in performance,

(iv)   the heat pump COP is frequently below the minimum of 2.5 which was suggested in paragraph 2.2,

(v)   the collection of a large proportion of free energy does not necessarily result in a low consumption of primary energy - for example, for load pattern 1 and the September 2% low weather condition, the primary energy is 150% even though the free energy is 49%.

The last of these conclusions is clearly a consequence of the frequent low values for the COP.   This would be partly offset if the system included a control for switching the heat pump off when the COP is likely to be too low.   As an example, the simulation has been re-run for load pattern 2 and September average weather conditions, but with the minimum COP set to 3 instead of zero.   The results for this calculation are:-

Free energy	45%
Auxiliary energy	33%
Heat pump energy	22%
Primary energy	113%

A comparison with the corresponding result in table 3 shows that although there has been a reduction in the free energy, there has also been a reduction in the primary energy.

6.   CONCLUSIONS

The results show that it is possible to use the roof of a building as an uncovered solar collector in a solar heat pump system.   However, although it is possible to collect a substantial amount of free energy, there is a danger that the system will tend towards a state in which the heat pump runs at a low coefficient of performance, resulting in a high consumption of primary energy.   Even so, the results show sufficient promise to suggest that by careful attention to the design of the system and its controls, the uncovered solar heat pump could be viable in the British climate.

7.   APPENDIX

The equation for the equivalent primary energy consumption can be written more generally as:

$$\text{Primary Energy} \quad E_{p} = \frac{E_{HP}}{\eta_e} + \frac{E_A}{\eta_f}$$

where η_e is the efficiency of electricity generation in terms of the quantity of electricity delivered to the final user, and η_f is the efficiency of a conventional fossil fuel fired boiler.   A report published in 1975 (ref.9) gives η_e as 0.27.   The higher value of 0.33 has been chosen on the grounds that the replacement of old inefficient power stations with better modern ones will create an upward trend in national electricity generating efficiency.

The value of 0.7 for η_f is typical for gas or oil fired boilers.

8. REFERENCES

1. **Bessler, W.F. and Hwang, B.C.** : "Solar assisted heat pumps for residential use".   ASHRAE Journal, $\underline{22}$, 9, Sept. 1980, pp 59-63.

2. **Terrell, R.E.** : "Performance and analysis of a series heat-pump assisted solar heated residence in Madison, Wisconsin."   Solar Energy, 23, 1979, pp 451-453.

3. **Hume, W.P.F.** : "Application of solar energy to a heat pump in a Northern climate."   Chartered mechanical engineer, $\underline{27}$, 8, Sept. 1980, pp 61-63.

4. **Lawaetz, H.** : "Solvarmsystem and varmepumpe."   (Solar heating systems with heat pumps).   VVS (Denmark), $\underline{14}$, 10, Oct. 1978, pp 25-28.   (In Danish).

5. **Wozniak, S.J.**: "Solar heating systems for the U.K. : design, installation and economic aspects."   London, Dept. of the Environment, H.M.S.O., 1979,

6. **Duminil, M.** : "Basic principles of thermodynamics as applied to heat pumps," in "Heat pumps and their contribution to energy conservation," edited by Camatini and Kester.   Proc. NATO Advanced Studies Inst. 1975.

7. **Freeman, T.L., Mitchell, J.W. and Audit, T.E.**: "Performance of combined solar-heat pump systems."   Solar Energy, 22, 1979, pp 125-135.

8. **Waters, J.R.** : "Modelling the performance of a series solar-heat pump for industrial buildings."   In preparation.

9.   "Energy conservation : a study of energy consumption in buildings and possible means of saving energy in housing."   Report of a BRE working party.   BRE Current Paper CP 56/75, Building Research Establishment, Watford, U.K., June 1975.

Table 1   Comparison of Solar Heat Pump Systems (after Freeman et.al.)

Type	Average COP of heat pump	Energy flows, per cent			
		E_F	E_A	E_{HP}	E_{PM}
Parallel	2.0	63	17	20	84
Dual Source	2.53	62	17	21	87
Series	2.84	57	31	12	80
Solar only	–	49	51	0	73
Heat Pump	2.07	35	35	30	140
Conventional Boiler	–	0	100	0	143

Table 2   Solar Heat Pump Performance, Light Load Pattern

Weather Type	Daily mean solar storage tank temp. ^{o}C	Daily mean heat pump COP	Energy flows, per cent			
			E_F	E_A	E_{HP}	E_{PM}
June 2% hottest	46.1	6.2	84	0	16	48
June average	27.3	3.6	72	0	28	84
June 2% coldest	8.3	2.1	51	0	49	147
September 2% hottest	34.7	4.5	77	0	23	69
September average	19.5	2.9	65	0	35	105
September 2% coldest	7.1	2.1	49	2	49	150
December 2% hottest	10.0	2.2	54	0	46	138
December average	5.0	2.3	11	80	9	141
December 2% coldest	5.0	2.3	0	100	0	143

Table 3   Solar Heat Pump Performance, Medium Load Pattern

Weather Type	Daily mean solar storage tank temp. ^{o}C	Daily mean heat pump COP	Energy flows, per cent			
			E_F	E_A	E_{HP}	E_{PM}
June 2% hottest	32.7	4.2	77	0	23	69
June average	17.6	3.0	64	3	33	103
June 2% coldest	6.8	2.2	37	31	32	140
September 2% hottest	22.2	3.3	70	0	30	90
September average	11.7	2.6	56	7	37	121
September 2% coldest	6.2	2.2	27	50	23	140
December 2% hottest	7.4	2.3	48	12	40	137
December average	5.0	2.4	5	92	3	140
December 2% coldest	5.0	2.5	0	100	0	143

Table 4   Solar Heat Pump Performance, Heavy Load Pattern

Weather Type	Daily mean solar storage tank temp. ^{o}C	Daily mean heat pump COP	Energy flows, per cent			
			E_F	E_A	E_{HP}	E_{PM}
June 2% hottest	19.3	3.5	60	14	26	98
June average	11.0	2.8	47	25	28	120
June 2% coldest	5.0	2.3	20	65	15	138
September 2% hottest	14.9	3.1	53	20	27	110
September average	8.2	2.5	40	33	27	128
September 2% coldest	5.0	2.5	14	76	10	139
December 2% hottest	5.5	2.3	29	48	23	138
December average	5.0	2.6	2	97	1	142
December 2% coldest	5.0	2.6	0	100	0	143

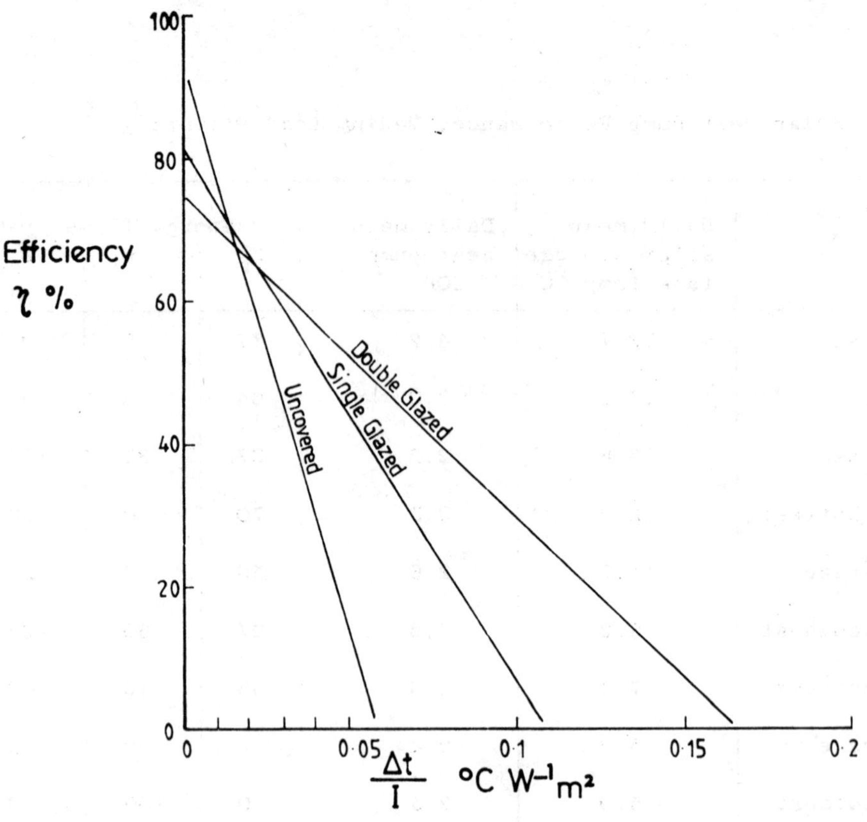

Fig. 1 Typical Solar Collector Efficiency

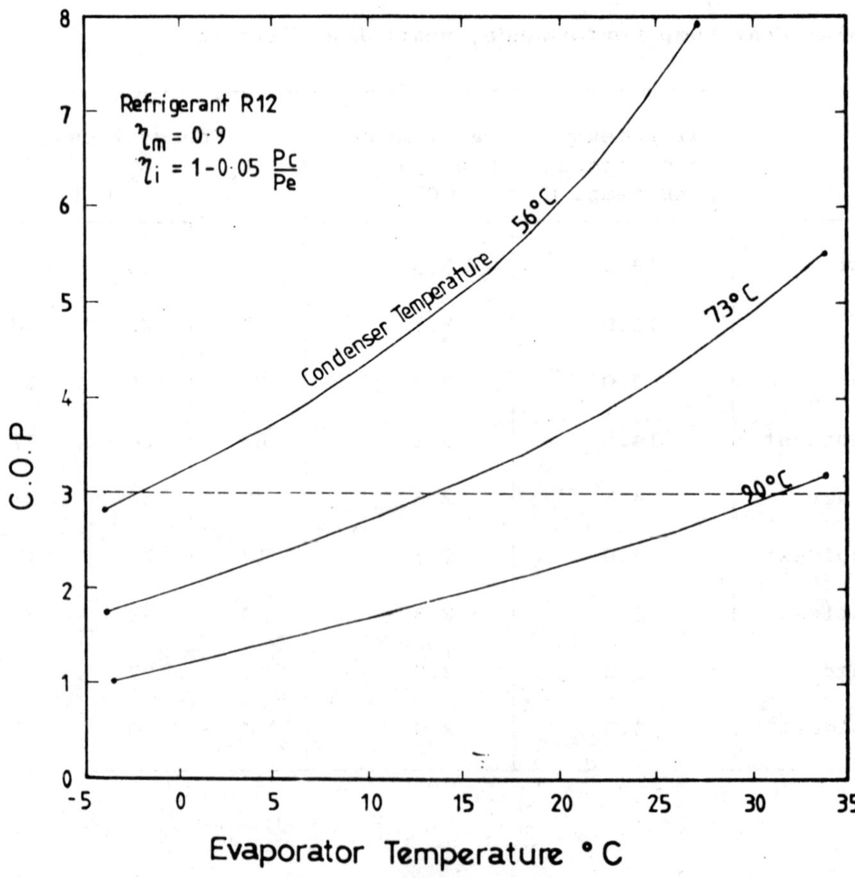

Fig. 2 Typical Heat Pump Coefficient of Performance

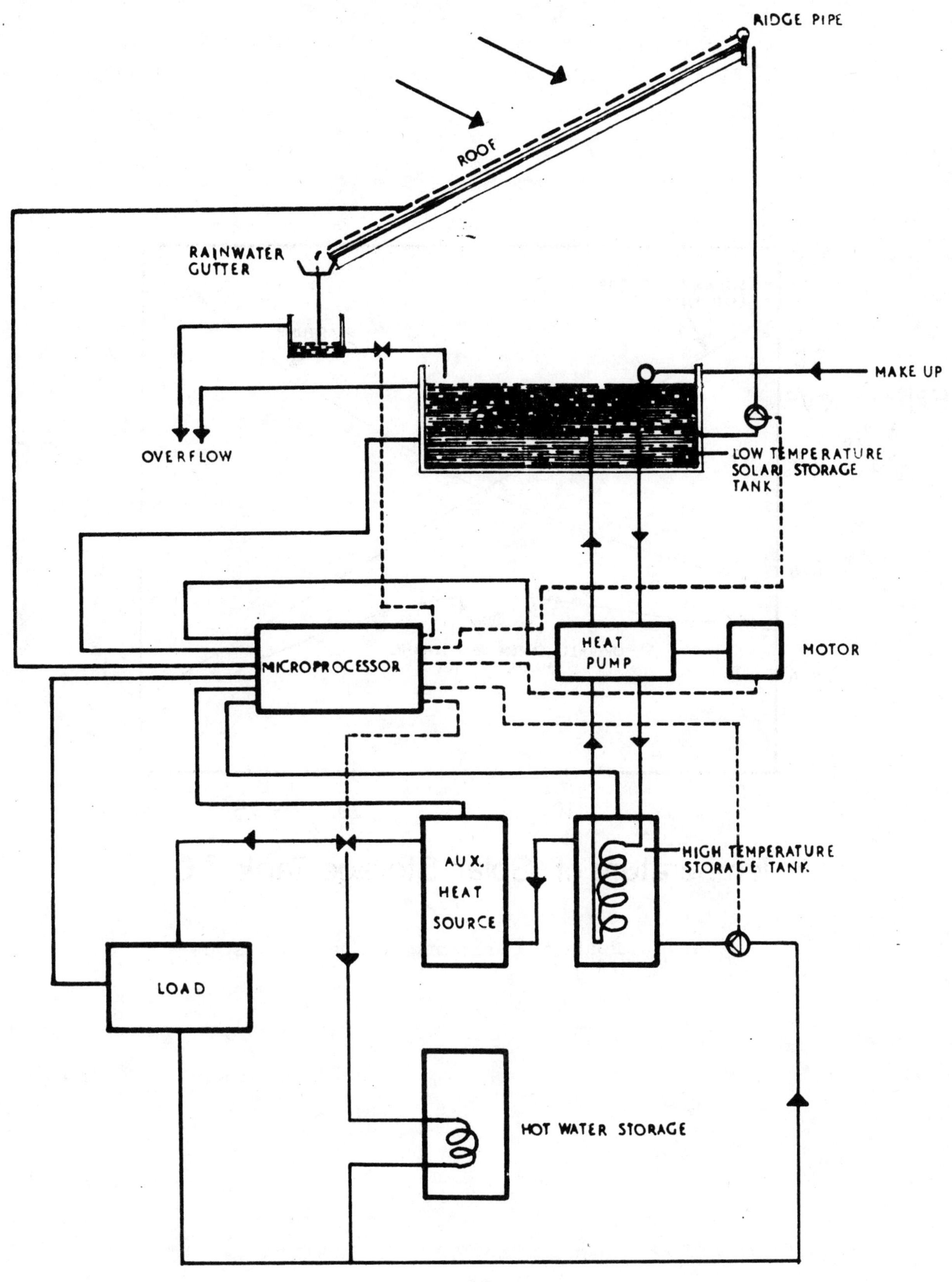

Fig. 3 The Industrial Solar heat pump

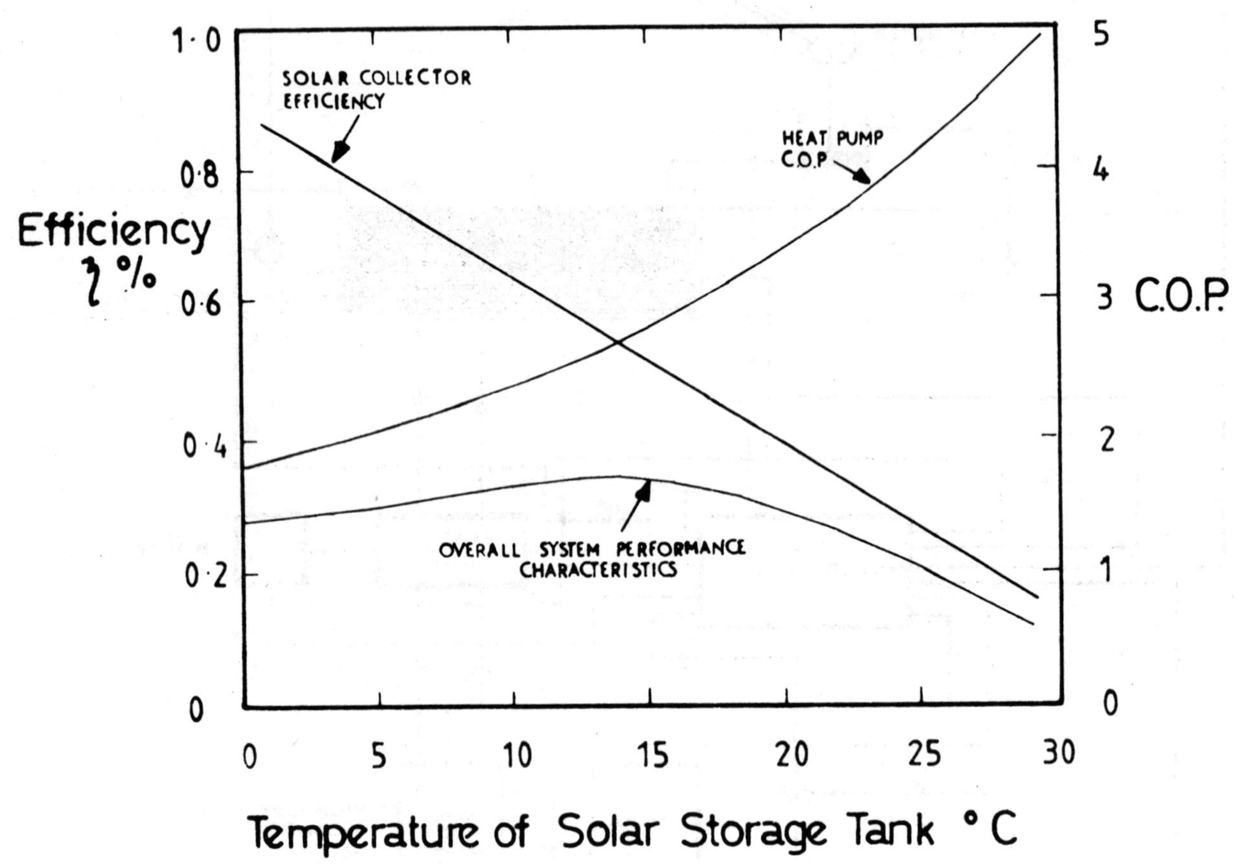

Fig. 4 Anticipated Performance Characteristics

USE OF METAL HYDRIDES IN HEAT PUMP CYCLES

O. Bernauer and H. Buchner

Daimler-Benz AG, F.R. Germany

Summary

The high hydrogen storage densities of metal hydrides with respect to weight and volume offers the possibility to use hydrides in heat pump cycles. On the base of some examples the working mode of such devices and their applications for vehicles are discussed and demonstrated.

Held at the University of Warwick, U.K.
Symposium organised and sponsored by
BHRA Fluid Engineering
©BHRA Fluid Engineering, Cranfield, Bedford MK43 0AJ, England.

NOMENCLATURE

α-phase	=	phase of the metal
β-phase	=	hydride phase
w %	=	weight percent with respect to the weight of the alloy
ΔH	=	enthalpy change of the hydriding reaction
R	=	universal gas constant
ΔS	=	entropy change with the hydriding reaction
η_h	=	efficiency of the hydride heat pump in the heating mode
η_c	=	efficiency of the hydride heat pump in the cooling mode
$_A C_p$, $_B C_p$	=	heat capacity of the hydride plus container
m_A, m_B	=	mass of the hydride
m_1, m_2	=	mass of the container
$_D Q$	=	heat amount to desorb hydrogen out of the hydride
$_A Q$	=	heat which is caused by the absorption of hydrogen

1. INTRODUCTION

The physical basis of the heat storage methods are in general:

- energy storage by heterogenous evaporation (Ref. 1, 2),
- heat storage and heat transformation by the chemical reaction of fluids with solids or gases with solids or fluids (Ref. 3, 4, 5),
- heat storage by using the enthalpy of solution, and
- heat storage by using adsorption and absorption processes.

Some of the promising heat storage systems are reversible chemical reactions like

$$NaOH\ (s) + n \cdot H_2O\ (\iota) \rightleftharpoons NaOH \cdot n\ H_2O\ (s)$$
$$CaO\ (s) + H_2O\ (\iota) \rightleftharpoons Ca(OH)_2\ (s)$$
$$FeCl_2 \cdot 2\ NH_3 + 4\ NH_3 \rightleftharpoons FeCl_2 \cdot 6\ NH_3$$

as well as the heat storage by molecular adsorption processes in Zeolites (Ref. 6).

Such systems are used in sorption heat pumps which are able to convert low grade into high grade energy without using mechanical or electrical energy.

Beside these well known sorption heat pump systems a lot of work (Ref. 7, 8, 9) has been done on hydride heat pump systems. These periodically working hydrides can be used as energy storage and recovery devices or as cooling and heating systems with high energy and power densities. The reaction of the hydrogen with the metals has the following features:

- high kinetics, when hydrogen reacts with the metals,
- high energy densities with respect to weight and volume,
- good heat conductivity of the metallic body and good heat transfer,
- no toxical elements.

In this report the physical properties of the hydride systems are discussed with respect to the applications.

2. PROPERTIES OF METAL HYDRIDES

Some metals react with hydrogen reversibly to form a metal hydride:

$$Me + x\ H_2 \rightleftharpoons MeH_{x.2} \qquad where\ \ 0 < x < 1$$

As this hydride formation or absorption process is exothermic, the heat from the reaction can be used for heating purpose. On the other hand the desorption process is endothermic, so that heat for the reaction must be supplied during the desorption. As some hydrides release hydrogen at low temperatures ($<0\ ^{\circ}C$) with pressures of more than 1 bar such systems can be used for cooling devices.

The formation of various exothermic hydrides as a function of temperature and their stoichiometry can be established with the aid of pressure/concentration isotherms (Fig. 1). In some cases there still exist several hydride phases which lead to different steps in the pressure/concentration isotherms (Fig. 2).

The physical properties of hydrides can be summarized as follows: The H_2-storage capacities of these hydrides are in the range between 1.3 and 2 w % H_2 with respect to the weight of the metal alloy for so called low temperature hydrides (they release hydrogen in a temperature range between -40 °C and +100 °C).

High-temperature hydrides (hydrides which release hydrogen above temperatures of 100 °C) reach storage capacities with respect to the weight of the alloy between 7 and 8 w % H_2 (Ref. 10). The plateau pressure p of the hydrides with respect to the absolute temperature can be described by the van't Hoff equation

$$\ln p = \Delta H/RT - \Delta S/R$$

where H is the heat of formation of the hydriding reaction, R is the universal gas constant and ΔS is the entropy change with the hydriding reaction. The van't Hoff curves for some hydrides are shown in Fig. 3. The picture demonstrates the wide range of pressure and temperature in which the hydrides can be used.

Beside the storage capacity and the temperature dependence of the plateau pressure the rate of the decomposition of hydrides at operating temperature is also important. Measurements with TiFe-hydrides and Mg_2Ni-hydrides carried out by J. J. Reilly (Ref. 11) show that the reaction kinetics of the absorption and desorption process is always quick enough to charge or discharge the hydrides in a few minutes when the temperature of the hydride material is high enough to guarantee a hydrogen pressure of more than 1 bar. Fig. 4 shows the decomposition rate of a TiFe-hydride at different temperatures measured by Reilly and Co. (Ref. 11).

This example makes it clear that the efficiency of hydrogen hydride systems depends on the ability to transfer heat between the hydride bed and the heat exchanging medium. In most applications the heat transfer of the pulverized metal hydrides is a restriction on the design and construction of hydrogen hydride systems.

The integral thermal conductivity of a total system depends on:

- the thermal conductivity of the hydride or metal powder,

- the thermal conductivity of the hydrogen gas, and

- the thermal conductivity of the container material.

To improve a high thermal conductivity the hydride powder is embedded in a stable matrix of aluminum, magnesium, copper and other metals with a high thermal conductivity and a relatively low melting point. The preparation process of the matrix body is described in Ref. 12 and 13. Fig. 5 shows the microstructure of such a TiFe-hydride - aluminum - matrix body which was hydrided and dehydrided several times.

The advantages of such a matrix technique, besides the high thermal conductivity, are the safety and construction aspects because the hydride powder cannot get into the open should the container break but is held in the matrix, and the storage body can be of any shape whatsoever (plates, cylinders etc.).

Table 1 gives the physical properties of metal hydrides which are useful for heat pump applications.

3. GENERAL CONSIDERATION REGARDING HEAT STORAGE AND RELEASE WITH METAL-HYDROGEN SYSTEMS

The use of thermal energy at a high temperature level to pump heat from a low-temperature source to an environment at an intermediate temperature is possible with the combination of metal-hydrogen systems when the absorption and desorption processes involve thermal energy. These processes then can be used for heating and cooling purposes.

Fig. 6 illustrates a typical heat pump cycle for a hydride combination A and B where the hydride B is the more stable one and $T_1 > T_2 > T_3$. It can be seen that hydride A has a slightly higher and slightly lower hydrogen equilibrium pressure at temperatures T_1 and T_2, respectively, than hydride B has at temperatures T_2 and T_3.

In order to have a continuous cooling or heating effect it is at least necessary to have four tanks which run on two alternating cooling or heating cycles (Ref. 7). Tanks 1 and 2 (Fig. 7) are filled with the metal hydride A, while tanks 3 and 4 are filled with the hydride B. The hydride B in tank 4 absorbs heat from the outside at a temperature level T_3, and desorbs hydrogen which flows to tank 2 where it reacts to form hydride A (A' is the metal alloy). The heat from this reaction at a temperature level T_2 can be used when the system is working as a heat pump. At the same time, the metal hydride A in tanks 1 decomposes by absorbing energy at a low-temperature level into metal A' and hydrogen. The hydrogen flows to tank 3 to form in combination with metal B' the hydride B. If the system is working as a heat pump, the heat of formation on the temperature level T_2 can be used again. The overall efficiency of such a system depends on the heat of formation of the hydriding processes, the heat capacities of the hydride and container materials and on the temperature differences between hydriding and dehydriding modes as already shown by Gruen et.al. (Ref. 7), and van Mal (Ref. 9).

If such a system is working as a heat pump the efficiency can be expressed by the following equation:

$$\eta_h = \frac{\Delta H_B + \Delta H_A - C_p \cdot (m_A + m_1) \cdot (T_2 - T_1)}{\Delta H_B + C_p \cdot (m_B + m_2) \cdot (T_3 - T_2)}$$

$$+ \frac{C_p \cdot (m_B + m_2) \cdot (T_3 - T_2)}{\Delta H_B + C_p \cdot (m_B + m_2) \cdot (T_3 - T_2)}$$

In practice the efficiencies reach values between 1.3 and 1.8 and, therefore, are comparable to standard absorption heat pump cycles.

If such a metal hydrogen system is working in a cooling mode the thermodynamic efficiency can be expressed by the following equation:

$$\eta_c = \frac{\Delta H_A - C_p (m_A + m_1) \cdot (T_2 - T_1)}{\Delta H_B + C_p (m_B + m_2) \cdot (T_3 - T_2)}$$

and in a concrete case it can reach values again comparable to standard absorption refrigeration cycles (~0.4 - 0.6).

4. APPLICATIONS FOR METAL HYDROGEN SYSTEMS

Hydride storage units can be used in industrial processes to store and regain waste heat by the desorption and the absorption of hydrogen. To demonstrate such a system a storage unit which is able to store between 2,200 and 2,400 Nm^3 at a hydride weight of about 10 t will be constructed and built 1982 under the sponsorship of the European Community. It is planned to use a semitrailer which, in principle, can be parked anywhere. Moreover this has the advantage that a comparison with transportation of hydrogen in pressurized bottles will be possible as a by-product of the project. It is suggested that the hydride storage system will be connected with an existing hydrogen line (hydrogen supply-storage-consumption) in an industrial process. Therefore the results should be evaluated considerably higher than in a purely experimental test.

Applications where the principles of the hydride systems can be used to particular advantage is in air conditioning and the preheating of vehicles and their engines (Ref. 14 and 15). The typical automobile air conditioner requires in general 5 to 10 kW during normal operation depending on the cooling demand. Consequently, a significant part of the energy consumed goes to operate the air conditioner. This problem can be solved by employing an air conditioning system which uses the waste heat from the exhaust to operate the air conditioner. In this way no mechanical energy is drained from the automobile engine; and, in comparison to an automobile equipped with a conventional air conditioner, a 10 to 20 % reduction (Ref. 16) in fuel consumption can be obtained while the air conditioner is working.

The same advantage can be obtained by employing a hydride based preheating system which reutilizes the waste heat from the exhaust gas or coolant water for operation. The effect is, that there are no cold start problems with the engines, resulting in a reduction in consumption and an extension of the life of the engine.

5. CHARACTERISTICS OF HYDRIDE HEAT PUMP CYCLES FOR COOLING PURPOSES

One important objective of hydride heat pumps and cooling devices resides in the provision of systems which are capable of efficiently utilizing low-grade energy while retaining the capacity for utilizing high-grade thermal energy, and which in other cases are capable of storing energy and making it available upon demand for heating or cooling or both heating and cooling. As shown in chapter 4, the coefficient of performance η_c for cooling for hydride heat pump cycles is determined by the enthalpies of the reactions in the heat pump and is in the range of 0.4 - 0.6. In comparison to this fact for a mechanical compressor cooling apparatus, we have a coefficient of performance of ~1.5, but to operate it we need mechanical energy from the engine. This energy can therefore not be used to drive the engine. An alternative to this may be hydride systems. The system $LaNi_5-H/Ti_{0.8}Zr_{0.2}CrMnH$ can operate at the same temperature levels as contempory mechanical refrigerators, with the difference that in this case the heat at the low-temperature level (0 $^\circ$C) is pumped to the medium temperature level (45 $^\circ$C) by heat on a temperature level of ~120 $^\circ$C (Fig. 8). If we presume that one hydriding-dehydriding cycle as shown in Fig. 9 will last for 5 minutes altogether (ab/desorption time 2 minutes; heating up/cooling down 30 seconds each) then for 1 kW cooling power ($\dot{Q}_c$) we need about 3 - 3.5 kg of $Ti_{0.8}Zr_{0.2}CrMn-H/LaNi_5-H$ material in the four containers which have nearly the same weight again (see Table 1). This means that a 1 kW unit will have a total weight of 6 - 7 kg without additional items like heat exchanging medium, fan etc. The total volume will reach values between 3 and 3.5 litres. The coefficient of performance η_c (chapter 3) can be calculated by the equation:

$$\eta_c = \frac{\Delta H_A - C_p \ (m_A + m_1) \cdot \Delta T_{2-1}}{\Delta H_B + C_p \ (m_B + m_2) \cdot \Delta T_{3-2}}$$

where

ΔH_A	= 28 kJ/mol H_2	($Ti_{0.8}Zr_{0.2}CrMnH_3$)
ΔH_B	= 31 kJ/mol H_2	($LaNi_5H_6$
$_A C_p$	= 0.54 J/g grd	(hydride plus container)
$_B C_p$	= 0.42 J/g grd	(hydride plus container)
ΔT_{2-1}	= 45 °C	
ΔT_{3-2}	= 75 °C	
$(m_A + m_1)$	= 333 g	
$(m_B + m_2)$	= 400 g	
η_c	= 0.46	

Hydride based cooling systems now have a power density of about 6 - 7 kg/kW. The general cooling power of contemporary compressor cooling devices is in the range of 5 - 10 kW. This means that equivalent hydride systems will have a weight between 30 and 70 kg with the advantage that no extra fuel consumption for the operation of the cooling unit is necessary. Whereby we have to point out that a further development in the hydride field is necessary in order to find hydrides with higher storage capacities than those of the $LaNi_5$- or MNi_5-based alloys (1 - 1.2 w %) because then the weight of the systems can be reduced. Hydrides with high storage capacities are for instance based on TiCrMn-H and reach values of 1.8 w %, so that the weight of the hydride pair in comparison to hydrides based on $LaNi_5$ and/or MNi_5 can be reduced by a factor of 0.6 to 0.7. From this it follows that in future power densities of 5 kg/kW might be realized. In comparison to current air conditioning systems we again get the advantage that no fuel will be consumed to operate the system with the result that a saving of fuel is possible.

6. PREHEATING OF ENGINES OR PASSENGER COMPARTMENTS IN CARS AND BUSES

Research work in the field of fuel consumption has shown that a remarkable reduction in consumption is possible, if one can dispense with cold starts in that the engine is preheated with an auxiliary heater. This effect becomes still more advantageous when the energy to heat the system comes from a heat storage system and not from fuel. The preheating times are in general 10 - 20 minutes so that the storage systems have to store an energy amount for these times without having too much weight. The hydride pair $MgH_2/Ti_{0.8}Zr_{0.2}CrMnH_3$ for example shows that the total weight of such heating systems is low enough to guarantee a distinct saving in fuel. The mode in which such a system works can be explained by Fig. 10. During operation, the engine heat of the exhaust gas is used ($_D Q_2$ in Fig. 10) to desorb hydrogen gas out of the MgH_2. This hydrogen gas is absorbed in a container with $Ti_{0.8}Zr_{0.2}CrMn$ to form $Ti_{0.8}Zr_{0.2}CrMn$-hydride when the heat $_A Q_1$ is diverted at a temperature level of about 16 °C. When the engine is stopped and it cools down to ambient temperature, for instance lower than -20 °C one has the possibility of preheating the engine before starting by desorbing hydrogen out of $Ti_{0.8}Zr_{0.2}CrMnH_3$

with heat supply Q_1 from the ambient air and by using the heat of formation Q_2 which is caused by the absorption of hydrogen in Mg. During engine operation, the hydrogen is refilled again from the Mg-container into the $Ti_{0.8}Zr_{0.2}CrMn$-container and the cycle is closed. With this hydride pair it is possible to preheat at temperatures as low as $-20\ ^{o}C$. If one needs to preheat at lower temperatures, one can use other hydride pairs like $Mg_2NiH_4/Ti_{0.9}Zr_{0.1}CrMnH_3$ with which preheating can be accomplished at temperatures as low as $-40\ ^{o}C$. The weight per kW and the operating time of such systems is determined by the

- heat of formation of the magnesium hydride

$$\Delta H = 77.4 \text{ kJ mol}^{-1} H_2$$

- H_2 storage capacity of the magnesium

$$H_2^C Mg_{0.95}Ni_{0.05} = 7 \text{ w \%}$$

- H_2 storage capacity of the low temperature hydride

$$H_2^C Ti_{0.8}Zr_{0.2}CrMn = 1.8 \text{ w \%}$$

- heat capacity of the high temperature hydride + container in the temperature range which cannot be used for preheating. In the case of engine coolant $< 100\ ^{o}C$.

$$Q_{C_p} = (1.04 \text{ J g}^{-1} \text{ grd}^{-1} \cdot m_{Mg} + 0.51 \text{ J g}^{-1} \text{ grd}^{-1}) \cdot 100\ ^{o}C$$

- weight of the container; for high temperature hydride the container amounts to 50 percent of the total weight, and in the case of low temperature hydrides the container amounts to 30 percent of the total weight.

    If all these factors have been taken into consideration the usable energy content of such systems runs up to 300 kJ/kg. From this ist follows that an auxiliary heating system (as shown in Fig. 11) for the engine (2.3 litres) which is able to heat up the engine from $0\ ^{o}C$ to $80\ ^{o}C$ (we need about 6,700 kJ), weighs 22 - 27 kg.

    In cooperation with the Benteler AG a prototype of such a system will be constructed and tested in 1982. If the results are positive an air conditioning device will be built afterwards.

7. CONCLUSION

    The technical feasibility of hydride based refrigerators and preheating systems especially for mobile application has been shown.

    The high energy densities with respect to weight and volume and the rapid kinetics of the hydriding and dehydriding processes mean that such devices can be installed in cars, buses and other vehicles to help save primary energy.

    The work in the field of hydride development has shown, that with respect to the storage capacity the maximum is reached. Further development is to be done on the following items:

- low costs of the materials
- hysteresis between adsorption and desorption
- heat exchanger technique
- container weight
- heat conductivity

    An innovation in this will open a wide spread field of further applications for metal hydride systems.

8. REFERENCES

1. Alefeld, G.: "Energy Storage by Heterogeneous Vaporization I. Physical-Technical Principles". (Energiespeicherung durch Heterogenverdampfung I, Physikalisch-Technische Grundlagen). Wärme, 81, 1975, pp. 89 - 93. (In German).

2. Alefeld, G.: "Energy Storage by Heterogeneous Vaporization II. A Method for Storing Heat in District Heating and for Generation of Peak Electricity". (Energiespeicherung durch Heterogenverdampfung II, Ein Verfahren zur Fernwärmespeicherung und Spitzenstromerzeugung). Energie, 27, 1975, pp. 180 - 183. (In German).

3. Alefeld, G.: "Heat Storage and Heat Transformation by Chemical Reactions". (Wärmespeicherung und Wärmetransformation durch chemische Reaktionen). VDI-Bericht, No. 288, 1977, pp. 111 - 114. (In German).

4. Schmidt, E. W.: "Reversible Chemical Reactions for Heat Storage". (Reversible chemische Reaktionen zur Wärmespeicherung). VDI-Bericht, No. 288, 1977, pp. 115 - 125. (In German).

5. Ervin, G.: "Method of Storing and Releasing Thermal Energy". United States Patent 3, Aug. 10, 1976, pp. 973, 552.

6. Alefeld, G., Bauer, H. C., Maier-Laxhuber, P. and Rothmeyer, M.: "A Zeolite Heat Pump, Heat Transformer and Heat Accumulator". In: Proc. Int. Conference on Energy Storage, Brighton, U. K., April 29 - May 1, 1981, pp. 61 - 72.

7. Gruen, D. M. and Sheft, I.: Metal Hydride Systems for Solar Energy Conversion". In: Proc. NSF-ERDA, Workshop on Solar Heating and Cooling of Buildings, Charlottesville, Vancouver, 1975.

8. Gruen, D. M., Mendelsohn, M. H. and Sheft, I.: "Proc. Solar Energy, Vol. 21, Pergamon Press, 1978, pp. 153 - 156.

9. van Mal, H. H.: "Stability of Ternary Hydrides and some Applications". Philips Res. Repts. Suppl., No. 1, 1976.

10. Reilly, J. J. and Wiswall, R. H., Jr.: Inorganic Chemistry 7, p. 2254, 1968.

11. Proc. Int. Symp. Hydrides for Energy Storage, Geilo, Norway, 1977.

12. Gutjahr, M. A., Buchner, H., Beccu, K. D. and Säufferer, H.: "A New Type of Reversible Electrode for Alkaline Storage Batteries Based on Metal Alloy Hydrides". In: Proc. 8th Int. Power Sources Conference, Brighton, U. K., Sept. 1972.

13. Bernauer, O. and Buchner, H.: "Special Problems of Metal Hydride Storage Tanks". (Spezielle Probleme der Metallhydridspeicher). In: Proc. Mobile Stromversorgung im Felde, Deutsche Gesellschaft für Wehrtechnik e. V., May 1980. (In German).

14 Buchner, H.: "The Hydrogen Hydride Energy Concept". (Das Wasserstoff-Hydrid-Energiekonzept). Chemie-Technik 9, 7. Jahrgang, 1978. (In German).

15. Toepler, J., Bernauer, O. and Buchner, H.: Journal of the Less Common Metals, 74, 1980, pp. 385 - 399.

16. Duffy, T. E., Rohy, D. A.: "Methods of and Apparatus for Energy Storage and Utilization". United States Patent 4, 161, 211, July 17, 1979.

TABLE 1   PHYSICAL PROPERTIES FOR METAL HYDRIDES

Alloy composition	ΔH (kJ mol^{-1} H$_2$)	A (K)*	B*	capacity (w %)	reference
TiFe	28.03	−3383	12.76	1.8	3
LaNi$_5$	30.96	−3712	12.96	1.4	1
MNi$_5$	20.92	−2539	11.64	1.4	2
CaNi$_5$	31.80	−3838	12.17	1.4	2
Ti$_{0.8}$Zr$_{0.2}$CrMn	24.66	−2968	13.07	1.8	−
Mg$_2$Ni	64.43	−7736	14.71	3.8	7
Mg$_2$Cu	72.80	−8771	17.12	2.0	7
Mg$_{0.95}$Ni$_{0.05}$	75.86	−9128	16.34	7	−
Mg	77.40	−9314	16.63	7	8
TiVMn	38.52	−4640	13.81	2.6 − 3	−

* $\ln p = \dfrac{A}{T} + B$ (plateau pressure (bar) taken in the middle of the plateau)

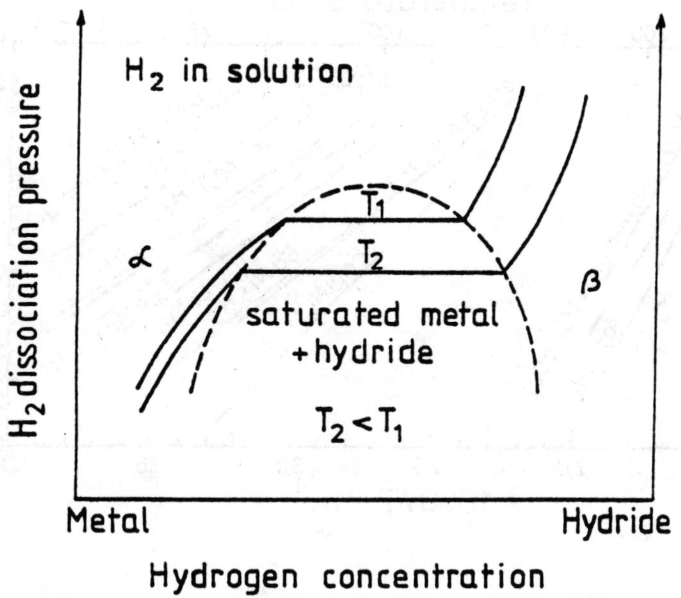

fig.1 Schematic CPI diagram for a metal-hydrogen system forming an hydride

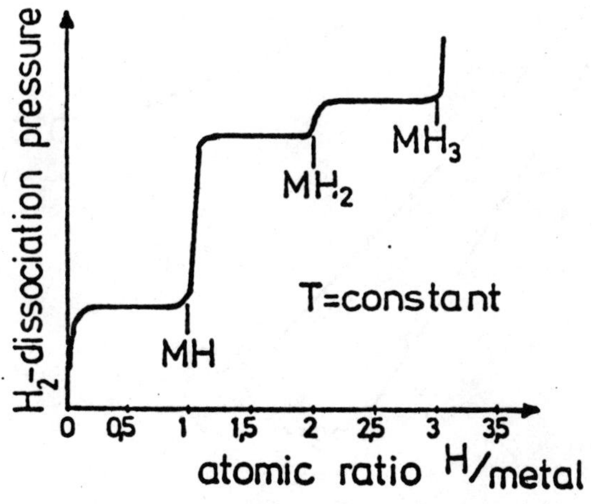

fig. 2 Pressure Concentration Isothermes for Hydrides with Different Hydride Phases

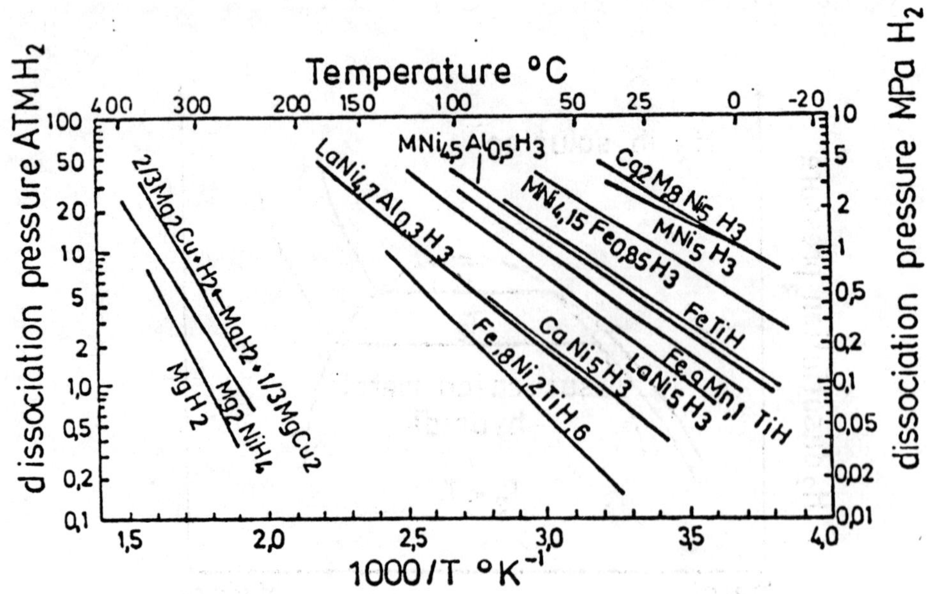

fig. 3 van't Hoff plots (descrption for various hydrides)

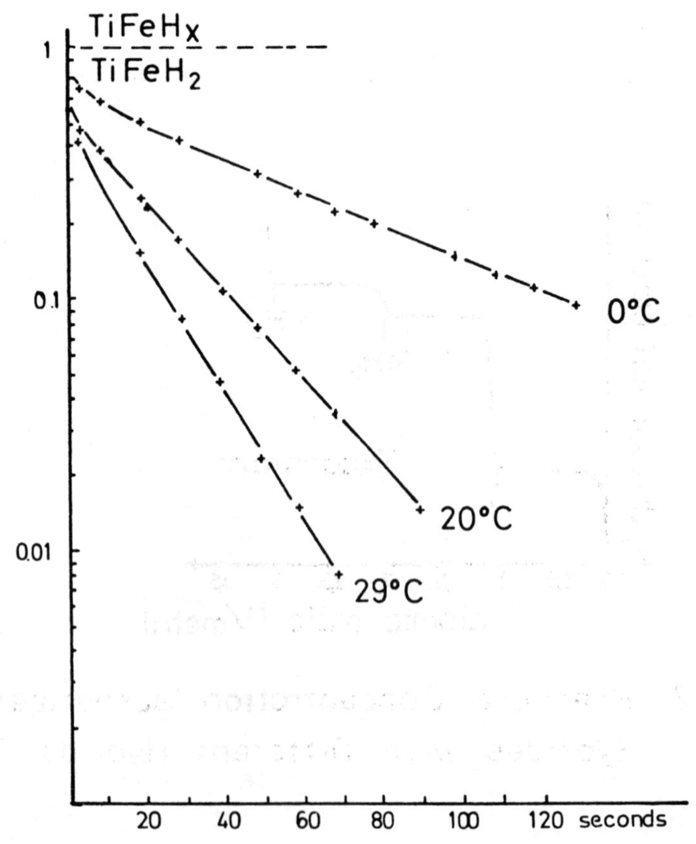

fig.4 H₂-desorption rate out of TiFe-hydride (from 11)

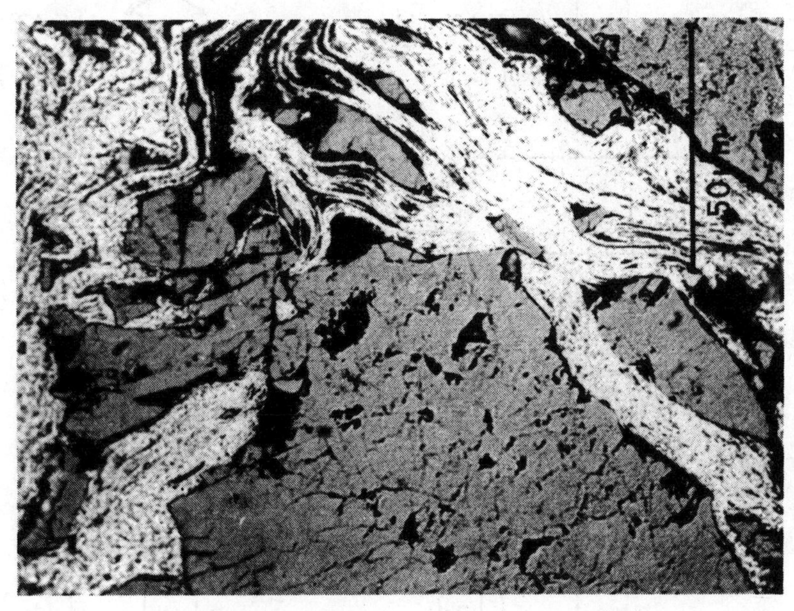

fig. 5   Ti Fe + 5w% Al
         (several times hydrided and dehydrided)

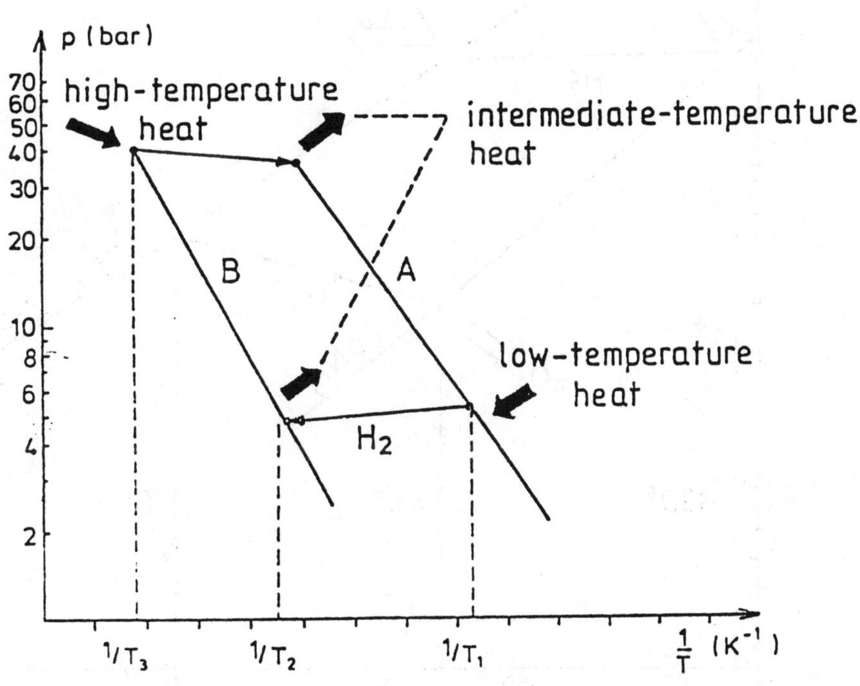

fig. 6 Hydride heat pump cycle with two different hydrides

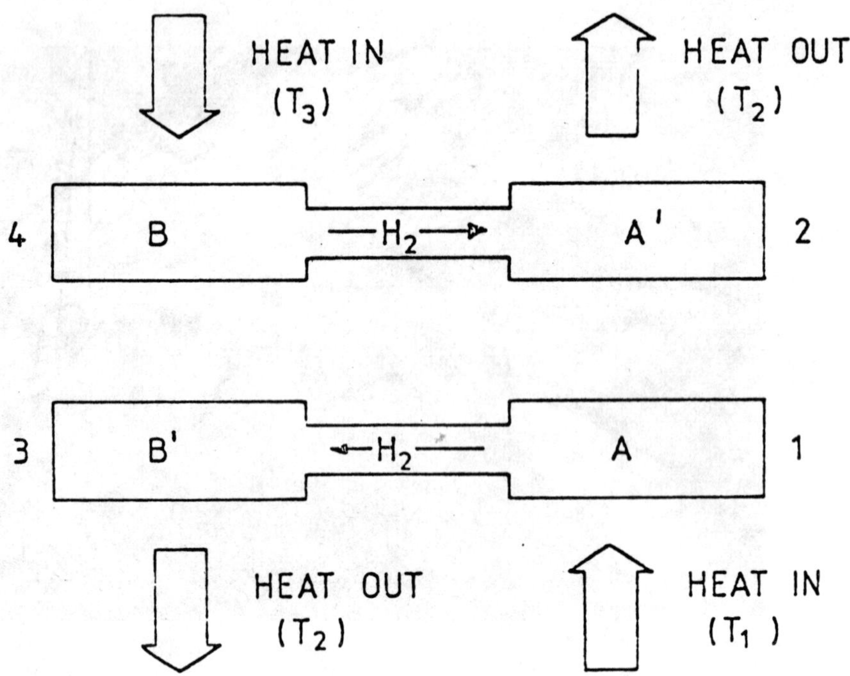

fig.7 A schematic diagram of a hydride heat pump

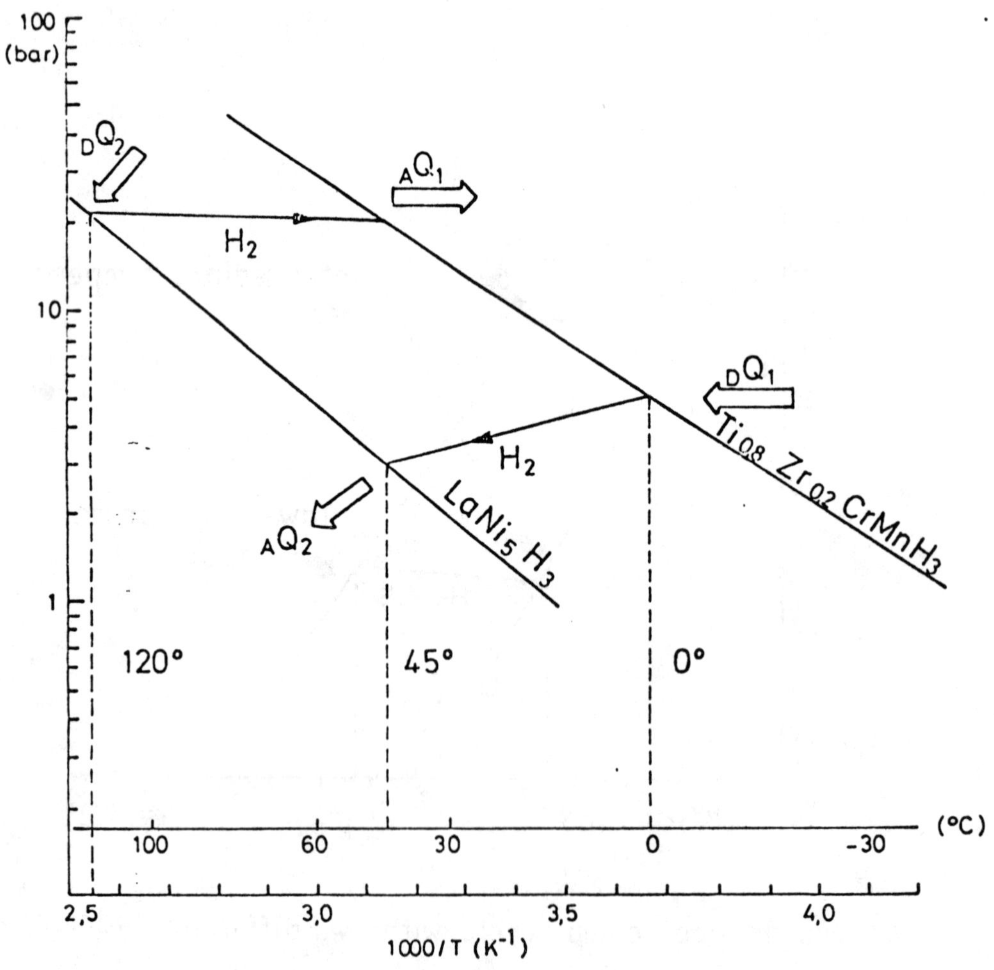

fig.8 Plateau pressure vs 1/T for $LaNi_5H_3$ and $Ti_{0,8}Zr_{0,2}CrMnH_3$

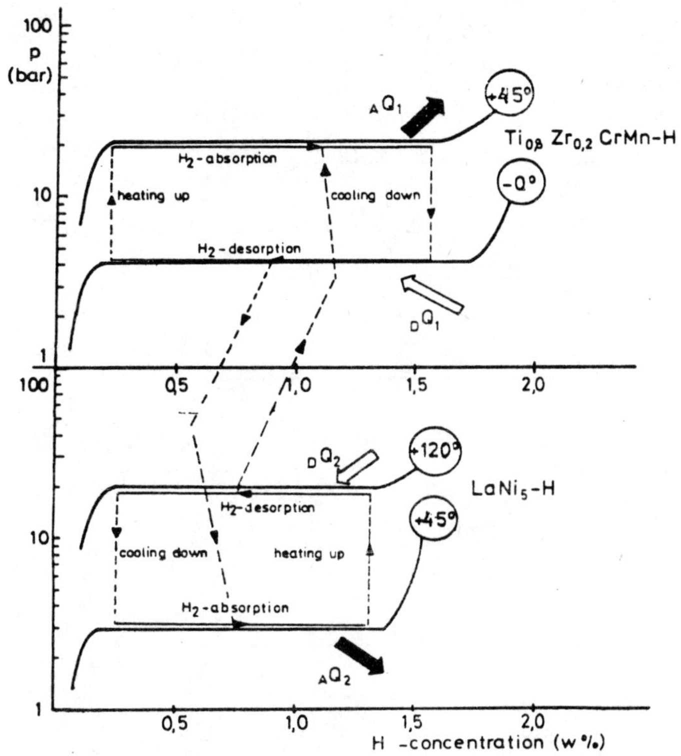

fig.9 Hydride heat pump cycle for the system LaNi$_5$-H / Ti$_{0,8}$Zr$_{0,2}$CrMn-H

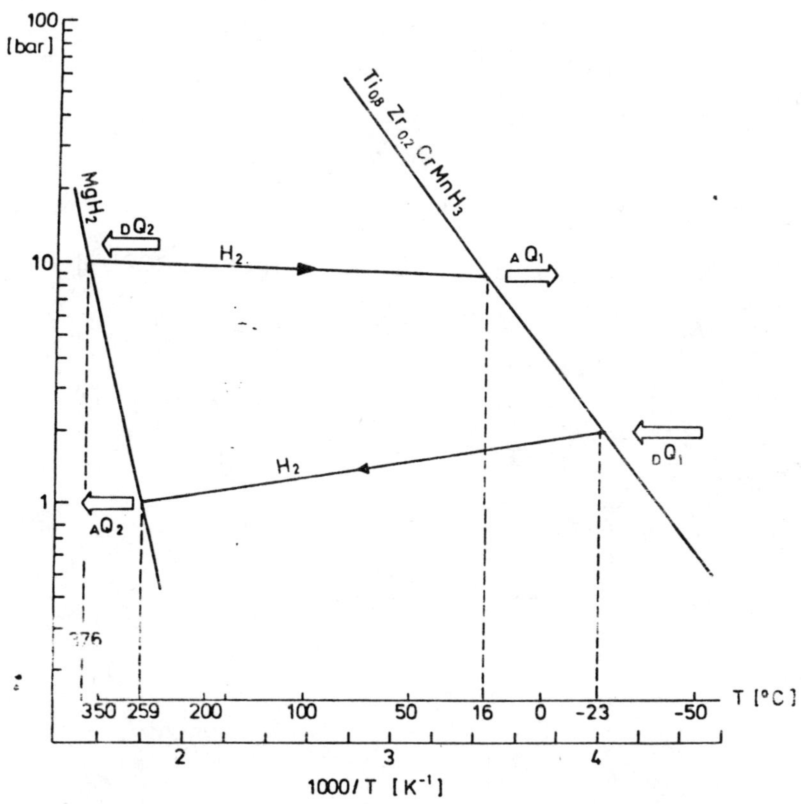

fig.10 Plateau pressure vs 1/T for MgH$_2$ and Ti$_{0,8}$Zr$_{0,2}$CrMn H$_3$

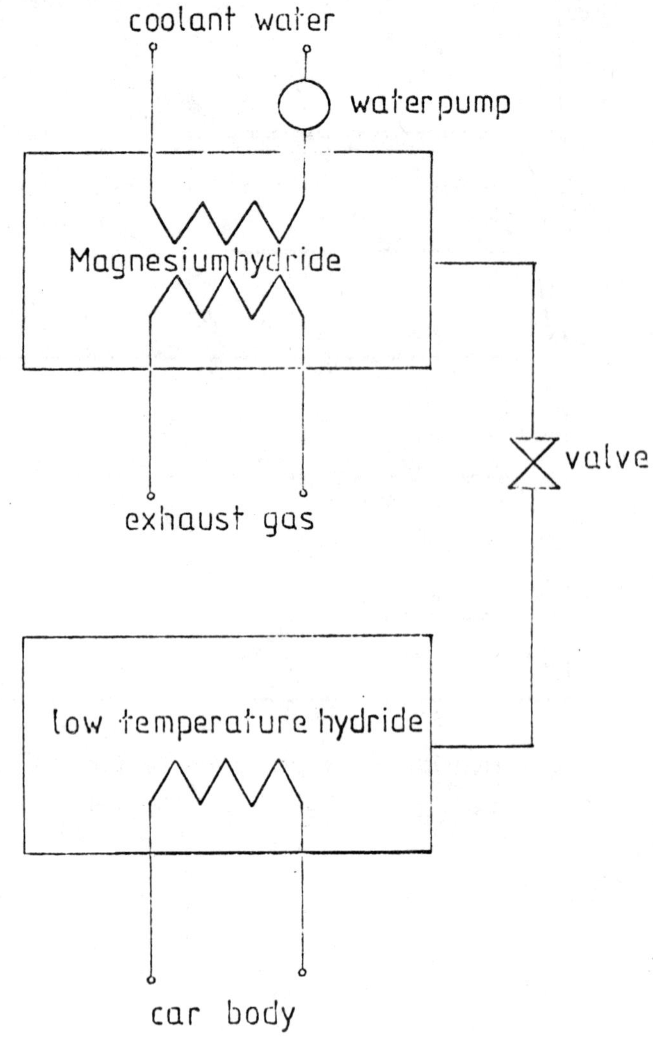

fig.11 Principle of a hydride auxiliary heating system

HEAT RECOVERY TECHNOLOGY FOR SWIMMING POOLS BY THE APPLICATION OF HEAT PUMPS

G.D. Braham

The Electricity Council London, U.K.

Summary

Most United Kingdom swimming pools are unnecessarily profligate in their energy consumption. Great opportunities therefore exist for the application of energy conservation techniques. Unfortunately the increasing number of technologies has added to the growing confusion as to their relative merits and suitability.

This paper proposes a simple but objective method of analysis, to allow the energy utilisation of each technology to be assessed. The method is based on the use of the energy ranking scale and the heating analysis chart.

Held at the University of Warwick, U.K.
Symposium organised and sponsored by
BHRA Fluid Engineering

Introduction

The past decade has provided ample evidence that some of our accepted sources of
primary energy, particularly oil and gas are being consumed at such a rate that
severe world shortages will occur well within the useful life span of the majority of
our existing swimming pools. Fortunately the future prospects of energy supply and
demand in the UK are secure until the end of the century. As well as substantial
reserves of coal, oil and gas and there is already a significant and growing
contribution from nuclear power.

Nevertheless, this does not eliminate the need to plan sensibly for the future and to
manage our energy resources as effectively as possible. It does not make economic or
political sense for the UK to isolate ourselves from the world energy situation and
the growing international pressures for conservation of energy resources.

The vast majority of the existing UK indoor swimming pools are unnecessarily
profligate in their energy consumption. It is a hard reality that leisure centres
and swimming pools have a low priority in the political hierarchy of local authority
buildings. Consequently, during periods of fuel scarcity or energy cost escalation,
all major fuel energy uses are examined and the continued operation or even existence
of these swimming pools become vulnerable!

In retrospect, it is rather surprising that the potential market for energy
conservation in UK swimming pools was not recognised earlier. The low energy cost of
the 1950s and 1960s prevailed too long, and supported a general reluctance by
authorities to recognise the changing energy environment. The technological changes
required for major improvements in energy conservation constituted too great a risk
for designers and operators. However more recent events have accelerated the rate of
technological change for both existing and new designs. As a consequence there are
now a number of competing energy conservation technologies. Unfortunately each
technology unwittingly contributes to a growing confusion about their relative merits
and suitability for the future energy scenario.

In order to aid the designer and pool manager, a simple but objective analysis is
required which may be clearly understood. For these reasons this paper proposes the
energy ranking scale and uses the heating analysis chart to explain the energy
utilisation of each technology.

Energy Ranking Scale

This scale (Fig 1) provides the professional and layperson alike, with a simple
method of depicting and ranking the energy saving potential of competing technologies
by listing their total annual bought energy consumptions. The total 'bought' energy
is inclusive of all the fuel and the electricity actually delivered to the building
and may be express in kWh/m^2 of pool area per annum. The energy ranking scale can
also depict the free energy component derived from solar heat, when this becomes
significant in the total annual energy consumption. To assist the calculation, for
location on the energy ranking scale of existing buildings, a simple procedure is
shown in Appendix 1. This allows the energy ranking scale value to be determined
from the known fuel and electricity consumptions.

HEATING ANALYSIS CHART

This chart (Fig 2) provides both designers and managers with a means of analysing,
calculating, and depicting both the energy demands and the annual heating energy
consumption for each competing energy conservation technology.

Heating Analysis Charts may be prepared for any part of the country where adequate meteorological data is available. Currently, the Electrical Supply Industry have 43 such locations covering England, Scotland, Wales and Northern Ireland. The charts can show the heating energy requirements for an average year (1967), and also for severe and mild years in each location.

The chart is subdivided into weather sensitive heating loads, such as ventilation and fabric heating loads, and the non weather sensitive loads, such as the domestic hot water heating and the pool water evaporation for dehumidifying systems. The horizontal axis represents the length of the heating season expressed either in days 0 - 365, or hours 0 - 8,760. The vertical axis represents the weather sensitive loads proportional to the temperature difference between the prevailing ambient temperature and the internal temperature. This axis also represents the heating demand or boiler power required for the prevailing temperature. The area under the curve is the energy requirement for space heating expressed in kilowatt hours (kWh). Non weather sensitive loads are shown below the horizontal axis. The total area encompassed by the weather sensitive curve and the non weather sensitive line represents the total annual energy requirement for the building and its environmental system.

Basis for Comparative Study
<u></u>

Consider a typical existing swimming pool complex (Fig 3) with a 33 m main pool and teaching pool, total pool area 515 sq.m. The pool hall has maximum glazing on the south elevation and generous provision of changing rooms, pre-cleanse facilities and toilets, entrance foyers, offices and centralised plant room for the heating and ventilating installation. The ventilation plant has a capacity of 0.018 m^3/sec/m^2 pool area. The average pool water temperature is 27°C and pool hall air temperature is 28°C. The electrical energy consumption of this pool is 1356 kWh/per m^2 pool area (although the average swimming pool consumption is 1000 kWh/per m^2 pool area).

Construction of the Energy Ranking Scale

The starting point on the energy ranking scale, is the conventional, perhaps even traditional heating system, with continuous operation of the heating and ventilating plant (Fig 4). The total fuel energy bought, to provide space ·heating is equivalent to 8795 kWh/m^2a, which together with the bought electrical energy of 1356 kWh/m^2a makes a total annual energy consumption of 10151 kWh/m^2a. This represents the starting point on the energy ranking scale (Fig 5). A realistic target on the energy ranking scale must include continuous heating of the building fabric, together with the minimum ventilation and water consumption for the occupants requirements.

The heating analysis chart for this system and building is shown in Fig 6.

Heat Conservation Technologies

Improved Thermal Insulation

The fundamental requirement of energy conservation studies is an examination of the building envelope to minimise the fabric heat loss. All such studies in both new and existing projects should be professionally checked for interstitial condensation. Moisture migration should be minimised between the humid pool hall and cold parts of the building. The techniques for additional insulation and vapour barriers in walls and roof spaces are well documented. Recently the introduction of infra-red thermography has provided a rapid and positive technique for identifying the 'rogue' heat losses and cold bridges which can cause surface condensation.

The minimum thermal insulation standards are now specified in the building
regulations part FF. These standards reduce the heat loss through the building
fabric.

Ventilation Heat Conservation

In all conventional indoor swimming pools, the ventilation and evaporation heating
loads are the main targets for energy conservation. There are two known conservation
techniques for reducing the annual ventilation heating requirement, controlled
ventilation and heat recovery, the combination of which will achieve an even lower
annual energy consumption. Before analysing this combination, it is of interest to
analyse their respective merits when used individually.

Controlled Ventilation

Manual control of ventilation fans is widely used in older UK swimming pools. Where
two speed motors have been installed on the supply fan, it is common practice to
operate for most of the year on the lower speed with a reduced number of roof exhaust
fans. This practice ensures that the structure is always kept warm, and minimises
condensation damage. The effectiveness of winter heating is dependent on the air
distribution pattern under reduced air flow. Unfortunately, in many cases, the
supply fan isswitched off during the unoccupied period thereby relying on the roof
exhaust fans for ventilation to remove the evaporated moisture. Obviously this
practice also cools the structure causing premature structural deterioration and
accelerated corrosion of the ventilating plant. Better protection of the structure
during the unoccupied period is achieved with automatic controls, particularly those
which continuously monitor the temperature of the coldest surfaces. These controls
achieve significant annual energy savings whilst protecting both the building and its
environmental services.

Controlled ventilation is achieved by varying the fresh air supply to the pool hall.
This may be achieved either by modulation of the motor speed on all fans, or by using
automatic dampers to vary the quantities of fresh and recirculated air.

With fan speed control, the internal humidity is maintained at a pre-set condition
for most of the year by electronically modulating the motor speed of ventilation
fans. Its principal advantage is the minimum installation disruption to either the
building structure or the air handling plant. The major disadvantage is the
reduction in the total air changes to transport the heat throughout the pool hall
during the coldest part of the year when maximum heating is required. Generally the
minimum winter air flow is restricted to 40% of the designed total air flow.

In contrast, the fresh air-recirculation system (Fig 7) allows the designed air
distribution pattern to be maintained in the pool hall at all times throughout the
winter period, whilst still modulating the fresh air quantities in accordance with
the internal relative humidity. This system has full modulation (ie 0 - 100%) of the
fresh air and no reduction in the heat transporting capability during the cold
weather periods. The principal disadvantages of the recirculation system for
existing swimming pools is the need to incorporate a return air system from the pool
hall, which may have higher capital cost implications.

In all respects controlled ventilation by recirculation is always preferable to
controlled ventilation by fan speed control. With both recirculation and fan speed
control the pool hall relative humidity is maintained by varying the fresh air
supply. This heating analysis chart for this system is shown in Fig 6.

The value on the energy ranking scale for RH controlled ventilation is
7870 kWh/per m^2 of pool area/annum.

Ventilation Heat Recovery (Heat Exchangers)

Three types of heat exchanges have been used for ventilation heat recovery in UK
swimming pools over the past decade. These are:

a   The Run Round Coil

This system figure 8 comprises two heat exchanger coils, one installed
in the exhaust air and the other in the fresh air inlet. A water and
anti-freeze solution is pumped between the two heat exchanger coils.
The warm exhaust air heats the fluid which then preheats the fresh
air. The main advantage of run round coils is their locational
flexibility, they may be used where the exhaust air discharge is
remote from the fresh air intake. Depending on the number of rows in
the heat exchanger coil and the air velocity through the coil matrix,
the heat transfer efficiency of run round coil systems ranges from 50%
to 65%.

b   Cross Flow Heat Exchangers

These are of plate construction comprising an arrangement of parallel
and separated air paths. It requires little maantenance but it is
relatively bulky and requires the supply and exhaust air flow ducts to
be adjacent. The heat transfer efficiency of plate heat exchangers is
generally in the order of 72% to 75%.

c   Thermal Wheels

This system uses a heat retentive honey comb matrix which allows air
to pass through it, whilst it is slowly rotating through both the
exhaust air and supply air. Thermal wheels are relatively bulky and
also require adjacent air flow ducts. The heat transfer efficiency of
thermal wheels is approximately 75%.

All three heat exchanger systems can be converted to the geared dehumidifier systems,
which are described later in this report. The heat analysis chart in Fig 10 assumes
an air to air heat exchanger with an overall thermal efficiency of 62%. The value on
the energy ranking scale for this system is 6280 kWh per m^2 pool area/annum.

Controlled Ventilation and Heat Recovery Systems

The combination of controlled ventilation and air to air heat exchangers offer a
simple and reasonably efficient method of reducing the heating energy consumption of
the ventilation system. The energy saving potential of a typical arrangement
comprising relative humidity controlled recirculation and run round coils (Fig 11),
is shown on the heat analysis chart (Fig 12). The value on the energy ranking scale
for this combination is 5750 kWh per m^2 pool area/annum (Fig 13).

Water Heat Conservation Technologies

Pool Covers

Without the benefit of pool covers, the evaporation rate in indoor swimming pools is
significantly reduced during the unoccupied period, due to the very low air movement
over the undisturbed pool surface. Depending on the type of air distribution in the
pool hall, the moisture evaporation rates during unoccupied periods can vary from
0.10 kg per m^2 per hour, 0.2 kg per m^2 per hour.

Regular use of manual or automatic pool covers will virtually eliminate even these low evaporation rates. Consequently, it is possible for two speed air handling plants incorporating recirculation to take full advantage of the minimal unoccupied evaporation, by operating on half speed and full recirculation whilst maintaining the building fabric heating requirement during the night period. A saving of 350 kWh m^2 pool area is possible with automatic pool covers when used in conjunction with other heat conserving systems. For example, the regular use of an automatic pool cover in conjunction with relative humidity controlled ventilation system using two speed fans and full recirculation, heat exchangers on the exhaust air/fresh air intakes, the annual energy consumption on the energy ranking scale would be 5400 kWh per m^2 pool area/annum.

Waste Water Heat Recovery

The heat loss in the water, currently discharged from UK swimming pools, is typically 300 kWh/m^2/pool area per annum. It has recently been suggested that pool water management throughout the UK should move towards bather related dilution. This minimises the obnoxious and increasingly suspect refractory compounds associated with chlorine and bromine disinfection as currently practiced, as well as reducing the total dissolved solid levels. Heat wastage through increased dilution based on continental recommended practice of 30 litres per person per day and assuming an annual bather density of 550 swimmers/m^2/pool area per annum the total heat loss in the discharge water would be approximately 1400 kWh/m^2 pool area per annum.

Obviously the energy cost implications in addition to the water cost would be a mator deterent against improving UK pool water management. Fortunately, both heat exchangers and heat pumps are available to recover heat from the waste water, which will allow the energy cost penalty to be significantly reduced or even eliminated in terms of annual operating costs. Parallel plate heat exchangers have been used for shower water heat recovery both on the continent and in the UK and have clearly demonstrated their practical value with operating efficiencies in the order of 75%. Recently new self cleaning heat exchangers have been used very successfully on the continent and their introduction to the UK should be seriously considered. When heat pumps are used for dehumidification or exhaust air heat recovery it is also possible to use the same refrigerant circuit to reclaim heat from the waste water. Heat pump heat recovery can easily reduce the discharge water temperature to equal the cold water supply temperature to the building thereby avoiding any heat loss, although, imposing the 300 kWh/m^2/pool area heat gain to the building. Combinations of heat exchangers and heat pumps would reduce this heat gain to less than 100 kWh/m^2/pool area per annum.

In addition to possible future public health recommendations, bather related dilution would in all non ozone pools materially improve the pool hall environment, thereby allowing full use of controlled ventilation and dehumidification techniques. Generally, heat exchangers may be used in conjunction with continuous pool dilution and would be advantageous for the majority of existing swimming pools and small new pools without ozone deozonisation. The heat pump - heat exchanger combinations are preferred for the new larger pool and leisure centres. All new designs and energy refurbishment schemes should allow for waste water heat recovery.

Solar Heating

Initially the direct use of solar heat to reduce the energy consumption of indoor swimming pools appears a very attractive concept. Unfortunately, the maximum solar heat is available in summer when the heat requirements are minimal. Due to the pool water temperatures of 27°C and higher, the potential for roof mounted solar collectors to contribute to the free heat component by heating the pool water, is very limited, although its use for preheating the domestic hot water deserves comparison with other techniques. The indirect use of solar-heat is described later in this paper.

The use of pool water direct in the solar collectors is only practical in the UK for outdoor pool applications used for summer only operation. For preheating of the cold water supply, the preferred technique is to use waste water heat recovery with heat exchangers as this heat source is available throughout the year.

The solar heat gained throughout the year from south facing windows makes a useful contribution to indoor swimming pools. In this example, it is equivalent to 258 kWh/m^2/pool area per annum.

<u>Heat Pumps</u>

In order to reduce the energy consumption of indoor swimming pools below 5500 kWh/m^2/pool area annum on the energy ranking scale, it is necessary to use a method of heat recovery which is independent of the prevailing ambient temperature. Heat pump technology provides this capability.

Electric heat pumps have been used in swimming pools for over a decade. Currently there are over 100 electric heat pump installations in UK pools. Normally, the electric motor and the compressor are built into the same casing, and the refrigerant gas is used to cool the electric motor, thereby reclaiming all the motor heat and increasing the heat pump heating performance. This type of compressor is called a semihermetic compressor and is to be preferred in all heat pump applications to the older 'open' compressor which uses a separate motor. Whilst the traditional problems associated with diesel or gas engine driven compressors such as noise, bulk and increased maintenance may be reduced by further development, all engine drives will require special site considerations to minimise noise and vibration nuisance. However, the comparative performance between the alternative heat pump drives deserves further analysis to improve the understanding of the principles before considering in detail the potential applications.

The electric heat pump of the semihermetic type and with an operating coefficient of performance of 5:1 is illustrated in Fig 17. It can be seen that for every kW of evaporator (ie cooling coil) duty, 0.25 kW of heat gain from the motor is contributed to the system.

The energy flow diagram for typical gas engine heat pump system is shown in Fig 18 from which it can be seen that every kW of evaporator (cooling coil) duty, produces $\frac{2.0}{1.12}$ $-1 = 0.786$ kW of heat gain from the motor to the system.

The ratio of the available heat for each kW of evaporator duty, between gas engine heat pumps and electric heat pumps is $\frac{0.786}{0.25} = 3.144$ to 1.

For heat pump applications where lower refrigerant temperatures are required at the cooling coil (evaporator) the coefficient of performance is reduced. For example, when heat pumps are used as external source heat pumps, which take heat from the ambient air, the typical operating coefficient of performance is 3 to 1. However, this improves as the outside temperature rises. The heat from heat pumps may be used to heat the ventilation air, the pool water, the shower water, and also to provide space heating. The heat pump is also capable of providing heat for all these functions simultaneously.

<u>Heat Pump Recovery</u>

Electric heat pump systems using exhaust air heat recovery were the first systems using heat pumps, to be installed in UK swimming pools. Six installations have been working with exhaust air heat recovery since the mid-1970's. Subsequently, the introduction of recirculation and controlled ventilation into UK swimming pools have made exhaust air systems of this type obsolete. All installations using exhaust air heat recovery with electric heat pumps should now be converted to geared dehumidificationsystems, which will be discussed later in the paper. The energy consumption on the energy ranking scale of the exhaust air heat recovery system using electric heat pumps is 4000 kWh/m^2 pool area per annum.

Controlled ventilation systems with exhaust air heat reccovery using both diesel and gas engine driven heat pumps are widely used on the Continent. In these installations the recirculation air quantity is controlled by the relative humidity in order to limit the winter ventilation air. All the exhaust air which is not recirculated goes through the heat recovery cooling coil, which is the heat source for the gas engine driven heat pump. This system allows full advantage to be taken of the higher heat production per kW of evaporator load, which is a characteristic of engine driven heat pumps. A gas engine driven heat pump with controlled ventilation and optional heat exchanger heat recovery is illustrated in Fig 19. The associated heating analysis chart is shown in Fig 20.

On the energy ranking scale the energy consumption for gas engine heat pump operating in conjunction with control ventilation is 3660 kWh/m^2/pool area per annum. When the optional heat exchanger is used in this arrangement, the energy consumption reduces to 2900 kWh/m^2/pool area per annum.

Heat Pump Dehumidification Systems

Compared to exhaust air heat recovery, when heat pumps are used as dehumidifiers, it is possible to achieve larger savings in the annual energy consumption. Heat pump dehumidification is essential if energy consumption significantly lower than 3000 kWh/m^2/pool area per annum on the energy ranking scale is required.

What is dehumidification? It is an air drier, where mot air is passed through a cooling coil (evaporator) to remove the excess moisture before the air is reheated by the heating coil (condenser). Fig 16 will illustrate the concept if it is considered that the energy flow is also the air flow.

All the compressor energy is usefully used, and the air going through the cooling coil is cooled below its dew point, and is therefore dried. It is subsequently reheated by the condenser before being discharged to the space.

The use of dehumidification allows maximum use of recirculation. During the occupied period, the fresh air required for ventilation can be matched to the occupancy, or alternatively set at the minimum fresh air proportions (say 10% of the total air flow). At night when the pool is unoccupied, the fresh air requirement is zero. Therefore, the pool can operate on 100% recirculation, and the evaporated moisture is removed in the dehumidification cycle. All the energy used to drive the compressor is then available as useful heat gain to the building. In fact it is used to heat the domestic shower water, the ventilation air, the pool water and even the building fabric.

Dehumidification allows the internal relative humidity to be stabilised which improves environmental comfort and ensures better structural protection. All dehumidification systems should be primarily controlled by the internal relative humidity to take full advantage of the reduced evaporation that occurs during periods of low occupancy.

<u>Controlled Ventilation Systems and Packaged Dehumidifier Units</u>

When controlled ventilation systems are used with or without air to air heat
exchanger recovery, the use of one or more packaged dehumidifiers in the pool hall is
a very simple and cost effective way to further improve the performance on the energy
ranking scale. Packaged units now commercially available can provide full
dehumidification for pools ranging from 14 sq metres to 140 sq meters for single
units.

Using several units in conjunction with a simple controlled ventilation system it is
possible to provide full dehumidification for public pools with a gross area of 450
to 600 sq metres. Fig 14 indicates the typical arrangement of a controlled
ventilation system employing recirculation and packaged dehumidifier units. The
heating analysis chart for this arrangement is shown on Fig 16.

On the energy ranking scale (Fig 15) the energy consumption of controlled ventilation
systems with two packaged dehumidifier units is 2680 kWh/m^2/pool area per annum.

<u>Centralised Plant for Direct Dehumidification</u>

The majority of electric heat pump conversions of UK swimming pools, employ this
principle (Fig 23), which is essentially the larger central plant version of the
smaller packaged unit. These larger units are now also packaged and offer greater
utilisation of the recovered heat than the smaller non centralised units. Central
plant versions are more robust, more flexible and use less energy. The energy
consumption on the energy ranking scale for direct acting central station
dehumidification plant is 2150 kWh/m^2/pool area per annum.

<u>Geared Dehumidification</u>

 a <u>Electric Heat Pumps</u>

   To achieve the maximum energy economy it is important to allow the
   system to operate on full dehumidification load as long as possible
   throughout the year. Whenever the fresh air exhaust dampers are
   opened unnecessarily the 'bought' heat is wasted. This energy wastage
   takes the form of heated ventilation air and uncondensed pool
   moisture.

   The geared dehumification principle allows the dehumidification
   operation to be maximised throughout the year whilst at the same time
   reducing the size and capital cost of the refrigeration plant.
   Essentially, this comprises an air to air heat exchanger arranged in
   the exhaust air before the refrigerant evaporator or cooling oil. The
   heat from the heat exchanger is used to preheat the recirculated air
   and fresh air mixture before being reheated by the heat pump system.
   The arrangement is shown in Fig 24, the heating analysis chart for
   this system is illustrated in Fig 25. Central station geared
   dehumidification heat pump systems are now commercially available from
   UK manufacturers ,for installations in both new and existing swimming
   pools from 50 sq metre pools up to the largest 2000 sq metre pools.
   These units come complete with automatic controls and are fully tested
   prior to leaving the factory.

   On the energy ranking scale the energy consumption for geared
   dehumidification systems using electric semihermetic compressors is
   2400 kWh/m^2/pool area per annum.

b   <u>Engine Driven Heat Pumps</u>

Dehumidification systems using either gas or diesel engine heat pumps are possible although the efficiency of their bought energy utilisation is both unproven and uncertain. Fig 22 illustrates that the gas engine driven heat pumps have surplus heat available. This is a distinct advantage with the exhaust air heat recovery systems but with the dehumidification mode of operation, the additional heat gained from the motor will overheat the environment of the average pool for the majority of the year. This will open the fresh air dampers to balance the heat and dehumidify the pool hall. Consequently ventilation air and moisture heat wastage will occur.

It can be seen from Figs 17 and 18 that for direct dehumidification, 1 kW of evaporator load produces 0.76 kW of heat gain per kW of dehumidification, whilst in a geared dehumidifier mode the energy surplus is 0.514 kW per kW of evaporator load. Compared to gas engine dehumidification plants, the electrically driven compressors only provide one third of heat gain to space on direct dehumidification, and one quarter of the heat gain to the space on geared dehumidification. Therefore, what initially seems to be an advantage for engined driven heat pumps tends to become a major disadvantage when integrated into the conventional swimming pool environment. The effect of this may be seen on Fig 22 which illustrates the heating analysis chart for gas engine heat pumps on both direct and geared dehumidification systems.

On the energy ranking scale the energy consumption for gas engine heat pumps working in the geared dehumidification mode is 3680 kWh/m^2/pool area per annum.

<u>Ambient Source Heat Pumps</u>

All ambient energy is derived from the sun. Solar energy is stored in the air, the ground and the water. This stored solar heat may be collected by the heat pump to provide the heating requirement of swimming pools.

External source heat pumps operating as air to water heat pumps are commonly used on the Continent for heating outdoor swimming pools. In the UK, air to water heat pumps are on sale for heating private outdoor pools. In this country, both river water and sea water are used to heat outdoor swimming pools through electric heat pumps. Ground source heat pumps have not been used here but are widely used in Germany.

Recent developments have extended the use of the geared dehumidifier system using semihermetic electric compressors to function both as a full dehumidifier and as an ambient air source heat pump. This is possible as the reduced evaporation at night from the indoor swimming pool ensures a surplus capacity during that period on the heat pump compressor. Electric heat pump systems using dehumidification should be now designed to take full advantage of the available ambient heat during the unoccupied period and the seven hours of lower cost night electricity tariff. Using the existing heat pump as an external source heat pump during the lower cost seven hour night tariff period will provide the cheapest heat currently available, irrespective of fuel source. The heat obtained during the night is stored in the pool and used to reduce, or if necessary eliminate, the need for supplementary heating during the occupied period from the fossil fuel boilers. A typical sequence is shown in Figs 26 and 27.

The method of operation during the unoccupied period is as follows:

During the unoccupied periods ambient air is admitted to the exhaust air handling plant through a separate fresh air intake and damper arrangement. It is filtered, passed through the run round coil and the main cooling system (evaporator). The refrigerant working at a lower temperature, extracts heat from the ambient air before it is discharged through the normal exhaust grill. The pool hall air bypasses the exhaust air plant, it is filtered, dehumidified by the run round coil during the cold weather and reheated at the condenser. Sufficient heat is admitted to the recirculated air from the refrigeration compressor, to maintain pool hall temperatures.

All surplus heat from the heat pump is directed to the pool water condenser and boosts the pool water temperature overnight in excess of the minimum set point. Generally this will be 1°C or 2°C higher than the normal pool water temperature. Automatic controls will ensure that the heat pump works fully loaded during the seven hour low tariff period. In mild weather when the pool water temperature has achieved its boosted temperature setting, the compressor will unload and revert to normal nightime dehumidifier operation on 100% recirculation.

When the pool is occupied during the day the fresh air dampers open to admit sufficient ventilation air, and the geared dehumidifier sequence operates in the normal way. As the pool water temperature is satisfied, all available compressor heat is directed to the air heater (condenser) to satisfy the fabric and ventilation loss during the day period. As the pool water cools during the occupied period, the evaporation will provide an additional energy source to the heat pump.

The heating analysis charts in Figs 27 and 29 show the operation of the electric heat pump system using both dehumidification and ambient heat for the unoccupied period. This is a multivalent system using ambient energy and electricity, with fossil fuel energy for extreme cold weather top up. All electric heat pump systems are now possible for both new and existing pools, by using a larger heat pump divided into two independent refrigerant circuits to provide additional ambient heat to satisfy the cold weather requirements.

Automatic Pool Covers compliment this concept since their regular use virtually eliminates the pool hall dehumidification during the unoccupied period and allows the heat pump to maximise the collection of ambient heat. Using waste water heat recovery both the hot water and the back wash heating loads can be considerably reduced from that shown on Fig 27.

On the energy ranking scale the total energy consumption for the multivalent electric heat pump system using ambient heat and dehumidification but without automatic pool covers is 1840 kWh/m^2/pool area per annum. With automatic pool covers the energy tanking scale value is 1500 kWh/m^2/pool area per annum.

For indoor swimming pools with general purpose electricity consumptions of 1000 kWh/m^2/pool area per annum the total energy consumption on the energy ranking scale for both the multivalent and the all electric ambient heat pum dehumidifier system with Automatic Pool Covers would be tween 1100 and 1200 kWh/m^2/pool area per annum. For both these ambient heat systems, the free energy use is 840 kWh/m^2/pool area per annum. The energy ranking scale showing electric heat pump dehumification systems is shown in Figs 30 and 31.

<u>TABLE 1 - ANNUAL ENERGY CONSUMPTIONS AND COSTS</u>
(costs based on 1981 prices)

System No	Type of System	Energy Ranking Scale Value kWh per m^2 Pool Area	Energy Cost per m^2 Pool Area
1	Conventional System	10,151	£128
2	<u>Controlled Ventilation Systems</u>		
2a	Controlled ventilation system using recirculation	7,870	£108
2b	Controlled ventilation system using recirculation with heat exchanger and monovalent heating system	5,750	£ 88
2c	ditto plus one packaged electric dehumidifier	3,470	£ 76
2d	ditto plus two packaged electric dehumidifier	2,680	£ 70
3	Electric Heat Pump Dehumidification System (non geared)	2,150	£ 68
4	<u>Geared Dehumidification Systems with Fossil Fuel Boilers for Supplementary Heating</u>		
4a	Gas engined heat pump with automotive engine and air conditioning compressor set (Life: 3 years or 30,000 hours)	3,660	£ 69
4b	Gas engined heat pump with industrial engine and industrial compressor set (Life: 100,000 hours plus)	3,660	£ 69
4c	Electric heat pump with semihermetic compressor (duplicate circuit) (Life: 100,000 hours plus)	2,400	£ 64
5	<u>Ambient Heat Pumps Using a Geared Dehumidification System and Electric Semhermetic Compressors</u> (Life: 100,000 hours plus)		
5a	Multivalent system with fossil fuel boilers for supplementary heating	1,840	£ 62
5b	ditto with Automatic Pool Cover	1,500	£ 55
5c	All electric heat pump system	1,800	£ 59
5d	ditto with Automatic Pool Cover	1,450	£ 52

<u>Comparative Cost Assessments</u>

The comparative annual energy consumptions for the principal systems outlined in this paper, are summarised in Table 1. This table also shows the comparative annual energy costs basedon 1981 prices. The comparative capital costs per square meter of pool area and crude payback periods with respect to the conventional system are shown in Tables 2 and 3 respectively.

The crude capital payback period for the majority of conservation systems including geared dehumidification and ambient heat pumps, range from 2.7 years to 4.41 years, with an average value of 3.45 years. When allowance is made for the cost inflation of fuel and energy over the crude payback period, this generally shortens the payback to approximately 3 years. Given the similarity in the paybackk periods the factors influencing selection are:

a   Capital Costs of the Installation

b   Future Energy Costs

c   Life Cycle Costs including Maintenance and Capital Amortisation

The comparative capital costs show a 1.5 to 1.0 ratio between the cost of the central plant systems using dehumidification heat pumps and the use of smaller packaged dehumidifiers. Clearly, the packaged dehumidifier used in conjunction with control ventilation and heat recovery is a very attractive proposition for the smaller swimming pool installation or the older swimming pool with limited useful life. In the short term, the annual energy costs for the smaller indoor swimming pool using packaged dehumidification, in conjunction with controlled ventilation is probably the best buy when gas or coal is the existing boiler fuel. However, when oil is the boiler fuel it is advantageous to use ambient source heat pumps in conjunction with dehumidification. Ambient source heat pumps minimise the 'bought' fuel component and provide greater stability for the annual energy costs for such installations in the short term. Over the medium term (8 years) and the long term, the use of ambient heat pumps in conjunction with dehumidification, secures the minimum value on the energy ranking scale; this stabilises and minimises the energy costs.

With regard to the maintenance costs and replacement costs for heat recovery and heat pump equipment, there is to date very little factual data available for the comparison of these systems when used in swimming pool applications. However, the UK experience with electric heat pump installations in swimming pools justifies the recommendation for use of semihermetic or fully hermetic compressors for both efficient operation and minimum maintenance. Little comparative data is available for engine driven heat pumps or packaged dehumidifiers at this point in time.

It is important to judge new proposals on the basis of comparable quality and life expectancy. The UK experience obtained in instaaling over 100 electric heat pump swimming pool installations has clearly emphasised the need for simplicity of maintenance, flexibility in dealing with heat pump load fluctuations, low operating noise levels and an inherent capability of continuous reliable operation, when unattended for 6 month intervals. The considerations of low capital cost, high reliability and low energy cost tend to favour the simple electric drive in preference to the more complex engine driven heat pump alternative. The use of ambient heat to minimise the bought energy consumption is a consideration that is going to be increasingly important in the future.

<u>TABLE 2 –CAPITAL COSTS</u>
(for conversions of existing pools based on 1981 prices)

System No	Type of System	Capital Cost Cost per m^2 Pool Area
1	Conventional System	Taken as datum
2	<u>Controlled Ventilation Systems</u>	
2a	Controlled ventilation system using recirculation	£100
2b	Controlled ventilation system using recirculation with heat exchanger and monovalent heating system	£120
2c	ditto plus one packaged electric dehumidifier	£140
2d	ditto plus two packaged electric dehumidifier	£160
3	Electric Heat Pump Dehumidification System (non geared)	£210
4	<u>Geared Dehumidification Systems with Fossil Fuel Boilers for Supplementary Heating</u>	
4a	Gas engined heat pump with automotive engine and air conditioning compressor set (Life: 3 years or 30,000 hours)	£220
4b	Gas engined heat pump with industrial engine and industrial compressor set (Life: 100,000 hours plus)	£260
4c	Electric heat pump with semihermetic compressor (duplicate circuit) (Life: 100,000 hours plus)	£220
5	<u>Ambient Heat Pumps Using a Geared Dehumidification System and Electric Semhermetic Compressors</u> (Life: 100,000 hours plus)	
5a	Multivalent system with fossil fuel boilers for supplementary heating	£240
5b	ditto with Automatic Pool Cover	£246
5c	All electric heat pump system	£240
5d	ditto with Automatic Pool Cover	£246

Future Technology

The microprocessor will have a major influence on the future operation and energy management of indoor swimming pools. The microchip technology (mini-computers) - is now being applied to commercial office buildings, where it is being used to control and monitor the operation of the heating, ventilating and electrical services. This technology is also available to control and monitor electric heat pump heating systems for individual houses. It is evident that direct digital controls (DDC) using the microprocessor are now competitive in capital cost terms at both the commercial and the domestic level.

The implications of the microprocessor control technology applied to indoor swimming pools will have far reaching consequences on operational flexibility. As well as controlling all heating ventilation and electrical services the microprocessor will be used to monitor and control pool water quality, including back washing, bather related dilution and chemical treatment. Microprocessor control systems based on available microcomputers can be applied to all new and existing installations. Future development will use the local telephone network in conjunction with a central computer and remote mini-computers at each pool. Such a system has been installed in Hamburg, West Germany.

Management of Conservation Technologies

Single conservation technology imposes little change on the daily operation of indoor swimming pools, the management implications are minimal. For example, the use of controlled ventilation and heat exchangers alone, impose no great operational change since all the existing environmental services including fans, pumps and boilers are continuously operational throughout the year. However, lower 'bought' energy consumptions will impose greater changes on the established operating procedure for heating, ventilating and pool water systems. This will impose additional responsibility on the pool management staff, especially in the initial stages after the commissioning of new or refurbished installations.

After the installation of heat conservation systems, including heat pumps, it is particularly important that Pool Management closely monitor and question the operation of all conventional equipment. Whenever major items of plant are not required they should be switched off to avoid unnecessary energy wastage. For example, when heat pump systems are installed in existing swimming pools, it is frequently found that the boilers and the low pressure hot water heating systems are unnecessarily operated throughout the year. In all heat pump systems, it is important to ensure that boilers are only used for top-up or supplementary heating purposes and that all controls are programmed to minimise standing losses from the conventional heating system.

CONCLUSIONS

Reducing the energy independence of indoor swimming pools has now acquired the same importance as the disinfection of pool water. Within the working lives of the majority of designers and pool managers the energy factor will be of prime importance.

The past decade has seen conservation of energy particularly the reduction of the annual energy costs from the 'good housekeeping' phase. This relied on the installation of heat exchangers and exhaust air heat pumps which has a minimal value on the energy ranking scale of approximately 5000 kWh/m^2/pool area per annum. The use of electric heat pumps has pioneered and developed dehumidification techniques,

<u>TABLE 3 - ENERGY COST SAVINGS AND PAY-BACK PERIODS</u>
(with respect to the conventional system)

System No	Type of System	Energy Cost Savings per m2 pool per annum	Crude Pay-back Period for Capital
1	Conventional System	Taken as datum	–
2	<u>Controlled Ventilation Systems</u>		
2a	Controlled ventilation system using recirculation	£20	5 yrs
2b	Controlled ventilation system using recirculation with heat exchanger and monovalent heating system	£40	3 yrs
2c	ditto plus one packaged electric dehumidifier	£52	2.7 yrs
2d	ditto plus two packaged electric dehumidifier	£58	2.76 yrs
3	Electric Heat Pump Dehumidification System (non geared)	£60	3.5 yrs
4	<u>Geared Dehumidification Systems with Fossil Fuel Boilers for Supplementary Heating</u>		
4a	Gas engined heat pump with automotive engine and air conditioning compressor set (Life: 3 years or 30,000 hours)	£59	3.73 yrs
4b	Gas engined heat pump with industrial engine and industrial compressor set (Life: 100,000 hours plus)	£59	4.41 yrs
4c	Electric heat pump with semihermetic compressor (duplicate circuit) (Life: 100,000 hours plus)	£64	3.44 yrs
5	<u>Ambient Heat Pumps Using a Geared Dehumidification System and Electric Semhermetic Compressors</u> (Life: 100,000 hours plus)		
5a	Multivalent system with fossil fuel boilers for supplementary heating	£66	3.64 yrs
5b	ditto with Automatic Pool Cover	£73	3.37 yrs
5c	All electric heat pump system	£69	3.48 yrs
5d	ditto with Automatic Pool Cover	£76	3.24 yrs

permitting the minimum value on the energy ranking scale of 2000 kWh/m^2 for indoor swimming pool installations. Future dehumidification designs will expand the heat pump technique to use ambient heat since this further reduces the 'bought' energy consumption to the value of 1500 kWh/m^2/pool are on the energy ranking scale value of 1000 kWh/m^2 pool area per annum, withhout reducing the internal environmental standards, or penalising the design life of the building.

All existing indoor swimming pools must aim for a maximum bought energy consumption of 3000 kWh/m^2/pool area per annum. All new indoor swimming pools should have a 'bought' energy design target of 1500 kWh/m^2 per annum written into both their design brief and detail specifications. Only by achieving these targets will the indoor swimming pool avoid wasting energy, and minimise annual operating costs. The heat pump, particularly the electric heat pump, will during the next decade, become an essential in all cost conscious swimming pools.

Acknowledgement

I am grateful to the Electricity Council for the opportunity to prepare and research this paper, in particular to my colleagues, T P Perera and A J Johnson, for their valued assistance and commitment to energy conservation in swimming pools. I am also indebted to all members of the Institute of Baths and Recreation Management and to the members of the Chartered Institute of Building Services who have willingly supplied information and knowledge for this technology.

References

1   Conservation and Management of Energy by G D Braham

2   Evaporation in an Indoor Swimming Pool by Basin & Krumme   OA Trans 976

3   The design of Ventilating Plants for Indoor Swimming Pools
                    by Prof E Doering   OA Trans 232B

4   Guidelines - Water Treatment for Swimming Pool Water   OA Trans 2086

Appendix 1

CALCULATION OF ENERGY BANKING SCALE VALUE

Fuel Consumption per Annum

Oil Consumption: No of litres   ________   x 10.542 = ________   kWh

or

Coal Consumption: No of tonnes   ________   x 7431   = ________   kWh

or

Gas Consumption: No of therms   ________   x 29.3   = ________   kWh

Add

Electricity Consumption per annum   = ________   kWh

                        Total Energy Consumption   = ________   kWh

Total Energy Consumption   ________   kWh
Divided by the
Total pool area   ________ m^2

Equals

The Value on Energy Ranking
    Scale
                            ________   kWh/m2/pool area annum

ENERGY RANKING SCALE

'Bought' Energy in kWh/m² pool area. annum

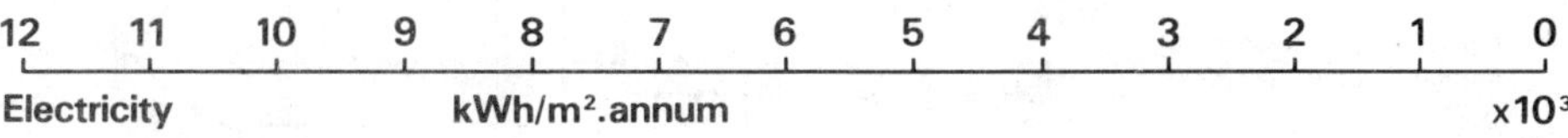

Fig. 1

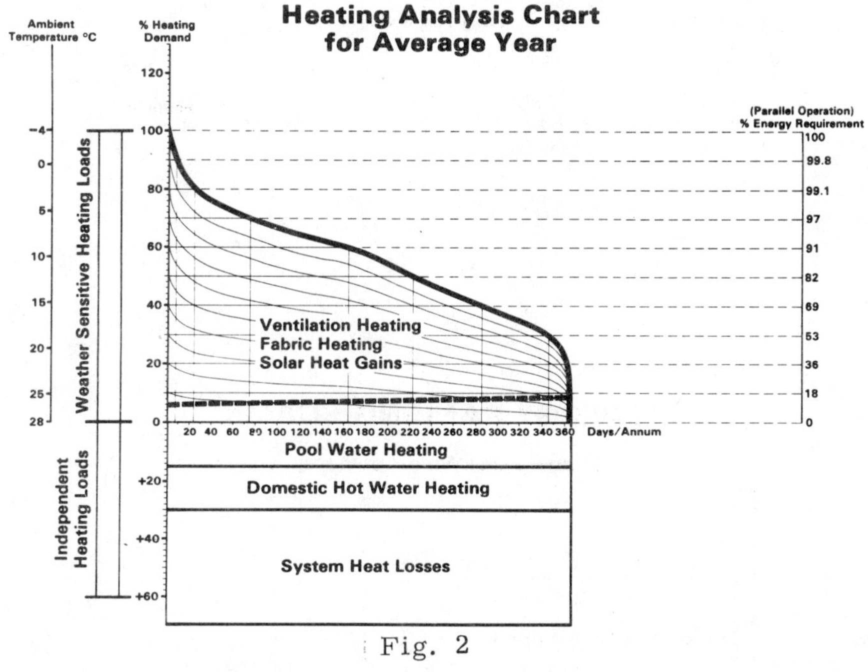

Fig. 2

SWIMMING POOL STUDY 1

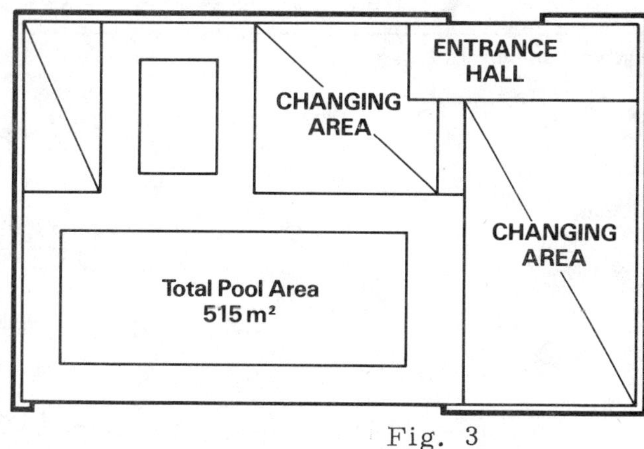

Pool Area 515 m²

Water 27°C/Air 28°C

Vent'n 0.018 $\frac{m^3}{m^2,s}$

Heat Losses @ −4°C

		kW
Vent'n Q_V	=	400
Fabric Q_F	=	71
Evap'n Q_E	=	55
Water Q_{HW}	=	17
Solar and Internal Heat Gains Q_S	=	22

Fig. 3

CONVENTIONAL
SYSTEM
Total Waste Air
Exhaust Discharge
28°C
Continuous Ventilation
Plenum Supply Operation
H.W.S.
Heating
27°C
1
2
3
Fossil Fuel Boilers
Pool Filtration
Plant

Fig. 4

ENERGY RANKING SCALE
'Bought' Energy in kWh/m² pool area. annum
Heating
Fuel
Consumption
8795 kWh/m².annum
12   11   10   9   8   7   6   5   4   3   2   1   0
Electricity   1356 kWh/m².annum
CONVENTIONAL
SYSTEM
Total 10 151 kWh/m².annum

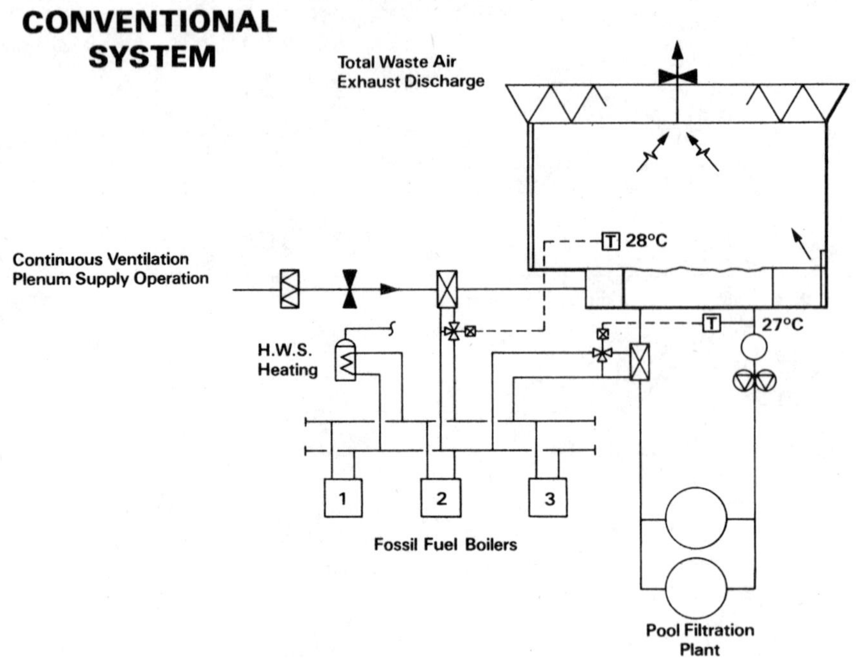

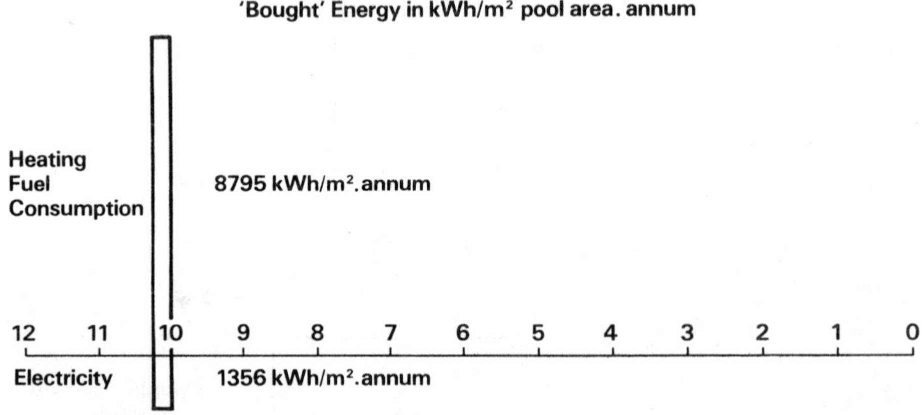

Fig. 5

HEATING ANALYSIS CHART
Conventional System
Total E.R.S. Consumption
10,151 kWh/m². annum
Ambient Temperature °C
% Heating Demand
(Parallel Operation)
% Energy Requirement
Ventilation QV
Fabric
471
QF
71
Boiler Load
Internal Gains
Solar Gains
Days/Annum
Pool Heating
HWS Heating
Heating System and Boiler Losses
Boiler Operating Period

Fig. 6

CONTROLLED VENTILATION
RH Controlled Ventilation for Occupied and Unoccupied periods
60%
RH
28°C
27°C
H.W.S. Heating
1
2
3
Fossil Fuel Boilers
Pool Filtration Plant

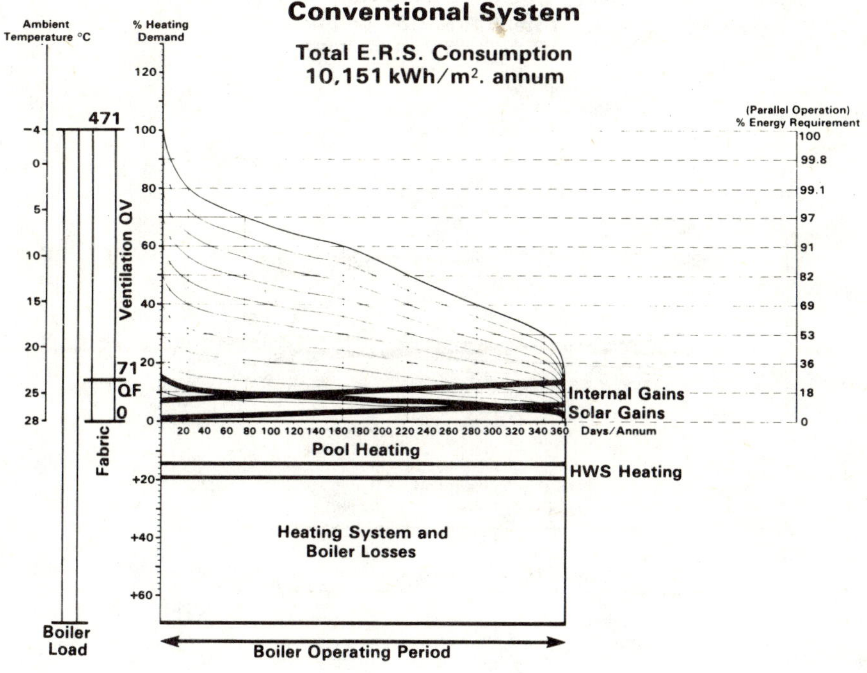

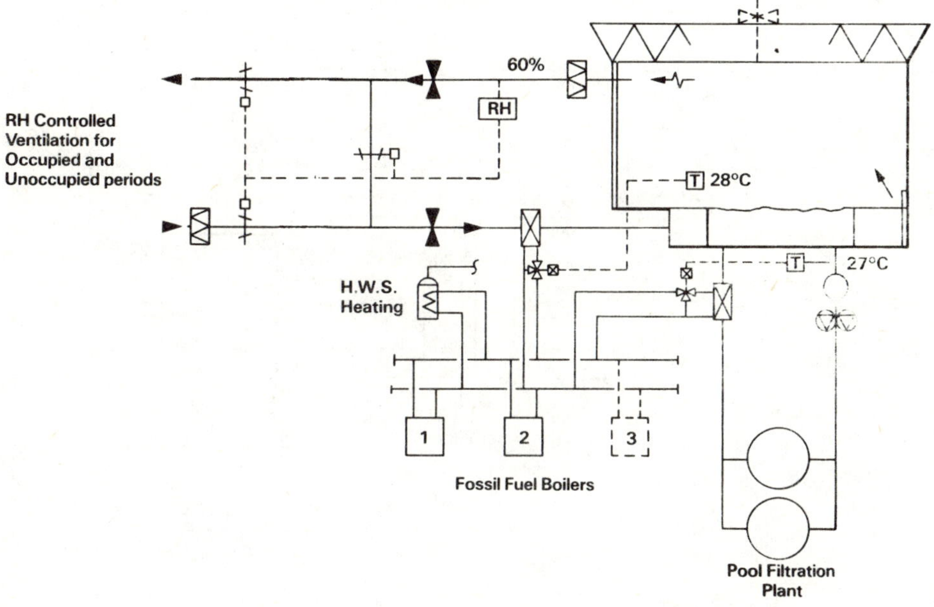

Fig. 7

HEATING ANALYSIS CHART
Controlled Ventilation System
Total E.R.S. Consumption
7,870 kWh/m². annum
Ambient Temperature °C
% Heating Demand
(Parallel Operation)
% Energy Requirement
120
100
80
60
40
20
0
QV
QF
Internal Gains
Solar Gains
Days/Annum
20 40 60 80 100 120 140 160 180 200 220 240 260 280 300 320 340 360
Pool and Water Heating
Heating System and Boiler Losses
Boiler Load
Boiler Operating Period
100
99.8
99.1
97
91
82
69
53
36
18
0
+20
+40
+60

Fig. 8

VENTILATION
HEAT RECOVERY SYSTEM
28°C
27°C
H.W.S. Heating
1
2
3
Fossil Fuel Boilers
Pool Filtration Plant

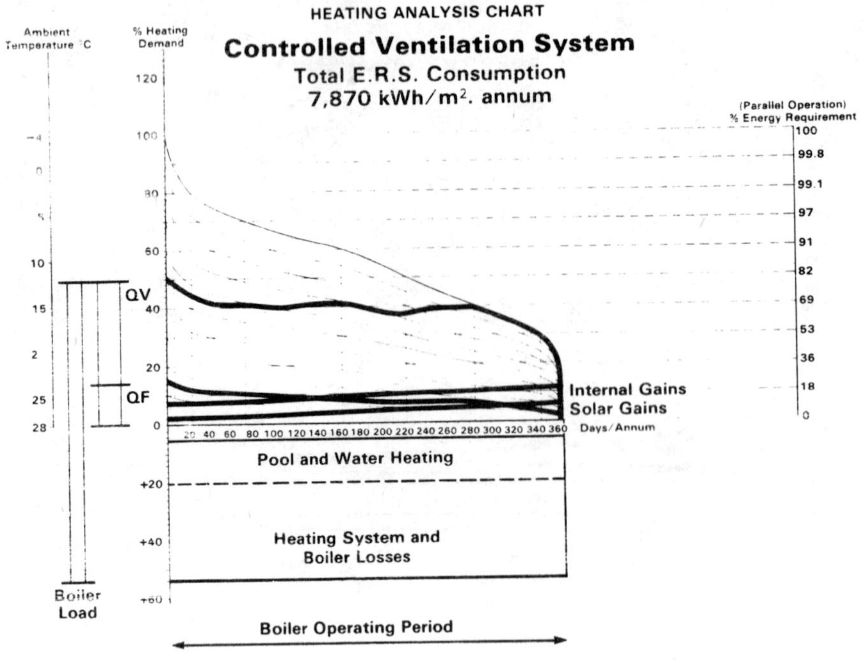

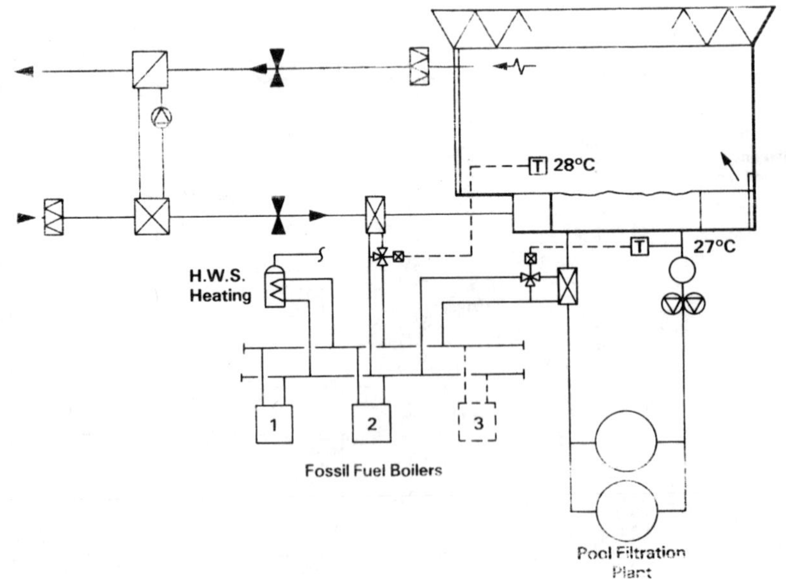

Fig. 9

HEATING ANALYSIS CHART
Ventilation Heat Recovery System
(with Air/Air Heat Exchanger, 62% η)
Total E.R.S. Consumption
6,280 kWh/m². annum
Ambient Temperature °C
% Heating Demand
(Parallel Operation)
% Energy Requirement
QV
QV'
QF
Internal Gains
Solar Gains
Days/Annum
Pool and Water Heating
Heating System and Boiler Losses
Boiler Load
Boiler Operating Period

Fig. 10

CONTROLLED VENTILATION
WITH HEAT RECOVERY SYSTEM
RH Controlled Ventilation for Occupied and Unoccupied periods
60%
RH
28°C
27°C
H.W.S. Heating
1
2
3
Fossil Fuel Boilers
Pool Filtration Plant

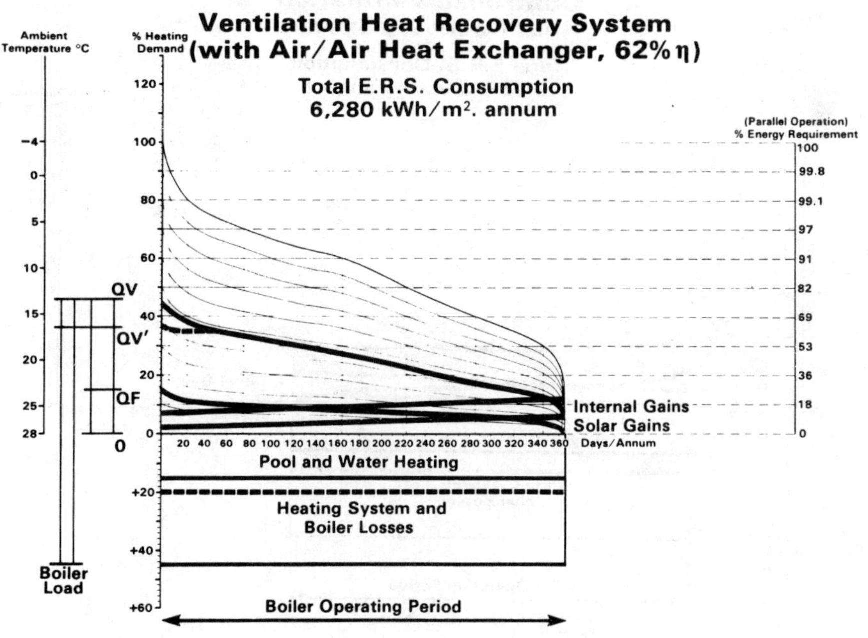

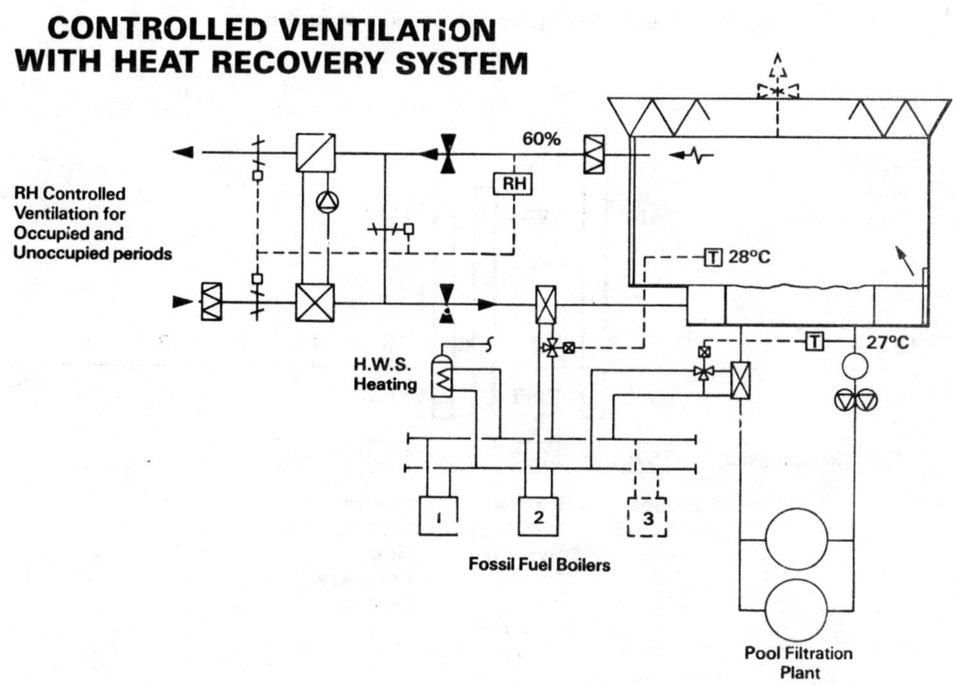

Fig. 11

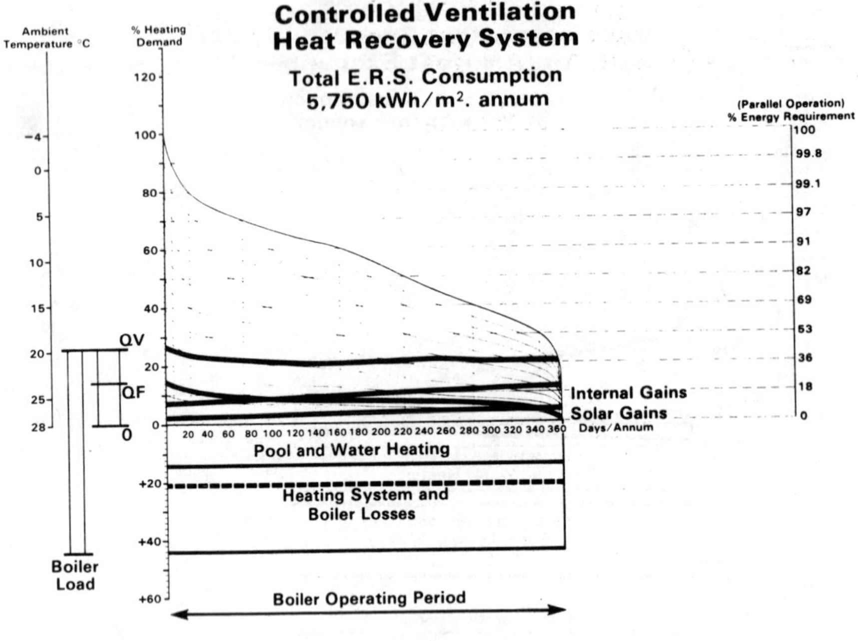

Fig. 12

ENERGY RANKING SCALE

'Bought' Energy in kWh/m² pool area. annum

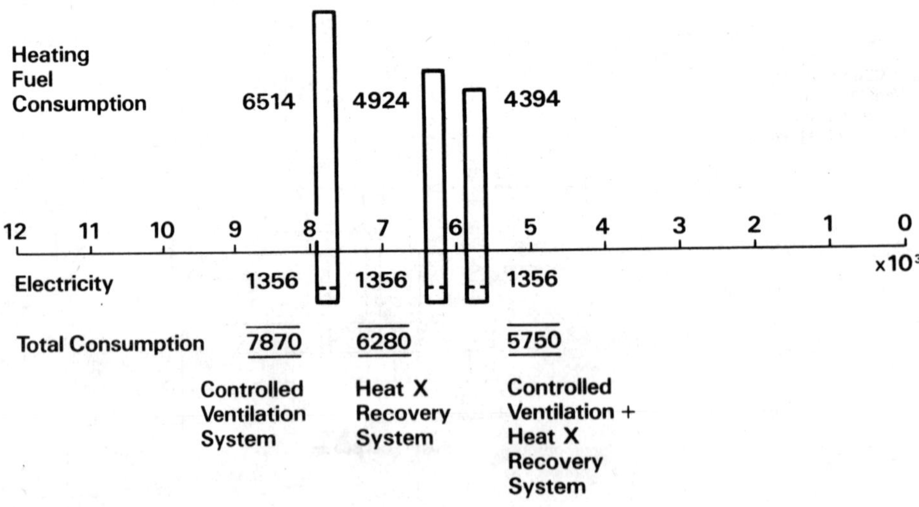

Fig. 13

CONTROLLED VENTILATION HEAT RECOVERY SYSTEM WITH DEHUMIDIFIER UNIT AND BIVALENT POOL HEATING

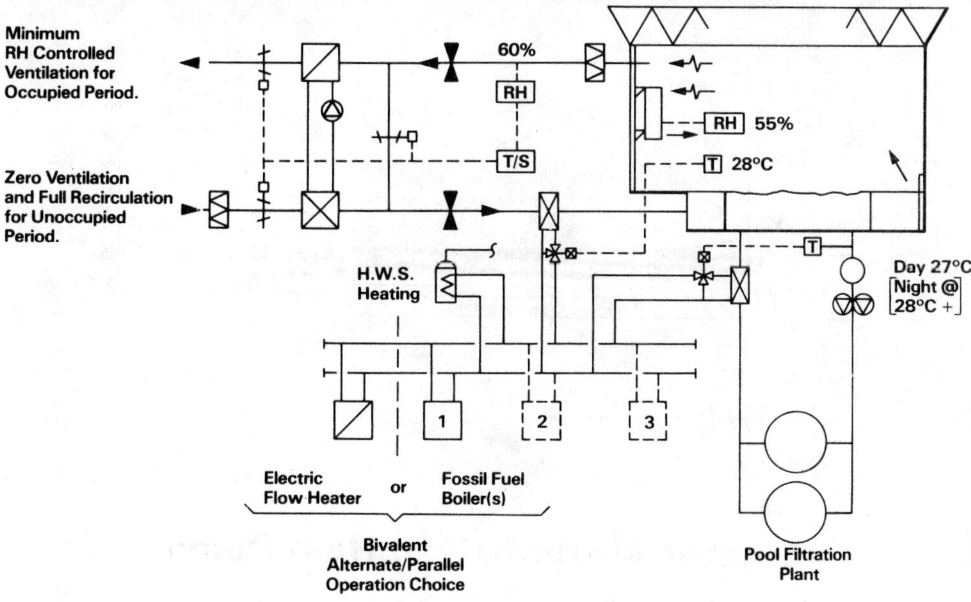

Fig. 14

ENERGY RANKING SCALE

'Bought' Energy in kWh/m² pool area. annum

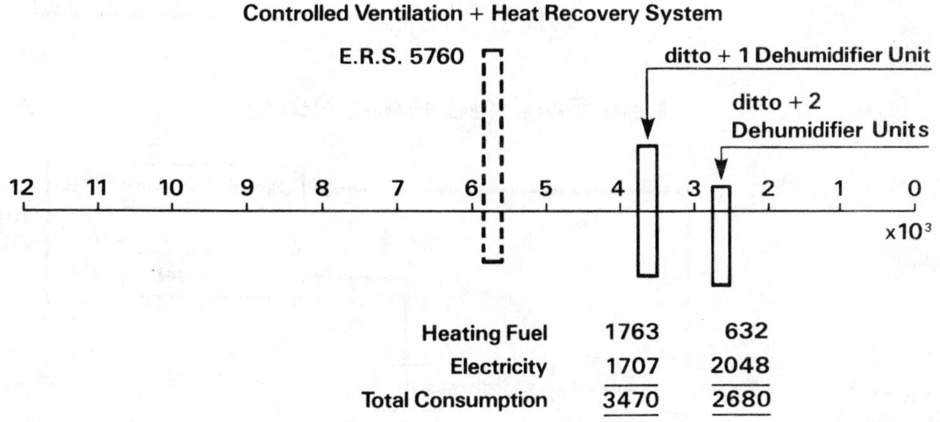

Fig. 15

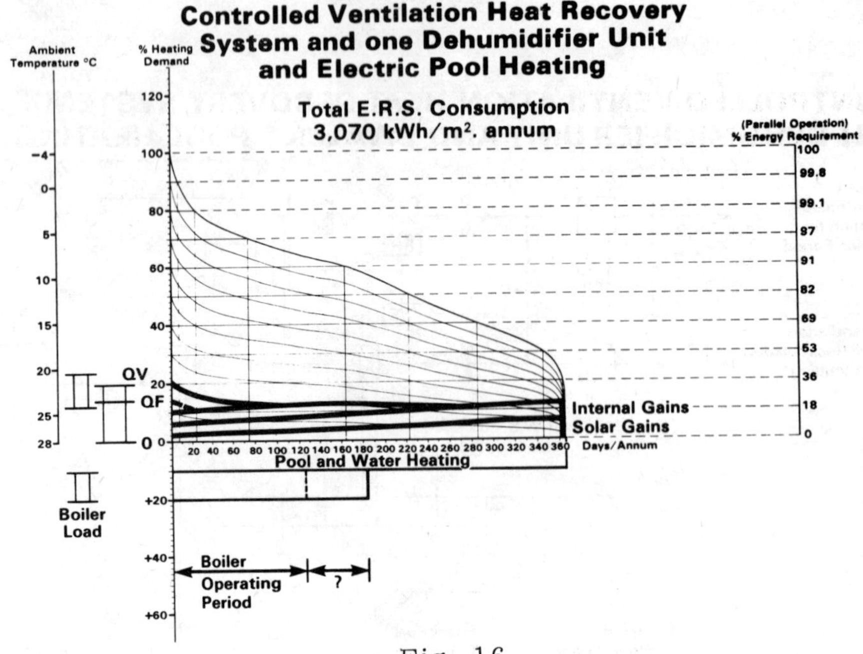

Fig. 16

Electric Motor Driven Heat Pump

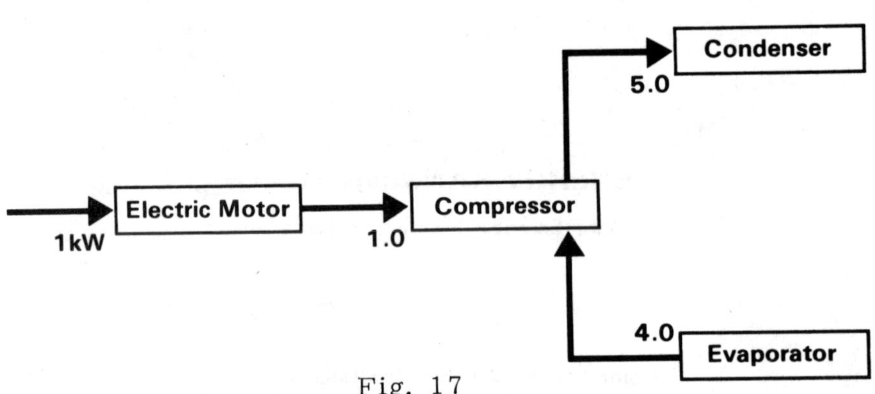

Fig. 17

Gas Engined Heat Pump

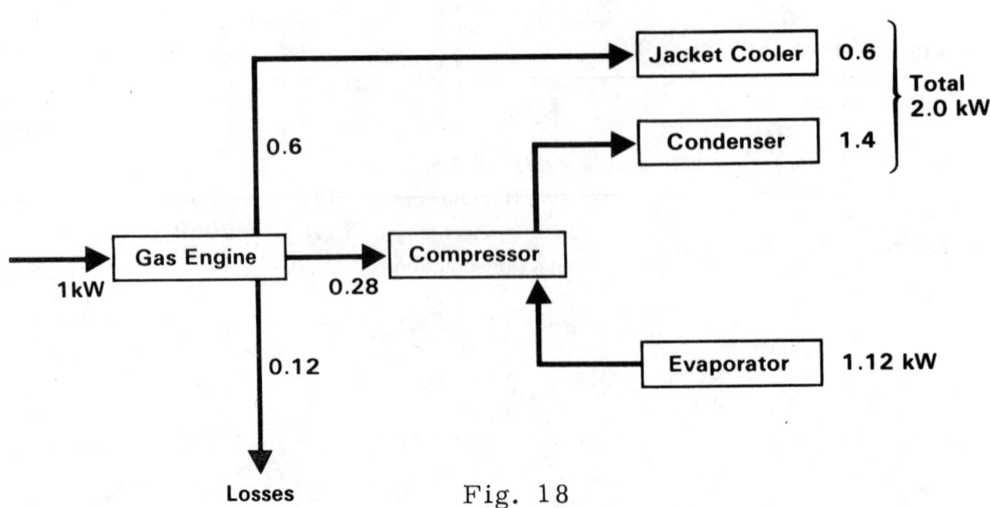

Fig. 18

GAS ENGINE DRIVEN HEAT PUMP WITH CONTROLLED VENTILATION AND OPTIONAL HEAT EXCHANGER RECOVERY SYSTEM

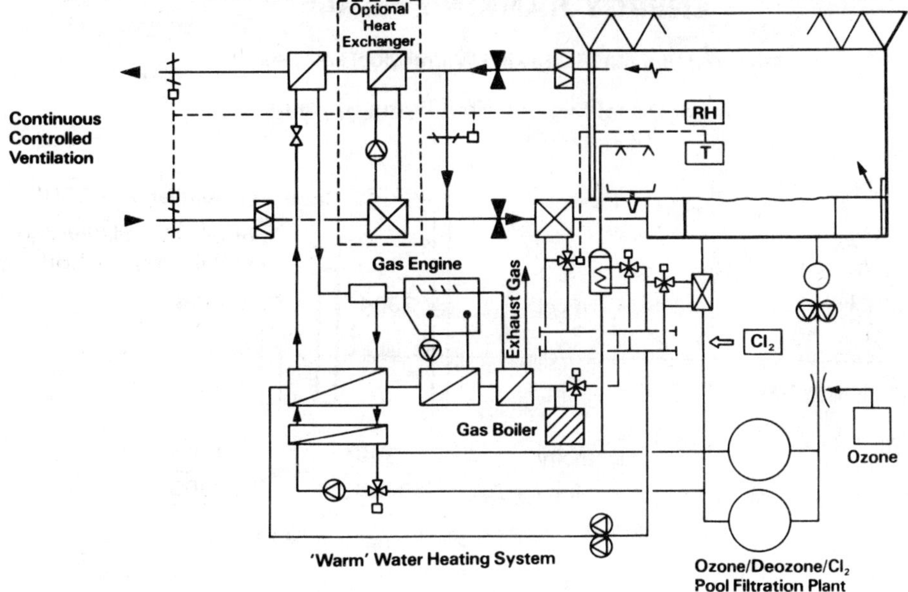

Fig. 19

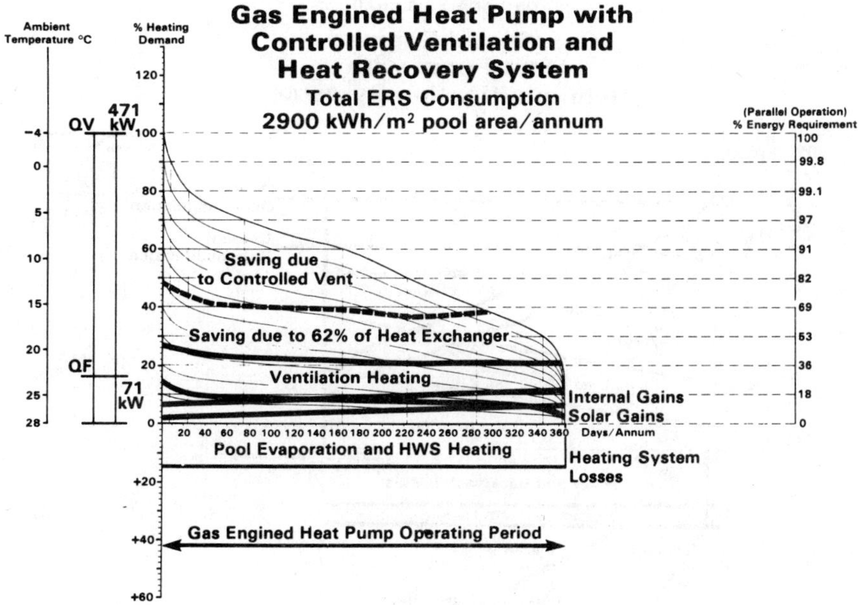

Fig. 20

ENERGY RANKING SCALE

'Free' and 'Bought' Energy in kWh/m² pool area. annum

Gas or Diesel Engined Heat Pump Systems

Controlled Ventilation + GEHP

Controlled Ventilation and
Heat Recovery + GEHP

Fuel
2304
1544

12 11 10 9 8 7 6 5 4 3 2 1 0

Bought Energy Use
x10³

Electricity   1356   1356
Total   3660   2900

Free Energy Use

Fig. 21

HEATING ANALYSIS CHART

Gas/Diesel Engined
Heat Pump with
Dehumidification Systems

Ambient
Temperature °C

% Heating
Demand

(Parallel Operation)
% Energy Requirement

QV 93 kW 100
QF 71 kW

Direct
Dehumidification

Geared
Dehumidification

Wasted Ventilation
Heat

Heat Gains
from Engine and
Compressor

Internal Heat Gains

Solar Heat Gains

Days/Annum

20 40 60 80 100 120 140 160 180 200 220 240 260 280 300 320 340 360 380

Hot Water and Backwash Losses

Heating System Losses

Direct   Geared

De-
humid-
ification
Operating
Period

Evaporation Wasted Heat

Exhaust Air Heat Recovery

100
99.8
99.1
97
91
82
69
53
36
18
0

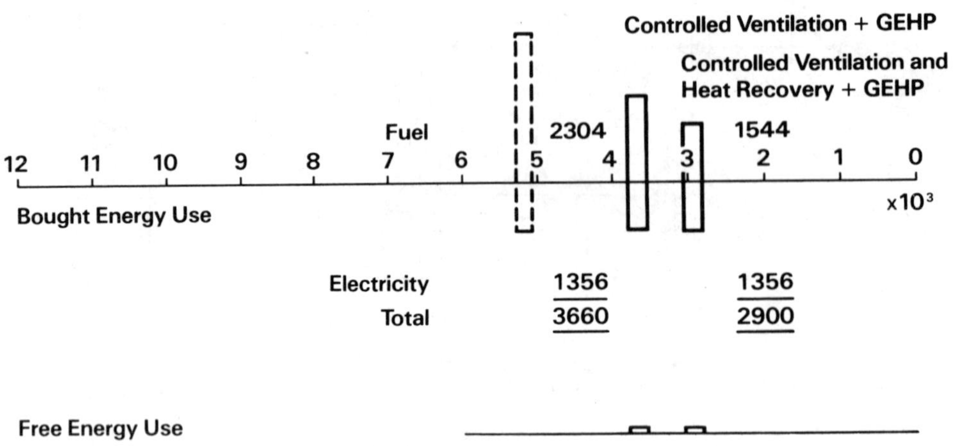

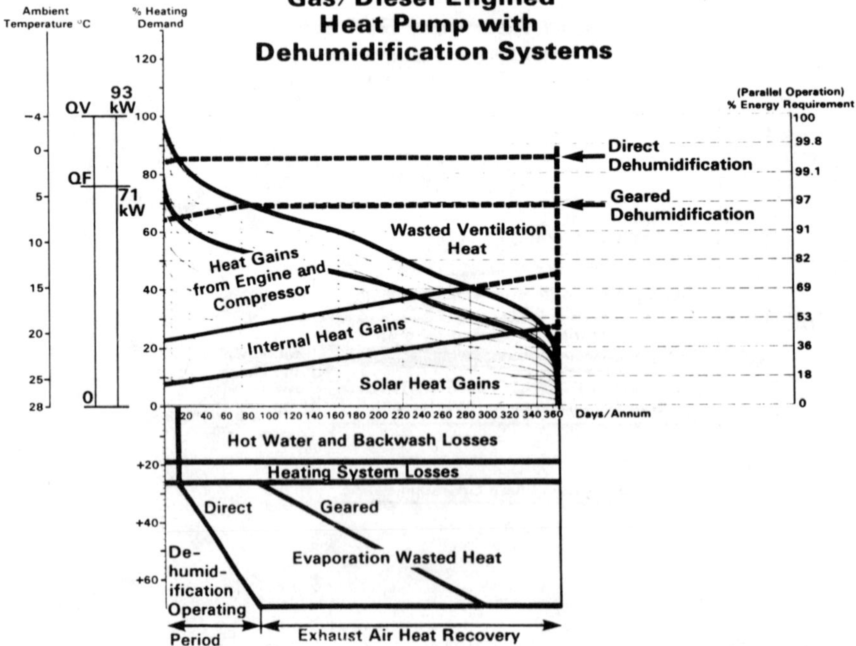

Fig. 22

DEHUMIDIFICATION SYSTEM WITH ELECTRIC HEAT PUMP

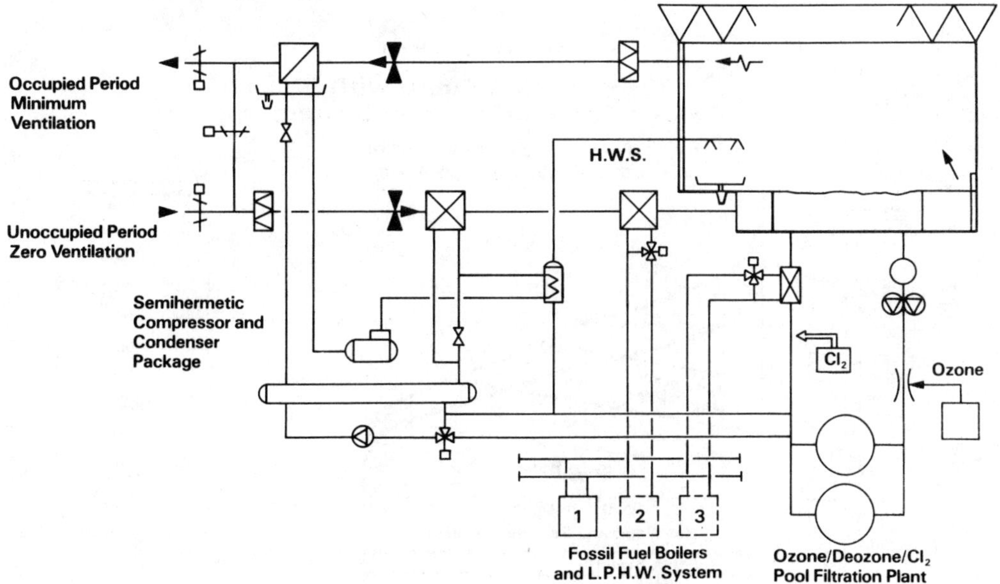

Fig. 23

GEARED DEHUMIDIFICATION SYSTEM WITH ELECTRIC HEAT PUMP

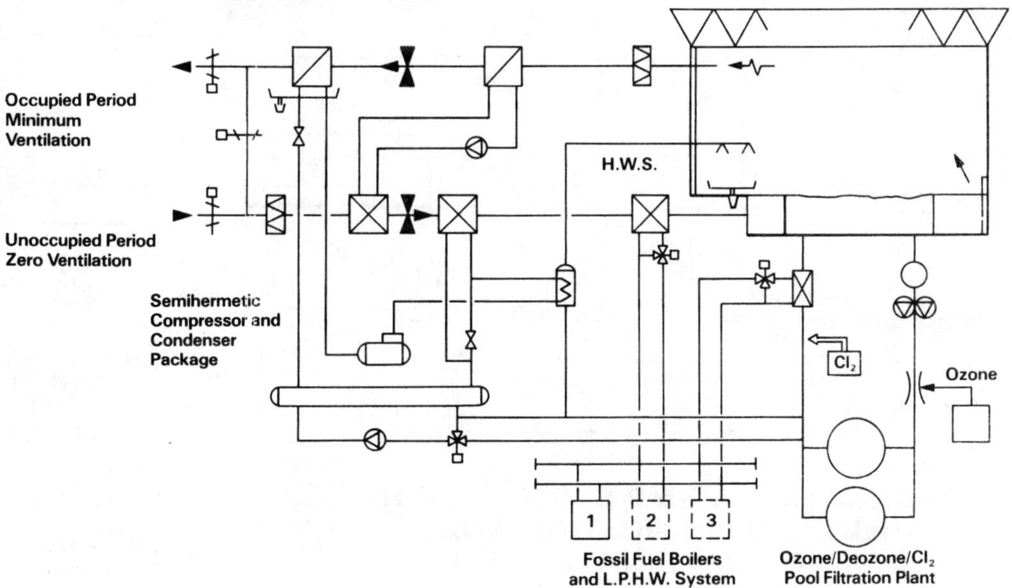

Fig. 24

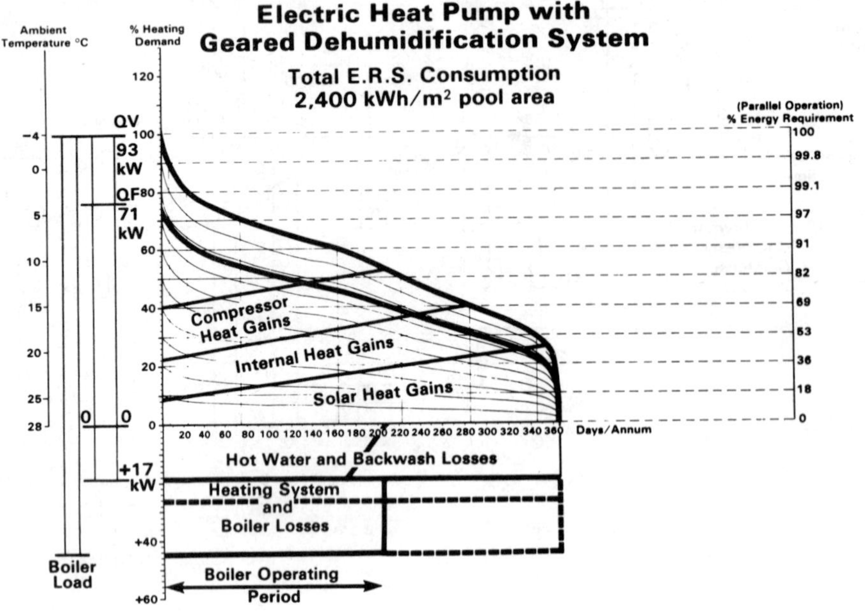

Fig. 25

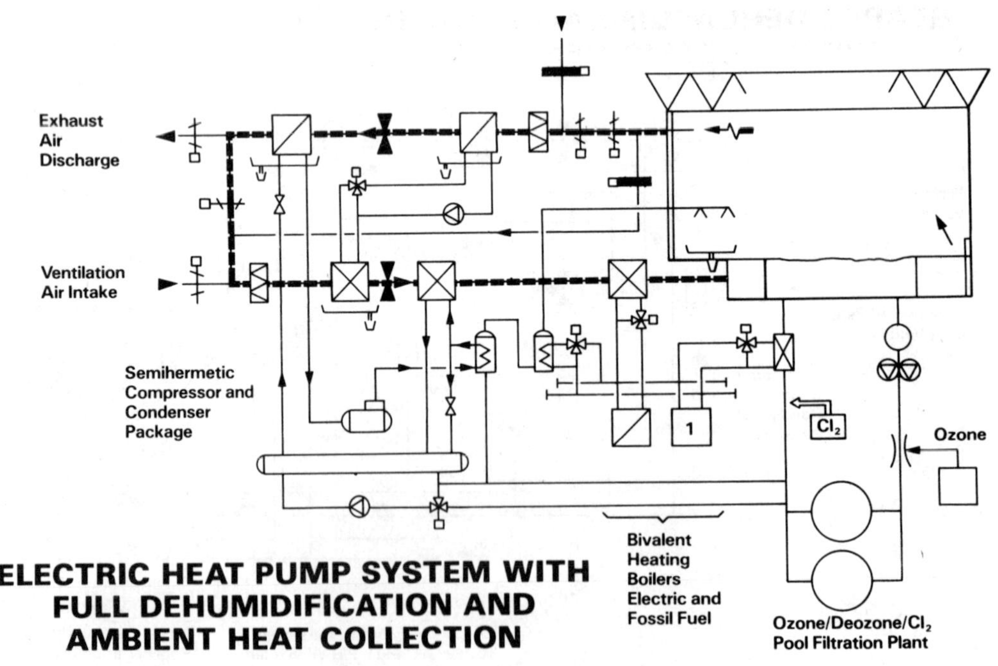

ELECTRIC HEAT PUMP SYSTEM WITH FULL DEHUMIDIFICATION AND AMBIENT HEAT COLLECTION

Occupied Period (13 hours)

Fig. 26

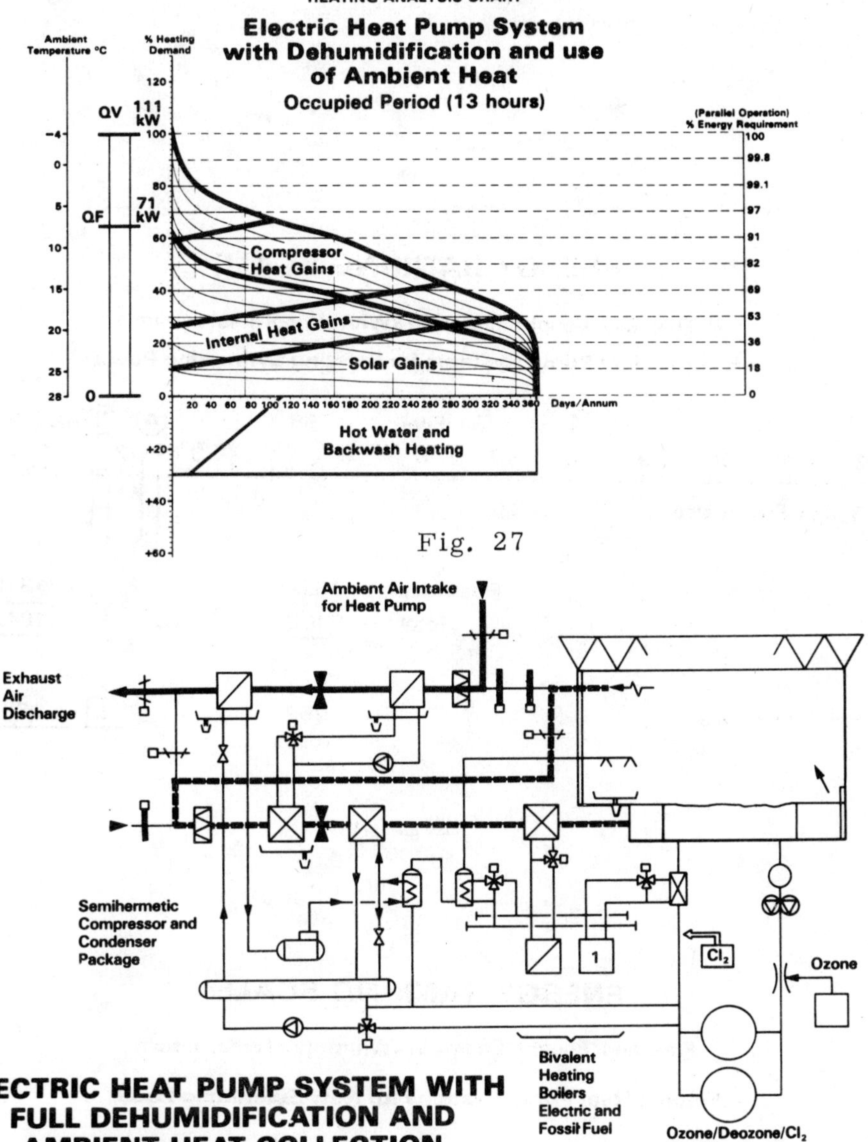

Fig. 27

Fig. 28

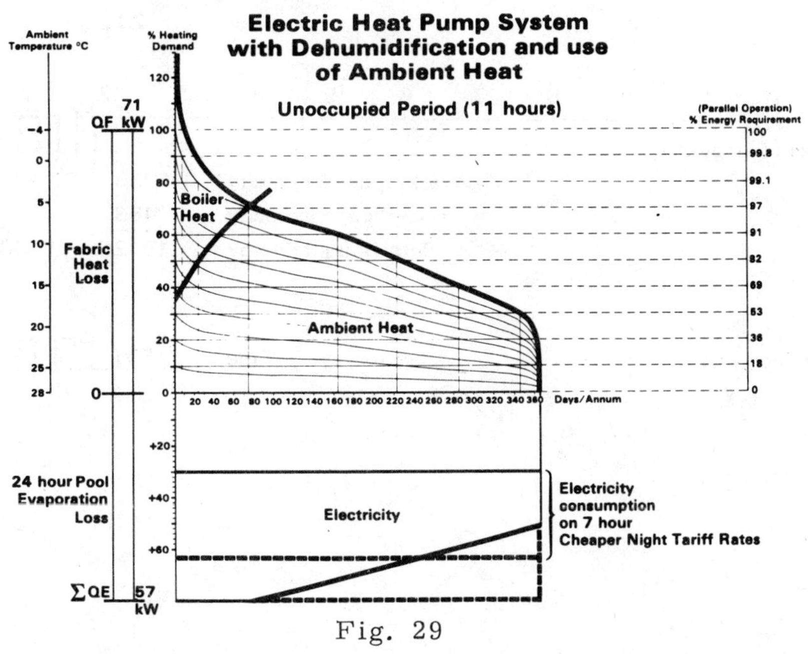

Fig. 29

ENERGY RANKING SCALE
'Free' and 'Bought' Energy in kWh/m² pool area. annum
Electric Heat Pump Systems for Existing Swimming Pools
Fuel
768
A B
23
12 11 10 9 8 7 6 5 4 3 2 1 0
Bought Energy Use
x10³
Electricity 1632 1819
Total 2400 1842
Free Energy Use 152 831

Fig. 30

ENERGY RANKING SCALE
'Free' and 'Bought' Energy in kWh/m² pool area. annum
Electric Heat Pump Systems for New Swimming Pools
Multivalent
All Electric
Fuel
23
12 11 10 9 8 7 6 5 4 3 2 1 0
Bought Energy Use
x10³
General Purpose Electricity 1356 1000
Heat Pump Electricity 463
Total Bought Energy 1842 1500
Free Energy Use Free 831 831

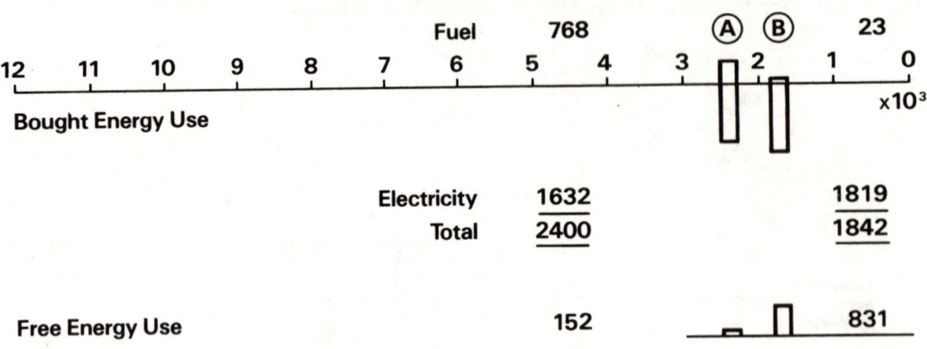

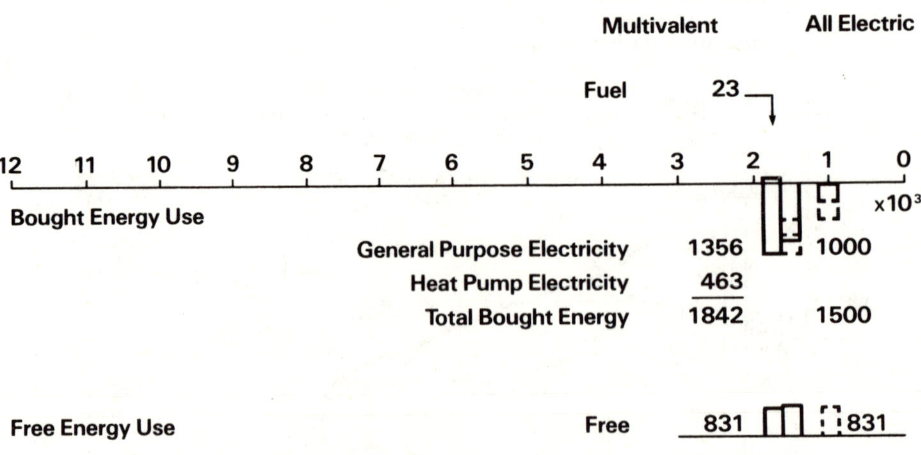

Fig. 31

DETERMINING THE CAPITAL COST OF INDUSTRIAL HEAT PUMPS BY CORRELATION

A. Van Looveren, J. Verelst, J. Berghmans and L. Gelders

Katholieke Universiteit Leuven, Belgium

Summary

Correlations are derived for the capital cost of electrically driven vapour compression heat pumps for industrial application. Two approaches have been taken. First, the component approach based upon an analysis of the cost of evaporator, condenser and compressor is presented. It leads to correlations which can be used in optimization studies related to heat pump component size and type. In a second approach the cost of the heat pumps is correlated to the power and the temperature levels. This method is particularly suited for optimization studies related to heat sink and heat source temperature levels. An example of such a study based upon the derived correlations is given.

Held at the University of Warwick, U.K.
Symposium organised and sponsored by
BHRA Fluid Engineering

NOMENCLATURE

C_{hp} = heat pump cost

C_{ev} = evaporator cost

C_{cd} = condenser cost

C_{cp} = compressor cost

C_{m} = cost of minor components, control equipment and assembly costs

X_{m} = variable which indicates the size of a heat pump

X_{ev} = evaporator surface area

X_{cd} = condenser surface area

X_{cp} = compressor volume

t_{0} = water temperature at the evaporator outlet

t_{1} = water temperature at the condenser outlet

R = working fluid

Q = condenser capacity

Q_{max} = maximum space heating load

E = heat delivered during the heating season

E_{max} = total heat required during the heating season

a, b, c, d, e, f, g, h
$\alpha, \beta, \gamma, \delta$ = constants

1. INTRODUCTION

It has been realized for quite a number of years now that presently there are heat pumps available which technically can be applied to a large number of industrial applications in order to conserve energy (see e.g. Reay (Ref. 1)). The reason why heat pumps are installed in only a few instances is to a large extent of an economical nature. Very often it is required that the expenses caused by the heat pump system (purchase cost, installment cost...) should be recovered in a period of no more than 2 years. This criterion drastically reduces the number of acceptable heat pump applications.

Very often the payback period, for a given application, will depend upon the size and therefore the cost (*) of the heat pump installed. An optimization directed towards a minimum payback period therefore requires the relationship "heat pump size-cost" to be known. It is the purpose of this paper to present the derivation of such a relation and to illustrate its application.

In most optimization studies the cost of the electrically driven vapor compression heat pump is simply taken to be linearly increasing with the compressor capacity or the condenser capacity (e.g. Refs. 2 and 3).

Josephsson (Ref. 4) considers the proportionality factor between heat pump cost and capacity to be a function of the evaporator and the condenser outlet temperature and the working fluid.

In this paper it is shown that even for given evaporator and condenser temperatures and a given working fluid the cost per kW is not a constant but a decreasing function of the capacity.

Urdaneta (Ref. 5) determines the cost of the major components to arrive at the heat pump cost.

Also in the present study the costs of the major heat pump components have been determined. These costs are related to the total cost. This allows one to estimate the importance of the cost of minor components, the cost for the control equipment and the assembly costs.

Finally the use of the correlations derived is illustrated.

2. HEAT PUMP COST : COMPONENT APPROACH

The cost of a heat pump can be considered to consist of the cost of the major components (evaporator, compressor, condenser). In addition to this there is the cost of minor components, the cost of the control equipment and the assembly costs.

The cost of each major component clearly will depend upon its size. Therefore one may put :

$$C_{ev} = a \, X_{ev}^{b} \tag{1}$$

$$C_{cd} = c \, X_{cd}^{d} \tag{2}$$

$$C_{cp} = e \, X_{cp}^{f} \tag{3}$$

in which C represent costs (in BF) and X_{ev} and X_{cd} are the surface area of evaporator and condenser respectively. X_{cp} is the compressor cylinder volume (in m^3). It was in fact found to be more efficient to correlate the compressor cost to its physical size than to its power. This is not so surprising because the same compressor can operate at different power levels (and temperature levels) by utilizing a different working fluid.

For the evaporator the price correlation is listed in Table I in which also the data base is described by the number of data points, the range for X_{ev} and the date. All data points refer to the same brand of evaporator.

For the condensor, data on three brands were available. For each brand a separate correlation was derived. The results are given in table II. The correlations derived for the compressor cost are listed in table III in which distinction was made

---

(*) in this paper the "cost" stands for the cost for the buyer, in other words, the selling price.

between semi-hermetic and open compressors. All evaporators and condensers considered
were of the shell and tube type.

The correlations listed in tables I to III were developed by means of a regression technique. The quality of the correlations is indicated by the R^2 value, the coefficient of determination.

The total heat pump cost can be expressed as :

$$C_{hp} = C_{cd} + C_{ev} + C_{cp} + C_m \tag{4}$$

where C_m represents the cost of the minor components, the assembly costs and the cost of the equipment.
Certainly C_m is difficult to estimate. However, it is obvious that it will depend upon the size of the heat pump. Therefore one may put :

$$C_m = g \; X_m^h \tag{5}$$

X_m is a variable which indicates the size of the installation (e.g. the horse power of the electric motor driving the compressor).
To confidently estimate the final relationship :

$$C_{hp} = a \; X_{cd}^b + c \; X_{ev}^d + e \; X_{cp}^f + g \; X_m^h \tag{6}$$

a lot of data points (for the same brand) are required. In this case only 13 data points were available which was insufficient to derive statistically significant results. However it was found that the ratio of the total cost to the cost of the major components was fairly constant so that one is allowed to put :

$$C_{hp} = A \cdot (C_{cd} + C_{ev} + C_{ep}) \tag{7}$$

The mean value of this ratio A for the 13 data points and for :

$$C_{cd} = 1.000 \; X_{cd}^{0.66}$$
$$C_{ev} = 1.000 \; X_{ev}^{0.66} \qquad \text{(See Tables I to III)}$$
$$C_{cp} = 10.000 \; X_{cp}^{0.66}$$

is 1.96. The relative error distribution is listed in table IV.
For this case, it may be concluded that the total cost of a heat pump is about 2 times the cost of the major components.

3. HEAT PUMP COST : TEMPERATURE-CAPACITY APPROACH

The component approach utilized above forces one to determine the physical size of the major components of the heat pump. This often is a cumbersome task requiring considerable design experience.

On the other hand, as is very often the case in industrial heat pump applications, the heat source and the heat sink temperature levels are known. In addition the thermal capacity of the heat pump is known due to a given heating requirement or a given waste heat power level.

The major heat pump components depend upon the temperature levels and the heat pump capacity. One more variable is the working fluid used. One may therefore put :

$$C_{hp} = F \; (t_0, t_1, R, Q) \tag{9}$$

in which : t_0 = water temperature at the evaporator outlet (C)

t_1 = water temperature at the condenser outlet (C)

R = type of working fluid used

Q = condenser output (kW)

First it was found that the relation between C_{hp} and Q, for a given set (t_0, t_1, R) and a given brand, is not a linear function but a power function :

$$C_{hp} = \Omega \; Q^{\delta} \tag{10}$$

with : $\delta < 1$

and $\quad \Omega = F (t_0, t_1, R)$

The cost of the heat pump can be expected to increase with increasing t_1, and to decrease with increasing t_0. By means of a regression analysis the following relationship was found :

$$\Omega = \alpha - \beta t_0 + \gamma t_1 \tag{11}$$

The cost of a heat pump therefore becomes :

$$C_{hp} = (\alpha - \beta t_0 + \gamma t_1) Q^{\delta} \tag{12}$$

A set of constants α, β, γ has to be derived for each brand and for each working fluid. The value of δ only depends upon the brand. The results of the analysis are listed in table V. The error spread for these correlations is listed in table VI.

It can be seen that for brand 1 at least a very satisfying correlation is obtained. The error between predicted and real heat pump cost almost never is larger than 10 %. In 75% of the cases it is smaller than 5 %. A less satisfying correlation is obtained for brand 2. Nevertheless the error is still less than ± 10 % in 65 % of the cases and less than ± 15 % in 90 % of the cases.

It may be concluded therefore that satisfactory correlations have been derived for two brands with t_0, t_1, Q and the freon type as input parameters.

The type of freon to be used is given by the temperature range between which the heat pump is operating. These are roughly :

$$R\ 22\quad : 0 < t_0 < 35\ ,\quad 30 < t_1 < 50$$
$$R\ 500\ : 0 < t_0 < 45\ ,\quad 40 < t_1 < 60$$
$$R\ 12\quad : 0 < t_0 < 45\ ,\quad 50 < t_1 < 70$$
$$R\ 12\text{-}114 : 25 < t_0 < 50\ ,\quad 60 < t_1 < 85$$
$$R\ 114\ : 30 < t_0 < 55\ ,\quad 70 < t_1 < 105$$

The influence of the working fluid used on the cost of a heat pump is very important. E.g. for $t_0 = 40$ °C and $t_1 = 70$°C one may use R 12-114 or R 114. However, the cost increases with 35 % if one uses R 114 instead of R 12-114.

4. APPLICATION

The usefulness of the above correlations can be illustrated by means of an example concerning recuperation of industrial waste heat to be used for space heating.

On the average the maximum space heating (Q_{max}) load occurs at an outside air temperature of -10°C. If a heating capacity Q of the heat pump recuperating the waste heat is less than Q_{max} then the heat E delivered during the heating season will be less than the total heat E_{max} required over the whole heating season. The relationship between $\dfrac{E}{E_{max}}$ and $\dfrac{Q}{Q_{max}}$ will depend upon the temporal variation of the outside air temperatures and thus will depend upon the local climate. This relationship is plotted in figure 1 for the case of central Belgium. It can be easily seen that even with a heating capacity of only 50 % of Q_{max} the heat pump can provide about 90 % of the total amount of heat required. The benefits, resulting from the "50 % heat pump" will be almost the same as if one had installed a "100 % heat pump" but the cost will be much reduced.

It is obvious that in such a situation optimization of the heat pump size is a necessity.

The results of such an optimization study are shown in figures 2 to 7. Table VII summarizes the assumptions on which the example is based.
Figures 2 and 4 show the net present value (NPV), the internal rate of return (IRR) and the payback period (PB) for a maximum load of 200 kW and condenser capacities ranging from 0 kW to 200 kW. Different curves correspond with different savings per kWh.

The best results are obtained for a 40 % - 50 % heat pump. If one installs e.g.
a 40 % heat pump, the payback period is 3,48 years (for a saving of 0.50 F/kWh), if
one installs a 100 % heat pump, the payback period amounts to 5,64 years.

Figures 5 to 7 show that the installation of a heat pump becomes more attractive
as the maximum load increases. The results are shown for a 50 % heat pump.

REFERENCES

1. Reay, D.A. : "Industrial Energy Conservation". Pergamon Press, Oxford, 1977, pp. 358.

2. Le Goff, P. : "Sur-investissement permettant d'économiser l'énergie, Optimisation et temps de récupération". EPE, Vol. XIV, n° 3-4, 1979, pp. 5-24.

3. Bolle, L. : "La production simultanée de puissance calorifique et frigorifique". EPE, Vol. XIV, n° 3-4, 1979, pp. 39-50.

4. Josephsson, H. : "Pompes à chaleur et industrie : les compresseurs à vis utilisés en pompe à chaleur". Revue pratique du Froid et du conditionnement d'air, nov. 1980, pp. 49-53.

5. Urdaneta, A.H. : "Energy and economic analysis of industrial process heat recovery with heat pumps". Ph. D. thesis of the University of Texas, 1978, pp. 265.

Table I

Correlation	R^2	range X_{ev}	No data points	date
$C_{ev} = 13.617\ X_{ev}^{0.79}$	0.98	$30 - 215\ m^2$	18	1980

Table I : the evaporator price correlation.

Table II

Correlation	R^2	range X_{cd}	No data points	date	brand
$C_{cd} = 11.683\ X_{cd}^{0.64}$	0.98	$0.18 - 25.66\ m^2$	13	1980	1
$C_{cd} = 15.965\ X_{cd}^{0.72}$	0.98	$6.60 - 289\ m^2$	54	1980	2
$C_d = 2.024\ X_{cd}^{0.72}$	0.99	$12 - 1500\ kW$	16	1980	3

Table II : the condenser price correlations.

Table III

Correlation	R^2	range X_{cp}	No data points	date	brand
Semi-hermetic					
$C_{cp} = 118.108\ X_{cp}^{0.70}$	0.98	$0.075 - 2.914\ 1$	39	1980	1
$C_{cp} = 134.264\ X_{cp}^{0.66}$	0.93	$0.044 - 1.458\ 1$	46	1981	2
Open					
$C_{cp} = 87.004\ X_{cp}^{0.71}$	0.99	$0.35 - 1.458\ 1$	8	1981	2 type A
$C_{cp} = 39.105\ X_{cp}^{0.48}$	0.99	$0.056 - 0.292\ 1$	4	1981	2 type B
$C_{cp} = 83.191\ X_{cp}^{0.75}$	0.97	$0.539 - 1.458\ 1$	6	1980	1 type A
$C_{cp} = 32.547\ X_{cp}^{0.50}$	0.94	$0.028 - 1.917\ 1$	8	1980	1 type B
$C_{cp} = 447.476\ X_{cp}^{0.47}$	0.92	$1.661 - 26.6\ 1$	9	1980	3

Table III : the compressor price correlations.

Table IV

$-20/-15\ \%$	$-15/-10\ \%$	$-10/-5\ \%$	$-5/0\ \%$	$0/5\ \%$	$5/10\ \%$	$10/15\ \%$	average error
1	1	1	1	4	1	4	$8{,}9\ \%$

Table IV : the component approach : the relative error distribution.

Table V

Brand	Freon	Correlation	No data points	R^2
1	500	$C_{hp} = (32.427 - 579\ t_0 + 200\ t_1)Q^{0.62}$	192	0.87
	12	$C_{hp} = (35.435 - 590\ t_0 + 204\ t_1)Q^{0.62}$	240	0.87
	12 - 114	$C_{hp} = (43.617 - 607\ t_0 + 205\ t_1)Q^{0.62}$	192	0.84
	114	$C_{hp} = (59.570 - 1.100\ t_0 + 432\ t_1)Q^{0.62}$	210	0.92
2	22	$C_{hp} = (8.797 - 86\ t_0 + 68\ t_1)Q^{0.74}$	663	0.62
	500	$C_{hp} = (11.953 - 173\ t_0 + 70\ t_1)Q^{0.74}$	728	0.80
	12	$C_{hp} = (13.351 - 220\ t_0 + 86\ t_1)Q^{0.74}$	728	0.82

Table V : the temperature-capacity approach : the results of the regression analysis.

Table VI

Brand	Freon	-30/ -25%	-25/ -20%	-20/ -15%	-15/ -10%	-10/ -5%	-5/ 0%	0/5%	5/ 10%	10/ 15%	15/ 20%	20/ 25%	25/ 30%
1	500				-	10%	46%	27%	17%	-			
	12				-	10%	45%	28%	17%	-			
	12-114				-	10%	46%	28%	16%	-			
	114				1%	11%	44%	27%	17%	-			
2	22	-	-	1%	8,5%	17%	22%	23%	17%	8%	2%	1,5%	-
	500	-	0,5%	5%	11 %	15%	15%	20%	16%	12,5%	3%	1%	1%
	12	-	1%	5%	9 %	14%	16%	19%	17%	12%	4%	2%	1%

Table VI : the temperator-capacity approach - the relative error distribution.

Table VII

t_0	30 °C
t_1	70 °C
C_{hp}	$25\,000\ Q^{0.70}$ (mean function based upon the results of table V)
tax rate	48 %
discount rate (without inflation)	6 % per year
rate of increase of energy prices (without inflation)	3 % per year
duration of life	13 years
maintenance costs	5 % of C_{hp} per year
depreciation method	straight line depreciation

Table VII : assumptions on which the example of heading 4 is based.

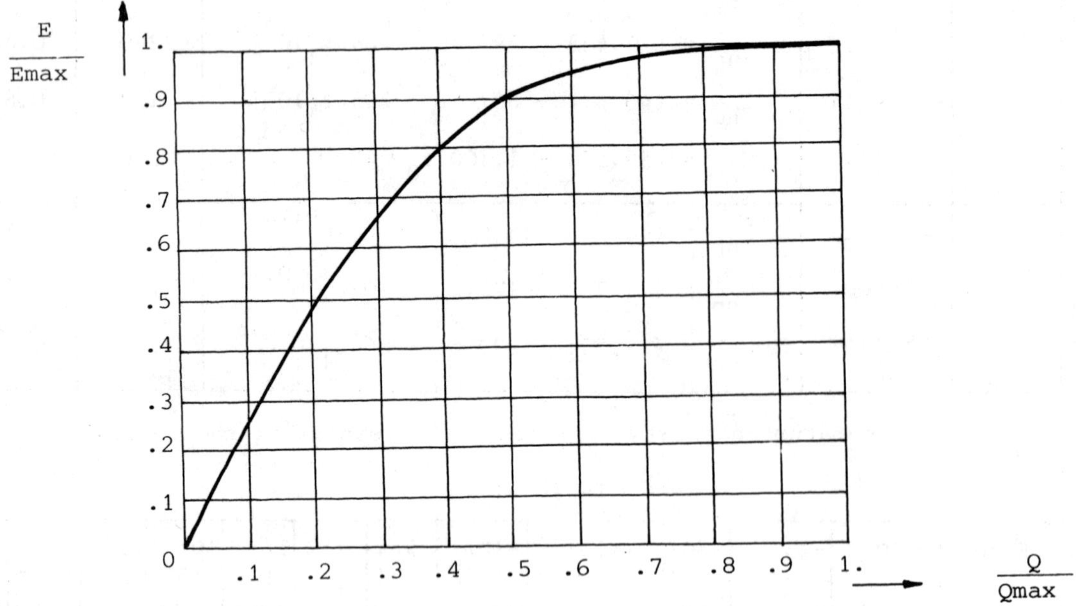

Figure 1 : the relationship between E/E_{max} and Q/Q_{max}.

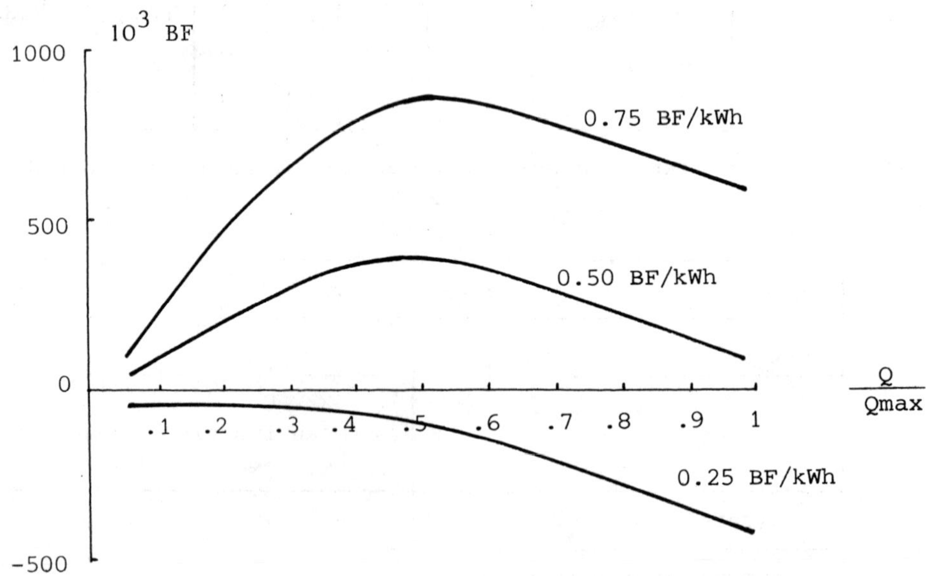

Figure 2 : the net present value for condenser capacities ranging from 0 kW to 200 kW.

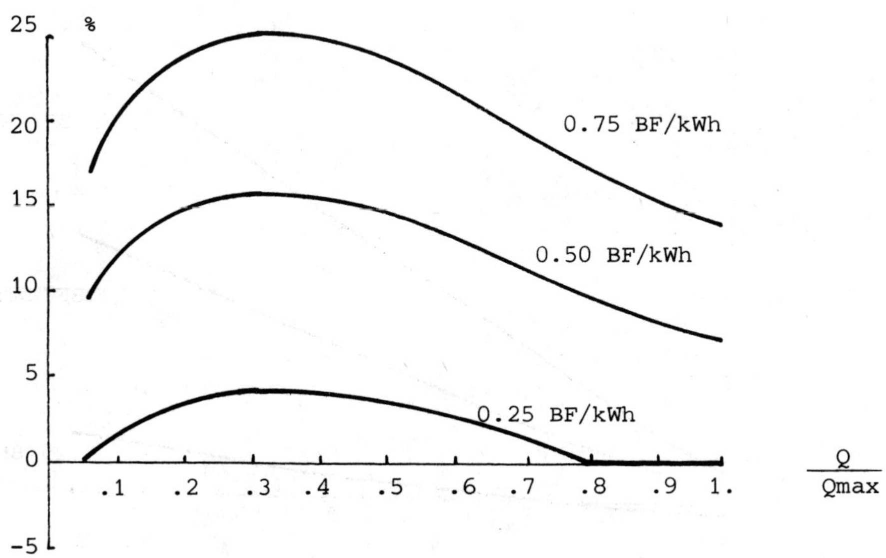

Figure 3 : the internal rate of return for condenser capacities ranging from 0 kW to 200 kW.

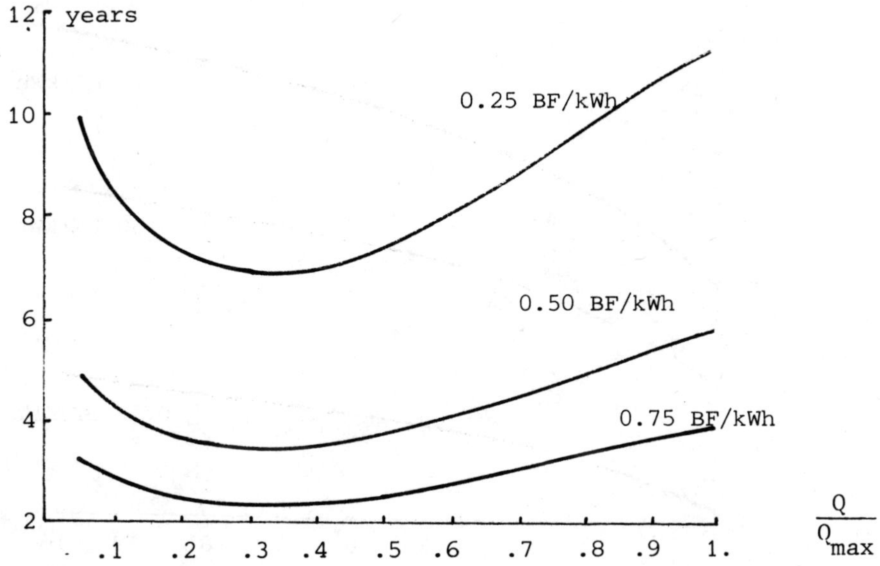

Figure 4 : the payback period for condenser capacities ranging from 0 kW to 200 kW.

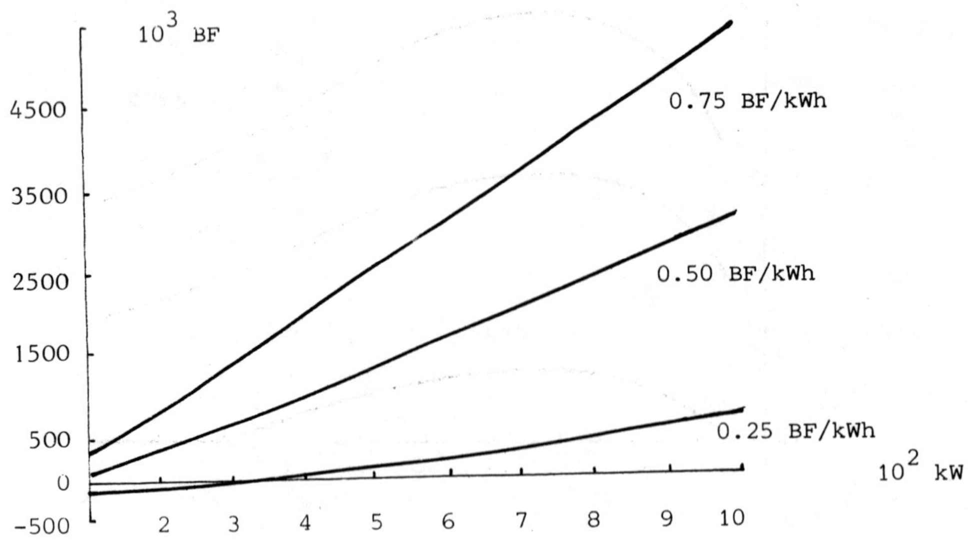

Figure 5 : the net present value for a maximum load ranging from 100 kW to 1000 kW (the condenser capacity is 50 % of the maximum load).

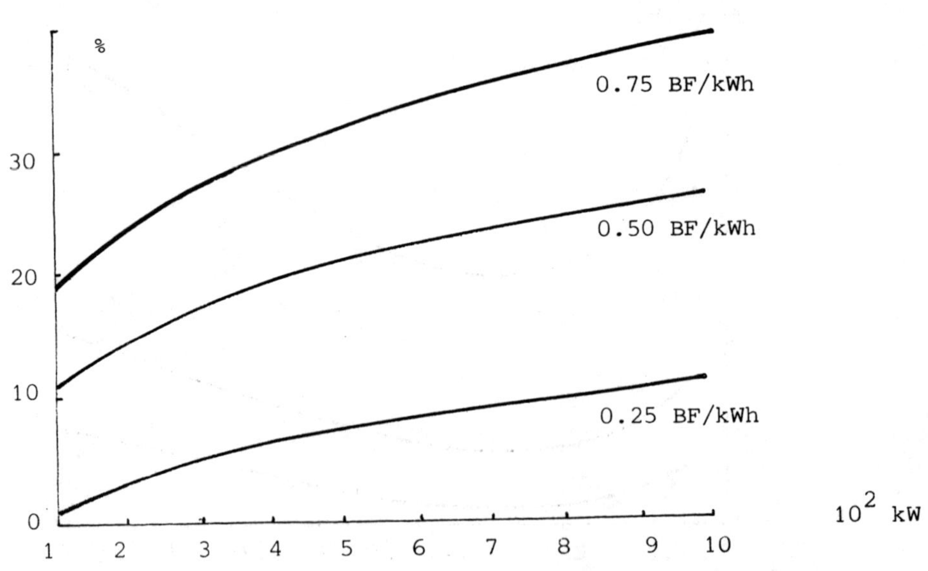

Figure 6 : the internal rate of return for a maximum load ranging from 100 kW to 1000 kW (the condenser capacity is 50 % of the maximum load).

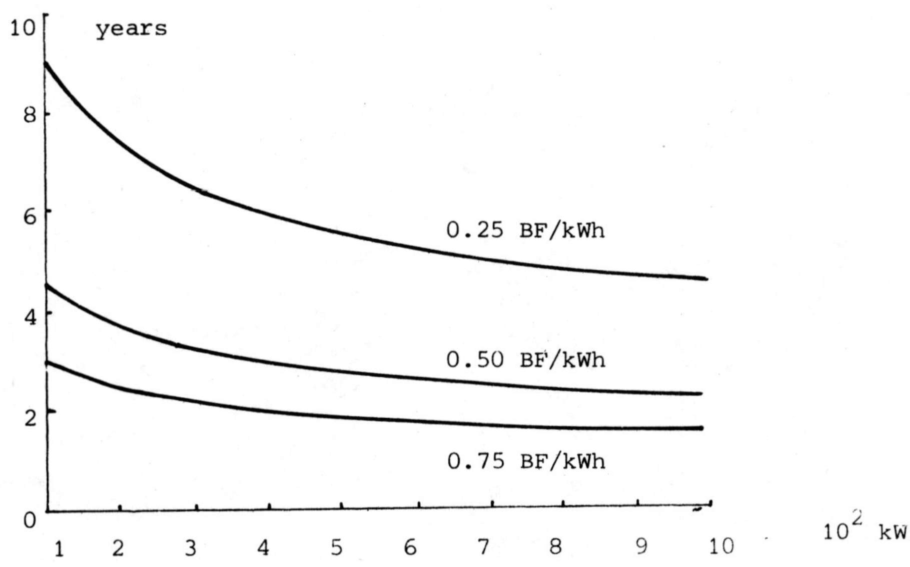

Figure 7 : the payback period for a maximum load ranging from 100 kW to 1000 kW (the condenser capacity is 50 % of the maximum load).